EXOBIOLOGY: MATTER, ENERGY, AND INFORMATION IN THE ORIGIN AND EVOLUTION OF LIFE IN THE UNIVERSE

EXOBIOLOGY: MATTER, ENERGY, AND INFORMATION IN THE ORIGIN AND EVOLUTION OF LIFE IN THE UNIVERSE

Proceedings of the Fifth Trieste Conference on Chemical Evolution: An Abdus Salam Memorial
Trieste, Italy, 22-26 September 1997

Edited by

JULIAN CHELA-FLORES

The Abdus Salam International Centre for Theoretical Physics, Trieste, Italy,
and
Instituto Internacional de Estudios Avanzados, (Universidad Simon Bolivar), Caracas, Venezuela

and

FRANÇOIS RAULIN

LISA, Universités Paris VII and Parix XII, Creteil, France

KLUWER ACADEMIC PUBLISHERS
DORDRECHT / BOSTON / LONDON

Library of Congress Cataloging-in-Publication Data

Trieste Conference on Chemical Evolution (5th : 1997)
Exobiology : matter, energy, and information in the origin and evolution of life in the universe : proceedings of the Fifth Trieste Conference on Chemical Evolution : an Abdus Salam memorial, Trieste, Italy, 22-26 September 1997 / edited by Julian Chela-Flores and François Raulin.
p. cm.
Includes index.
ISBN 0-7923-5172-X (alk. paper)
1. Life--Origin--Congresses. 2. Exobiology--Congresses. 3. Life on other planets--Congresses. 4. Molecular evolution--Congresses.
I. Chela Flores, Julián. II. Raulin, F. III. Title.
QH325.T74 1997
576.8'3--dc21 98-28195

ISBN 0-7923-5172-X

Published by Kluwer Academic Publishers,
P.O. Box 17, 3300 AA Dordrecht, The Netherlands.

Sold and distributed in North, Central and South America
by Kluwer Academic Publishers,
101 Philip Drive, Norwell, MA 02061, U.S.A.

In all other countries, sold and distributed
by Kluwer Academic Publishers,
P.O. Box 322, 3300 AH Dordrecht, The Netherlands.

Printed on acid-free paper

Coverpicture:
Courtesy Jet Propulsion Laboratory. Copyright © California Institute of Technology, Pasadena, CA. All rights reserved. Based on U.S. government - sponsored research under contract NAS7-1260.

Printed in the Netherlands

Dedicated to the memory of

ABDUS SALAM (1926-1996)

Advisory Committee

Herrick Baltscheffsky, Sweden,

Mohindra S. Chadha, India,

George V. Coyne, S.J. Italy,

Donald DeVincenzi, USA,

Frank Drake, USA,

J. Mayo Greenberg, The Netherlands,

Jean Heidmann, France,

Gerda Horneck, Germany,

Kensei Kobayashi, Japan,

Mikhail S. Kritsky, Russia,

Alexandra J. MacDermott, UK,

Christopher P. McKay, USA,

Alicia Negron-Mendoza, Mexico,

John Oro, USA,

Tairo Oshima, Japan,

Martino Rizzotti, Italy,

Giovanni Strazzulla, Italy,

Wang Wenqing, China.

Sponsors

International Centre for Theoretical Physics

International Centre for Genetic Engineering and Biotechnology

European Commission DG XII

Consiglio Nazionale delle Ricerche

UNESCO

The European Space Agency

National Aeronautics and Space Administration.

Weizmann Institute Genome Project

Universités Paris VII and Paris XII

Instituto de Ciencias Nucleares, Universidad Nacional Autonoma de Mexico

Assicurazioni Generali, and

Dipartimento di Biologia, Università degli Studi di Trieste.

1. R.O. Grubbs
2. L. Ianniti
3. R. Vikramsingh
4. G. Lemarchand
5. G. Strazzula
6. P. Stadler
7. M. Chadha
8. J. Oro
9. A. Pappelis
10. R. CardosoGuimaraes
11. P. Metrangolo
12. D. DeVincenzi
13. F. Drake
14 J. Tarter
15 A. Lazcano
16. R. D. Keynes
17. F. Raulin
18. J. Chela-Flores
19. S. W. Fox
20. H. Baltscheffsky
21. M. Baltscheffsky
22. J.M. Greenberg
23. Wang Wenqing
24. M.V. Ivanov
25. T. Owen
26. C. Cosmovici
27. H. Nakamura
28. V.C. Tewari
29. M.S. Krtitsky
30. V. Pironello
31. P. Davies
32. A. MacDermott
33. F. Westall
34. G.H. Schwehm
35. A.O.Kuku
36. S. Jain
37. R.I.M. Rashid
38. K. Kobayashi
39. J.A. Hogbom
40. K.R.K. Esawaran
41. Observer
42. P. Bruston
43. Observer
44. S. Moorbath
45. P. Schenkel
46. E. Chacon
47. D. Lancet
48. J. Seckbach
49. R. Vaas
50. M. Rizzotti
51. K.T. Shah
52. M.C. Guclu
53. D. Segre
54. R. Navarro Gonzalez
55. Mrs. M. Chadha
56. V. Basiuk
57. E.P. Whitehead

PREFACE

Exobiology was the main topic of the Fifth Trieste Conference on Chemical Evolution. The directors and the Advisory Committee, agreed to dedicate the present volume to the memory of Professor Abdus Salam, the first director of the International Centre for Theoretical Physics, who recently passed away. Due to his initiative the Trieste series on chemical evolution began in 1992, following an invitation he issued in the Spring of 1991 to Professor Cyril Ponnamperuma to become its first director.

Both this Centre and the ICGEB have been sponsors since we first started planning the series. We were honoured and privileged to have 12 sponsors. The following institutions have contributed generously to make this event possible:

The International Centre for Theoretical Physics

The International Centre for Genetic Engineering and Biotechnology

European Commission DG XII

Consiglio Nazionale delle Ricerche

UNESCO (through its regional offices in Latin America, India, China and its Headquarters in Paris)

The European Space Agency

National Aeronautics and Space Administration

Weizmann Institute Genome Project

Universités Paris VII and Paris XII

Instituto de Ciencias Nucleares,
Universidad Nacional Autonoma de Mexico

Assicurazioni Generali,
and
Dipartimento di Biologia,
Università degli Studi di Trieste.

We welcomed, once again, members from all the major space agencies. These included the National Aeronautics and Space Administration, the European Space Agency, the Russian Space Programme. We also had the opportunity to learn about the most recent points of view of Japanese efforts in exobiology.

We were particularly grateful to those scientists and engineers that allowed us to make a truly comprehensive review of Astrobiology. In the 68 presentations that we enjoyed we had contributions on many of the subjects that contribute to exobiology.

These included: historical perspective of biological evolution; philosophical and theological aspects of exobiology, cosmic, chemical and biological evolution, molecular biology, geochronology, biogeochemistry, biogeology, planetary protection, planetology.

Some of the current missions were reviewed, such as the Galileo, Cassini-Huygens and Rossetta Missions. To complete the programme, there were discussions of some of the proposed missions or missions that are only at the preliminary level of planning within working groups, such as the the Far InfraRed Space Telescope, Mars Express, the Europa Cryobot/Hydrobot mission and exobiology in earth orbit.

To complete the programme, there were discussions of some of the proposed missions that are only at the preliminary level of planning within working groups, such as the Europa Cryobot/Hydrobot mission and exobiology in earth orbit.

Last, but certainly not least, we were able to learn of the most exciting of all exobiological undertakings, such as the search for planets outside the solar system, and the possible manifestation of intelligence in such environments. Indeed the SETI research effort is well represented in this general survey of our subject.

Finally, we were particularly pleased with the wide interest that this conference has risen in such a world-wide representation form Africa, Asia, Australia, Latin America, Eastern Europe, the European Union, Switzerland, Japan, the United States of America and Canada. With the present group the Trieste series have had some 350 participants since they began in October 1992. Of the selected participants about 50% come from the Third World or East Europe, two geographical areas on which this Centre has put special emphasis.

Due to illness Professor F.W. Eirich was unable to present his paper, but we have included it in the proceedings.

We would like to acknowledge the collaboration of many staff members of ICTP, but particular Ms. Lisa Iannitti for her support throughout the conference.

Julian Chela-Flores,
Trieste, Italy
and

François Raulin,
Paris XII, Creteil, France.

CONTENTS

Section 1. General Overview

Section 2. Matter in the Origin and Evolution of Life in the Universe

Section 3. Energy from Inert to Living Matter

Section 4. Information

Section 5. Early evolution

Section 6. Exobiology: General Perspectives

Section 7. Exobiology on Mars and Europa

Section 8. The Interstellar Medium, Comets and Chemical Evolution

Section 9. Exobiology on Titan

Section 10. Extrasolar Planets

Section 11. Search for Extraterrestrial Intelligence

OPENING ADDRESS

F. K. A. Allotey,
Member of the Scientific Council,
International Centre for Theoretical Physics

Distinguished guests, conference directors and participants, ladies and gentlemen:

First I wish to apologise on behalf of the Director of the International Centre for Theoretical Physics ICTP) who is out if town on another assignment.

I am happy and privileged to be at this inaugural ceremony of the Fifth Trieste Conference on Chemical Evolution. The first of this series of conferences on Chemical Evolution was in 1992. From its small beginning the series has grown to become a major activity of ICTP.

For those new to ICTP, I should be permitted to say some few words about the Centre. ICTP was conceived and created by the late Professor Abdus Salarn (Nobel Laureate) and transformed into reality with the help of another visionary, Professor Paolo Budinich and the generosity of the Italian Government. It operates under the aegis of two United Nations Agencies namely the UNESCO and the International Atomic Energy Agency (IAEA). It has been in existence since 1964. It is regularised by a seat agreement with the Government of Italy which provides the major part of the Centre's funding.

The main objectives of the ICTP are:

(a) to help in fostering the growth of advanced studies and research in physical and mathematical sciences, especially in developing countries.

(b) to provide an international forum for scientific contacts among scientists from all countries

(c) to furnish facilities to conduct original research to its visitors, associates and fellows, principally from developing countries.

The Director of the ICTP is Professor Miguel Angel Virasoro from Argentina and the Deputy Director is Professor Luciano Bertocchi from Italy. In managing the scientific activity of the Centre, the Director is assisted by the Academic Board which consists of the Scientific Staff and Consultants of the ICTP.

There is also a scientific Council which has twelve distinguished scientists as members. At present three of the Council Members are Nobel Laureates. The present chairman of the Council is the Nobel Laureate Professor R. J. Shrieffer. The Council meets normally once a year to review and recommend programmes.

At ICTP research is carried out throughout the year in the following areas: Condensed Matter Physics, High Energy Physics and Astrophysics, Mathematics, Microprocessor Laboratory, Physics and Environment, Aeronomy and Radiopropagation Laboratory, Physics of the Weather and Climate Group, Structure and Non-linear Dynamics of the Earth, Training and System Development on Networking and Radiocommunication.

ICTP has External Activities Programme which supports activities initiated by scientists in the developing countries.

There are several scientific organisations which are headquartered at ICTP. Among them, may be mentioned, The Third World Academy of Sciences, International School for Advanced Studies and International Centre for Science and High Technology.

Mr. Chairman,

Scientists living and working in developing countries are faced with the difficulty of obtaining scientific literature and adequate laboratory facilities. For this reason the ICTP has initiated a scheme of providing science libraries in developing countries with books, journals and equipment through the channels of its "Book and Equipment Donation Programme". As a result, the ICTP is distributing an average of fifty thousand journals, proceedings and books every year among libraries in developing countries. At the same time the ICTP has sent to laboratories in the developing countries a large number of different items of equipment which has been donated by various European Laboratories.

Mr. Chairman,

Every year several thousand scientists, the majority of whom come from the developing countries, visit the Centre. In addition to the regular research activity, high-level training courses, workshops, conferences and topical meetings take place at ICTP. I wish the participants a successful conference. Thank you.

Section 1: General Overview

ABDUS SALAM
FROM FUNDAMENTAL INTERACTIONS TO THE ORIGIN OF LIFE

ABDUS SALAM
FROM FUNDAMENTAL INTERACTIONS TO THE ORIGIN OF LIFE

JULIAN CHELA-FLORES
The Abdus Salam International Centre for Theoretical Physics, Miramare P.O. Box 586; 34100 Trieste, Italy and Instituto de Estudios Avanzados (Universidad Simon Bolivar), Apartado 17606 Parque Central, Caracas 1015A, Venezuela.

Abdus Salam made many contributions to the science of the elementary interactions. He contributed, in particular, to the creation of the theory that unifies the electromagnetic and weak nuclear interactions. For this achievement he was awarded a Nobel Prize in Physics in 1979. I first met Salam in the second week of June 1968, during the inauguration of what is now known as the 'old wing' of the Main Building of the International Centre for Theoretical Physics (ICTP), which now bears his name. Earlier still, during the spring of 1966, while I was a graduate student at the University of London, I attended a series of two lectures at Imperial College. Salam delivered at that time an extended summary of his original work on the gauge theory of the weak nuclear interaction.

Salam's generosity was evident to all those near to him. In spite of the many unexpected commitments that must have arisen in October of the year when the news from the Nobel Committee was made public, Salam did not cancel a visit to Caracas, which I had promoted early in 1979. A visit to Universidad Simon Bolivar (USB) had been suggested, on the occasion of its 10th anniversary in January 1980. Salam came to us only five weeks after he had received his award in Stockholm. He thoroughly enjoyed his visit to Caracas; it was evident that he was particularly pleased to meet so many young Venezuelan students and the academic staff from our University.

J. Chela-Flores and F. Raulin (eds.),
Exobiology: Matter, Energy, and Information in the Origin and Evolution of Life in the Universe, 5-8.

I would like to refer to a significant contribution of Salam, which has influenced a large cross-section of those of us who are actively doing research in the field of astrobiology, which includes the sciences of the origin, evolution and distribution of life in the universe. Salam called me to his office at the beginning of February 1991 to introduce me to Cyril Ponnamperuma. It was not a surprise for me. I had collaborated with Salam in his efforts to get deeper insights in Chemical Evolution. I also knew Ponnamperuma's work well from my early research activity.

In that meeting Salam asked me to be a Local Organiser of a conference on the origin of life, which was to take place later on in the year under the direction of Ponnamperuma. However, Ponnamperuma and I took almost two years to complete the organisation of the First Trieste Conference, which eventually took place in October 1992 [1]. I still keep a copy of my first message to Ponnamperuma's office at the University of Maryland. It was a *telex,* rather than an e-mail, which is the means that has helped Francois Raulin and I to keep close contact during the Fourth and Fifth Conferences. Raulin joined us following Cyril Ponnamperuma's unexpected passing away in December 1994. Ponnamperuma and I were in almost daily telephone contact then preparing the Fourth Trieste Conference.

Salam himself passed away two years later, in November 1996. I was greatly honoured to have shared his interest in questions regarding astrobiology. Our common interest in the origin of the genetic code had begun a few years earlier, during my sabbatical leave in 1986. We started our discussions on a daily basis from late in 1990 till 1993, after my retirement from USB in 1990; it was the year when Salam [2] would take definite steps in his own original approach to astrobiology.

He would say on 29 January 1991 [3]:

"I feel particularly proud of my last paper on the role of chirality in the origin of life".

This statement was made only a few weeks before Salam went ahead with the suggestion, mentioned above, to hold the First Trieste Conference on Chemical Evolution at ICTP. Salam suggested that we should publish separate simultaneous papers on the question of the origin of chirality, a request to which I happily agreed. The papers appeared in the Journal of Molecular Evolution [2] and Chirality[4], respectively, both in 1991. Several of the participants in the Fifth Trieste Conference still remembered the First Trieste Conference. In that event Salam presented his personal contribution to the question of the origin of the homochirality [1]. This work led to an exciting debate chaired by Professor Laurence Barron, form the University of Glasgow. Salam's legacy, which Cyril Ponnamperuma contributed significantly to develop in the Trieste series, has been with us throughout the 1990s [1, 5-8].

Today almost 400 participants from all over the world, as Salam would have wished, have left a significant record of their research in five volumes. In Trieste we have witnessed discussions that span the whole range of astrobiology. We have debated topics in cosmic, chemical, prebiotic and biological evolution on Earth, and elsewhere. Efforts to find environments suitable for life in our own and other solar systems, have been presented. Finally, we have also learnt about the search for signs of extraterrestrial intelligence.

John Oro graciously accepted to deliver the *Abdus Salam Lecture.* Raulin and I had the honour to ask Oro, form his unique point of view - the fruit of one of the most successful careers in astrobiology - to attempt to put into perspective for us the fragmented contributions that make up our wonderful field of research. It seems fitting that the present *Abdus Salam Memorial* volume should begin with Oro's lecture.

References

1. Ponnamperuma, C. and Chela-Flores, J. *Chemical Evolution: Origin of Life.* Eds. A. Deepak Publishing: Hampton, Virginia, USA (1993).

2. Salam, Abdus, *The role of chirality in the origin of life.* J. Mol. Evol., **33,** 105-113 (1991).

3. *Ceremony in Honour of Abdus Salam on his 65th Birthday.* ICTP: Trieste (1991). p. 79.

4. Chela-Flores, J. , *Comments on a Novel Approach to the Role of Chirality in the Origin of Life.* Chirality **3,** 389-392 (1991).

5. Chela-Flores, J., M. Chadha, A. Negron-Mendoza, and T. Oshima. Eds. *Chemical Evolution: Self-Organization of the Macromolecules of Life.* A. Deepak Publishing: Hampton, Virginia, USA (1995).

6. Ponnamperuma, C. and Chela-Flores, J. Eds. *Chemical Evolution: The Structure and Model of the First Cell.* Kluwer Academic Publishers, Dordrecht, The Netherlands (1995).

7. Chela-Flores, J. and Raulin, F. Eds. *Chemical Evolution: Physics of the Origin and Evolution of Life.* Kluwer Academic Publishers, Dordrecht, The Netherlands (1996).

8. Chela-Flores, J. and Raulin, F. (Eds.). *Chemical Evolution: Exobiology: An Abdus Salam Memorial.* Kluwer Academic Publishers, Dordrecht, The Netherlands. (1998, the present volume.)

The Abdus Salam Lecture

COSMOCHEMICAL EVOLUTION
A unifying and creative process in the Universe

J. ORO
Department of Biochemical and Biophysical Sciences
University of Houston
Houston, TX 77204-5934, USA

ABSTRACT. This is a brief review paper. It starts with the (1) recognition of the fundamental work done by Abdus Salam, one of the greatest physicists of the 20th century. As is well known Salam, Weinberg, and Glashow independently found part of the unity in the physical forces of the Universe that Einstein had been searching for and was not able to find. After the introduction the paper covers briefly the creative processes of physical and chemical evolution and the origin and evolution of life on this Earth. This includes: (2) the physical events described by cosmologists on the origin of the Universe and Big Bang inflationary models explaining its evolution; (3) the nuclear synthesis of carbon and other biogenic elements in the core of stars; (4) the formation of circumstellar and interstellar biogenic molecules; (5) formation of the Solar System by the gravitational collapse of the solar nebula; (6) the massive impact that generated the Earth-Moon System; (7) the capture by the primitive Earth of water and other cometary molecules that led to the chemical synthesis of biochemical compounds; (8) comets and carbonaceous chondrites; (9) the processes of prebiological chemical evolution that eventually gave rise to the emergence of life on our planet; (10) some of the critical stages of the slow-pace and punctuated evolution of life such as the K-T boundary event, the evolution of hominids, and the appearance of *Homo sapiens sapiens* ; (11) the exploration of the Solar System in the twentieth century; (12) the possible existence of life on Mars, Europa, and beyond our planetary system; (13) the recent and important discovery of extrasolar planets; (14) the paper also touches briefly the possibilities of extraterrestrial intelligence in our galaxy. (15) Finally, some reflections are made on ethical unifying principles we may derive from a better knowledge of cosmic evolution and a biophysical view of the universe.

1. Abdus Salam and unification

In order to understand better the physical world around us, we look for the underlying forces responsible for the apparent unity of the observable Universe and for the creative processes and physical laws, that may offer an explanation of what we observe in the natural world. In other words, by our own nature, we are inclined to search for unity within diversity. Isaac Newton found this unity in the laws of gravitation applicable to planets, satellites, comets, and other cosmic bodies of the Solar System, and Charles Darwin found it in the law of natural selection which applies to the evolution of the many diverse and different forms of living systems on this Earth.

J. Chela-Flores and F. Raulin (eds.),
Exobiology: Matter, Energy, and Information in the Origin and Evolution of Life in the Universe, 11-32.

As I understand, it was presumably his own intuitive feeling that led Albert Einstein to make the statement somewhat as follows: "it is amazing how easy it is to comprehend the Universe". As it appeared for Einstein, harmony and unity were the first laws of nature. With this in mind, Einstein spent many decades working on a unified field theory but he failed. Nevertheless, the research efforts of other physicists continued in that direction. In the 1960's and 1970's, Abdus Salam in the United Kingdom and Steven Weinberg in the United States of America, working independently, developed a theory unifying the electromagnetic and the weak nuclear forces or fields, which was later elaborated by Sheldon Lee Glashow in the Glashow-Weinberg-Salam model of the electroweak interactions. For this accomplishment, the three scientists shared the 1979 Nobel Prize in physics. Eventually, the theoretically predicted new particles, weak positive, weak negative and neutral (W^+, W^-, Z^o) were detected by Carlo Rubbia and colleagues at CERN and the electroweak field theory became firmly established.

That gave a renewed impetus to further attempts at the unification of the other two fields or forces, the strong nuclear, and the gravitational. Some measure of success has been attained in the unification of the strong nuclear force or field with the electroweak field, the so called Grand Unified Theories (GUTs), in which Alan Guth has been a pioneering investigator. However, up until now, it has not yet been possible to include the gravitational forces, because, so far, we have not been able to detect the extremely small or weak gravitational waves and their corresponding particles, the so called "gravitons". So we have not yet attained the unification searched for by Einstein.

At any rate, these four different forms of energy, forces, fields, or interactions (strong nuclear, electromagnetic, weak nuclear and gravitational) are responsible for the fundamental laws and processes that govern the structure and evolution of our entire observable universe. It is amazing that the matter (and in the past antimatter) and energy of our cosmos ranges from the infinitesimally small particles and interactions of the microcosmos to the incomprehensively large dimensions of the beautiful nebular structures, such as the Trifid, Orion and Eagle nebulae, clusters of galaxies and galaxies, such as the M100 galaxy in the Virgo cluster, stars and planetary systems, such as the Solar System, and so many other cosmic bodies of the macrocosmos.

1.1. ABDUS SALAM: THE MAN

It is a great honor for me to write in the memory of Abdus Salam. I believe it is entirely appropriate to recognize that he not only developed the theory for unification of the electroweak force, and contributed in a major way to the advancement of knowledge about the physical world, but was also very conscious of the fraternity of human beings and of the need to make scientific knowledge available to the people of all countries. Abdus Salam was one of the ten most distinguished physicists of this century. In addition to his Nobel Prize, he was awarded the Einstein medal (UNESCO, 1979) and many other honors and distinctions from all over the world. He was dedicated to the cause of peace being the scientific secretary of the Geneva Conferences on Peaceful Uses of Atomic Energy, and the founder, first director, and president of the International Center for Theoretical Physics (ICTP) Trieste, Italy. As soon as ICTP was created, he was primarily responsible for initiating and carrying out a program, supported by the Italian government, the International Atomic Energy Commission, UNESCO, and other institutions, to advance the level of scientific knowledge in developing countries. This was done by allowing a large number of capable postdoctoral and senior scientists from these countries to spend time at the Trieste International Center for Theoretical Physics, so that they could contribute with their efforts to the development of science. He was also the President of the Third World Academy of Sciences. In this way, he significantly promoted the advancement and unification of knowledge in all countries.

He was a devoted man to his country of birth, Pakistan, as well as to his family and religion. Thus, he was the scientific advisor to the president of Pakistan. I have been told by a colleague of his, Dr. S. L. Glashow, that when Salam appeared at the ceremonies for the 1979 Nobel Prizes, Salam wore the traditional Pakistani pointed shoes and a huge white turban. This was done not for shock value, but because he was a deeply religious man, and devoted husband. I am also pleased to say that for his scientific and social contributions to the advancement of knowledge and peace he was awarded the "1990 Catalunya International Prize" in Barcelona by the Autonomous Government of Catalonia, Spain, where I had the good fortune to be with him and discuss problems on the physical aspects of the origin of homochirality and of the origin of life, in which he was very much interested. There is no question that by his remarkable scientific and social accomplishments he singularly contributed to the advance and spread of science and scientific cooperation all over the world for the betterment of humankind.

1.2. SCOPE OF THIS ARTICLE

In this paper, we will focus our attention on the origin and evolution of life and we will briefly cover the pertinent physical and chemical aspects of cosmic evolution which were responsible for the formation of the main biogenic elements in stars, which aside from H include C, N, O, S and P; and the circumstellar and interstellar organic and other molecules which constitute the dust and gas clouds of the interstellar medium. We will discuss the formation of both the Solar and the Earth-Moon systems, the role of comets in providing the water and the biogenic precursors necessary for the formation of biochemical compounds on the early Earth, the probable processes of chemical evolution that presumably preceded the emergence of Darwin's ancestral cell from which all other living systems have evolved on the Earth, including this chimerical and probably ephemeral creature we call *Homo sapiens sapiens*.

Moreover, after exploring some of the critical stages of biological evolution, such as the K-T boundary event responsible for the disappearance of dinosaurs and very probably, much later on, for the evolution of hominids, we will also consider the question of life beyond the Earth. including the possibilities of past and present life on Mars, the possible microbiological habitability in Europa, and beyond. Perhaps more importantly, we will discuss one of the major discoveries of this century, namely the recent findings of extrasolar planets around stars other than pulsars, situated less than sixty light years away from us. We will consider the possibility of development of technological space-faring civilizations on the basis of our present knowledge and the probability calculations of Frank Drake on the estimates of existence of extraterrestrial intelligence. Finally, we will end with some ethical unifying conclusions that can be derived from a better knowledge of the evolution of the cosmos and the origin and evolution of life. After all, we are all children of the Universe.

2. The physical Universe: Limitations of a chemist

I was giving a lecture at the New York Academy of Sciences in 1962 on some of the experiments we had carried out in our laboratory on the prebiotic synthesis of the fundamental building blocks of nucleic acids, such as adenine and other purines, as well as amino acids and other biochemical compounds, from simple molecules, like hydrogen cyanide, and others known to be present in comets. At the beginning of the talk I said "I don't know anything about the origin of the universe", and that I would be talking only about experiments that may be related to the origin of life [1].

After the dinner reception, an elegant couple approached me. The man was Edward P. Tryon, and he essentially said to me: "Dr. Oró, the problem you did not want to talk about, that is the origin of the universe, I have already solved it ". Obviously my only possible answer was: " I am all ears". And he said "yes, the birth of the Universe is simply a quantum-mechanical fluctuation of vacuum". I recall that I immediately answered back: "If that is the case, we should have an infinite number of universes". To which he responded: "Yes, that is the case". Under the title "Is the Universe a Vacuum Fluctuation?", E. P. Tryon published a paper about his quantum theory [2].

2.1. LIMITATIONS OF EARLY THEORISTS

By tracing back the theoretical question of the birth of the Universe, it is well known that early traditional systems of belief cosmologies assumed a beginning in the not very distant past and an eventual end of the physical Universe while retaining the concept of eternal existence from a human transcendental point of view. With the development of the theory of gravity, Newton realized that if there were a finite number of stars in the Universe they would eventually fall together by their mutual attraction. On the other hand, he argued, that if the Universe had an infinite number of stars this collapse would not take place, because the gravitational forces in opposite directions would cancel each other, and there would not be any central point for the stars to fall to. So Newton concluded that the Universe is infinite and static. Cosmology in the twentieth century began with Albert Einstein's theory of general relativity which is an improved theory of gravity since it describes gravity as a distortion of the geometry of cosmic space and time. However, Einstein was perplexed that it was impossible for him to build a static model of the Universe consistent with his theory of general relativity. Looking at the night sky and observing what appeared to him as fixed stars, he also concluded, very much against his own intuition, that the Universe is probably static. The concepts about the static nature of the Universe remained essentially philosophical and controversial. In the 1910s, V. M. Sipher, from the Lowell observatory, discovered that the spectral lines of many nebulae were primarily displaced towards the red and some slightly towards the blue, which was interpreted as a Doppler effect depending on their direction of movement. In 1917, the Dutch astronomer W. de Sitter found another solution for the modified Einstein theory that could predict the displacement towards the red proportional to the distance of the nebulae. Yet all this within the same Einstein concept of a static universe.

2.2. THE EXPANDING UNIVERSE

It was not until the Russian cosmologist A. A. Friedman precisely applying Einstein's field equations arrived at the conclusion in 1922 that the Universe was expanding.

Independently, and also using the theory of relativity the Belgian astronomer G. E. Lemaitre in 1927 arrived at the same conclusion. Yet he went further and proposed the idea that the Universe started from the explosion of a very small volume of matter. But it was not until 1929 that Edwin Hubble, after a long period of observation published the first observational evidence about the dynamic nature of the Universe [3]. Hubble showed that the galaxies were receding from each other in a process of universal expansion, as indicated by the red-shift of their light, which was proportional to their distance from us. Thus, the law of cosmic motion in a dynamic expanding universe became known as Hubble's law. The universal primordial explosion, which is used to explain the expansion of the cosmos as shown by the receding of galaxies, was given the name of Big Bang as the initial event or birth of the Universe by George Gamow in 1948. I understand that Fred Hoyle and colleagues who proposed an alternative steady-state theory of continuous creation of matter between the galaxies used to refer humorously to the proponents of the primordial explosive creation of the Universe as the "Big Bang " colleagues. Apparently the name has stuck since then.

It is generally accepted now that in addition to the many different fields, particles, and macroscopic cosmic bodies, the Universe consists primarily of clusters of galaxies and galaxies which are receding from each other. So if we go back in time, all the cosmic bodies must have existed as an extremely small pack of matter that exploded into the expanding Universe that we now observe. In 1965 Arno A. Penzias and Robert W. Wilson [4] discovered accidentally the microwave background radiation (CBR) received uniformly from all directions of space. The energy of this radiation was equivalent to a temperature of about 2.7 degrees Kelvin. This was taken as the temperature remaining today from the original flash of light and radiation from the primeval fireball, or primordial explosion of the Universe which has been cooling since its birth.

The radiation was found to be isotropic in all directions, as had been predicted almost simultaneously by Dicke and Peebles, and much earlier in 1948 by George Gamow and colleagues as the diffuse microwave after-glow, or relic fossil radiation, from the Big Bang. Some recent measurements carried out by a NASA satellite, the so called Cosmic Background Explorer or COBE, have given results that according to J. C. Mather and colleagues [5], have been interpreted as small nonuniformities in the primordial cosmos located probably very far away from us. They could indicate the first stages of condensation of matter in our Universe. It is of interest in this respect that also recently the Hubble Space Telescope found thin, apparently young, galaxies at a distance approximately fourteen thousand million light-years away from us. They could be interpreted as baby-galaxies formed a little after the Big Bang explosion took place.

2.3. STANDARD AND INFLATIONARY MODELS OF THE UNIVERSE

According to the standard Big Bang model about fifteen thousand million years ago the Universe existed condensed into an infinitesimally small ball of matter of an almost infinite mass and at an almost infinite temperature. In this condensed state of matter the four forms of energy, or fundamental forces, were unified, the strong nuclear, the weak nuclear, the electromagnetic and the gravitational. In this unique state, the accepted laws of physics would not apply. This is why the special event of the Big Bang explosion is referred to as a singularity. That is, one can not probe prior to its existence. But after the hot explosion occurred, the cosmos in which we are embedded began to evolve and it has been cooling down since then, producing a wonderful Universe.

Thus, as soon as the explosion began several things started to happen very fast, namely the four different forces became differentiated into the strong nuclear, the weak nuclear, the electromagnetic and the gravitational. Light chemical elements were formed, such as hydrogen, helium, deuterium and traces of lithium. The heat and the light of the explosion flash was irradiated throughout the Universe. And the first condensations of matter appeared to be formed as ripples in space in an otherwise perfectly uniform cosmos. As we said above they probably represent relics of the earliest condensation of matter into galaxies in our Universe.

Table 1. Big Bang inflationary theory model of the universe

	TIME	PROCESS
FORCES		
	10^{-43} s	Gravity is separated from other forces
	10^{-33} s	Electroweak, strong nuclear & gravity operate
	10^{-10} s	Weak nuclear and electromagnetic separate
MATTER		
	10^{-6} s	Quarks combine to form particles
	3 minutes	Light nuclei (H, He, Li, et cetera) form
	$3x10^{5}$ yr	Free nuclei & electrons condense to neutral atoms (Universe becomes transparent to CBR photons)
	$>10^{6}$ yr	Protogalaxies & stars form from H & He atoms

Where the standard Big Bang cosmological model and later inflationary models of the Grand Unified Theories, or GUTs, developed initially by Alan H. Guth [6] and Andrei Linde [7], differ is in the number of positive and better answers to the predictions which the latter theories seem to provide. For instance, the standard hot Big Bang cosmology can describe the history of the Universe back to one microsecond, (10^{-6} s) only, whereas the goal of GUTs models is to be able to extend the time back to an unimaginable short period of time. More specifically, Guth's scheme holds that the Universe began with a very brief but exceedingly rapid period of expansion called inflation. This so called inflation process would have lasted only for an unbelievably fleeting moment. Apparently this brief glitch of time inserted into the beginning of the hot Big Bang model is sufficient to solve essentially all the problems. To put it in more specific terms the events occurring after the Big Bang may be briefly described as shown in Table 1.

The GUTs inflationary models are claimed to provide plausible explanations to several major questions of our observable Universe, including: (1) the large number of particles; (2) the Hubble expansion: (3) its large-scale uniformity; (4) The nearness of its mass to a critical density; (5) the absence of magnetic monopoles; and (6) the nature of the spectrum of the microwave background fluctuations observed by the COBE satellite. Furthermore, (7) it is considered also possible that inflation be eternal and that there is an infinity of Universes aside from our own. See also E. P. Tryon [2], and G. V. Coyne [8] for interesting views on the birth and evolution of the Universe.

2.4. QUESTIONS ABOUT THE BIG BANG MODEL

Progress is being made, but not everything is rosy with the different Big Bang models proposed so far, either the early models, or the more recent ones derived by the application of quantum mechanics to the Universe as a whole. There are many questions that remain to be answered. Indeed two of the outstanding cosmologists S. W. Hawking and R. Penrose [9] agree in principle with a similar picture of the Universe although they differ in their vision of the application of quantum mechanics to the evolution the Universe. Also, while Hawking considers that a theory needs only to suggest predictions that agree with data, Penrose thinks that a simple comparison of experiments with predictions is not enough to explain reality. Even though the mainstream of cosmological theories are with the inflationary models, it should be pointed out that other views are possible. For instance, new solutions of Einstein's cosmological equations do not require a Big Bang model and a uniform Universe [10].

Leaving aside the details of cosmological evolution, it may be worth pausing for a moment and considering the basic experimental and observational principles of science. In ordinary phenomenology, objective knowledge is obtained by the application of the scientific method. This is based on the performance of an experiment or observation which can be measured, verified and confirmed by other scientists. Once a valid theory is developed from the established facts, predictions for future observations can be made and tested. Any theory is in turn subject to modification as more observations or experiments are made.

On the other hand when we are talking about the Universe at large we are supposing an initial "big bang" at the time of birth and perhaps a terminal "big crunch" at the time of death. Here we are dealing with a different kind of phenomenology, since these two events are singularities. The experiments can not be performed by anyone or repeated by anyone. However, it may be possible to eventually develop appropriate experimental procedures to simulate in a microcosmos the initial moments of the Big Bang, as indicated by the behavior of superfluid He-3 at a temperature of 100 microkelvin. In the meantime, and in accordance with Hawking, and Guth the best thing it can be done, is to make as many predictions as possible that agree with the theory.

2.5. THE KEY QUESTION ABOUT OUR UNIVERSE

As we mentioned earlier, and is well known, early cosmological models of the Universe were based on geocentrism, then heliocentrism and eventually universalism. Yet, the question of the birth of the Universe still remains one of the most difficult of science. One has to grasp something we are not accustomed to. For instance, one may wonder what happened one second before the birth of the Universe. One may disregard it and say that the question does not belong to the science of cosmology, or simply say that in reality we do not know. However, it is possible that there was nothing before, except perhaps the "magic" quantum fluctuations of vacuum about which E. P. Tryon theorized. Alternatively, one could say, like Weinberg, if we consider another physical property such as temperature we know that there is an absolute zero temperature. Therefore, we could also have a zero time. Indeed, this is the concept developed by Hawking about the history of time. The moment prior to the beginning is supposed to be a singularity point where our laws of physics do not apply.

Presumably, empty space is not entirely empty. The vacuum is filled with small quantum fluctuations. Some evidence for this, so-called "zero-point energy", has been recently measured by the Casimir-Polder effect, and shown to have a value of less than 10^{-6} newton. The vacuum fluctuations can lead to a limitless and eternal self-reproduction of inflationary Universes or space-time bubbles. We might be in one of these bubbles, and we call it our observable Universe. At any rate, leaving aside the difficult question about the birth of the Universe, the only thing that a chemist can do is to start at the last stage of the inflationary model of the Universe, when nucleosynthesis began in stars.

3. The nucleosyntheses of carbon and biogenic elements

From a chemist point of view one of the most creative processes in the evolution of the Universe is the nuclear synthesis of the chemical elements of the periodic system. That took place first during the Big Bang explosion, then in the core of the stars and later in supernova explosions. As shown in a previous section, three minutes after the Big Bang, the free nuclei of hydrogen and helium were generated. However, a few million years had to pass before protogalaxies and stars could be formed from the condensation of clouds of hydrogen and helium. First the light elements were synthesized in the core of stars. Most of the heavy elements were generated much later by supernova explosions. These two processes are continuously going on at the present time. As we know, the Universe is made of essentially hydrogen and helium, which total about 98%. The remainder is made of about 1% of C, N, O, S and P and about 1% of the rest of the elements.

The formation of carbon and of the biogenic elements (H, C, N, O, S and P) is most important. These elements are needed not only to make the organic compounds that are present in the circumstellar and interstellar medium, in comets and other cosmic bodies, but they also are necessary for the formation of the biochemical compounds of living systems.

The corresponding nuclear reactions for the synthesis of the biogenic elements have been briefly described in a paper from our laboratory [11]. It is worth to consider two of the first nuclear reactions which lead to the formation of helium and carbon nuclides. In addition to its synthesis during the Big Bang, the stellar formation of the helium nuclide takes place by the proton-proton chain. This occurs in the core of many ordinary stars and specifically in our Sun at a temperature of about 15 million degrees. The small deficiency of mass which results from the condensation of four protons into one nuclide of helium is converted into energy in accordance with Einstein's equation, $E = mc^2$. This large amount of energy in the form of heat and radiation has been partly responsible for the appearance of life on our planet and for its evolution during the past four thousand million years. The next most important nuclear reaction is the one that involves the condensation of three helium nuclides into one carbon-12 nuclide. This nuclear reaction is known as the triple-alpha process. It occurs inside carbon stars at temperatures of 100 million degrees. Obviously, without the triple-alpha process, we would not be able to talk about life in the Universe. In fact we would not be here.

Once carbon-12 is made by the low probability collision of three alpha particles, subsequent alpha-capture processes produced the oxygen and sulfur nuclides. The nitrogen nuclide is catalytically generated by the CNO cycle. However, the phosphorus nuclide requires the participation of many complex nuclear reactions. This explains the relatively lower cosmic abundance of this element [11].

4. Circumstellar and interstellar molecules

Carbon stars are a rich source of carbon compounds or organic molecules. From the core of these and other stars, the biogenic elements (H, C, N, O, S, P) migrate to the outer and cooler regions of the stars. There, ordinary chemical reactions give rise to the formation of diatomic and triatomic combinations that can be observed in stellar atmospheres. Among the most common species on finds C_2, CN, CO, CH, NH, OH, and H_2O that are present for instance in the atmospheres of ordinary stars such as our Sun, a third generation star.

At least one hundred chemical species have been identified so far in the interstellar medium (ISM) by their gas-phase molecular spectra [12]. All these molecules, ions, and radicals are relatively simple. The ones specifically identified are made from two, three, four, etc., up to about 11 atoms. About 75% of them are organic, or contain carbon.

Since the biogenic elements are the most abundant active elements in the Universe, and since most of the interstellar molecules contain carbon, one could say that the Universe is essentially organic, and prepared for life to emerge wherever and whenever the conditions are right. It should also be pointed out that these one hundred chemical species do not include many organic compounds that have not been individually detected in the interstellar medium but that probably form the bulk of the so called diffuse infrared bands (DIB). According to Allamandola and colleagues [13], polycyclic aromatic hydrocarbons, or PAHs, are one of the major components of interstellar dust and gas clouds. In a recent review, Allamandola shows that these compounds are not only abundant in the ISM, but in many other places of the cosmos such as meteorites, interplanetary dust particles, and comet Halley. Indeed, as he has said, PAHs are everywhere including the Martian meteorite ALH 84001. The Martian PAHs are not products of biological origin, since the same PAHs have also been found in the Murchison meteorite [14]. The most probable sources of cosmic PAHs are the circumstellar clouds. Eventually they become part of the ISM as well as comets, meteorites, and other bodies of the Solar System. On the basis of experimental work on the production of aromatic hydrocarbons cosmic PAHs probably result from the high temperature condensation of C_2 or C_2H_2 in the atmospheres of carbon stars and their circumstellar envelopes.

Among all these species, it is of interest to point out the following ten molecules: Hydrogen, ammonia, water, formaldehyde, hydrogen cyanide, cyanacetylene, carbon monoxide, hydrogen sulfide, cyanamide and phosphorus nitrile. With them one could synthesize in the laboratory under prebiological conditions, the amino acids, nucleic acid bases, sugars, lipids, and mononucleotides which are part of the biochemical compounds of all living systems.

5. The Solar System: The second Big Bang

Approximately 5 Ga ago the Solar System was formed by the gravitational collapse of a dusty gaseous nebula of interstellar matter, presumably triggered by the shock-wave of a supernova explosion. Hoppe et al. [15] have obtained evidence for the presence of SiC grains in the Murchison meteorite where several isotopic ratios indicate they are matter from a type II supernova. Subsequently, during its first several million years, the evolving Solar System was in a state of great upheaval, where the norm was the continuous collision of planetesimals or comets with other major bodies. Thus, at least two phases can be distinguished in the formation of the Solar System, first the proto-sun by gravitational collapse of the solar nebula, and then the planets, satellites, and other bodies by accretion of the differentiated nebular matter.

As pointed out by Delsemme, the precursors of the terrestrial planets were rocky planetesimals essentially devoid of light molecules due to the relatively high temperature prevailing around the orbit of the Sun within the inner part of the asteroid belt.

6. The third Big Bang: The Earth-Moon system

The impact record on the Moon is witness to the turbulent state of flux during the early stages of formation of the Solar System. The proto-Earth was obviously subject to many collisions from small and large planetesimals. A relatively recent model proposed by Cameron and other investigators, e.g., Cameron and Benz [16], has led to a new theory for the formation of the Earth-Moon system which apparently avoids the deficiencies of previous theories. According to this theory, a celestial body with a mass comparable to that of Mars collided with the proto-Earth, injected most of the iron into the nucleus of the Earth, and caused the fusion and ejection into orbit of portions of the mantle which eventually coalesced to form the Earth's only natural satellite, the largest moon of the terrestrial planets. This origin explains most of the similarities and slight differences between the composition of the Earth's mantle and the Moon, for instance the iron-poor nature of our satellite. In the dynamic aspects, it also explains the angular momentum of the Earth-Moon system. Thus, from a terrestrial point of view we could consider the formation of the Earth-Moon system as the third Big Bang.

7. Cometary matter captured by the early Earth

The above theory, simply expressed, means that the Earth-Moon system resulted from a single-impact massive collision which led to the loss of most of the volatiles into space. Also, as suggested by Cameron, comets and other small bodies of the evolving Solar System subsequently contributed by late accretion most of the water and biogenic compounds through smaller collisions which occurred during the first 600 million years of Earth's history. On the other hand, due to its low gravitational mass, the Moon could not retain any significant amount of volatile compounds.

We know that currently the atmosphere of the Moon is 10^{-10} torr. There is no doubt that, while some of the collisions contributed matter to the Earth, others removed water and volatile compounds, because of the dual aspect of impact capture and impact erosion. However, at the end of this bombardment period, as indicated by the Moon cratering record, the Earth came out with a significant increase of its mass, particularly in water, carbon, and other biogenic elements (Table 2), in a similar manner as it probably happened to the major terrestrial planets, to Titan and other bodies of the Solar System.

Table 2. Cometary matter trapped by Solar System bodies*

	Cometary matter (g)	Time-span (years)
Venus	4.0×10^{20}	2×10^{9}
Moon	2.0×10^{20}	late-accretion
Earth	$1.0 \times 10^{14-18}$	2×10^{9}years
	$1.0 \times 10^{25-26}$	late-accretion
	3.5×10^{21}	late-accretion
	7.0×10^{23}	4.5×10^{9}
	2.0×10^{22}	4.5×10^{9}
	1.0×10^{23}	2.0×10^{9}
	$1.0 \times 10^{24-25}$	1.0×10^{9}
	$6.0 \times 10^{24-25}$	1.0×10^{9}
	$1.0 \times 10^{23-26}$	4.5×10^{9}

* Calculated by several authors; listed in Oró, Mills and Lazcano [17]

8. Comets

Comets are aggregates of interstellar matter at the very low temperatures of interstellar space (Oort cloud) and outer regions of the Solar System (Kuiper belt). The dirty ice model that has prevailed with some modifications is that of Fred Whipple. The composition and many aspects of the relation of comets to the terrestrial planets has been studied in detail by Delsemme [18], and the model for the condensation of interstellar grains to larger and larger aggregates to generate cometesimals and planetesimals has been developed theoretically and experimentally by Greenberg [19]. Several interesting studies have been reported on Halley's comet, by Kissel and Kruger on Shoemaker-Levi's 9 comet, by Gautier and other investigators; and on Hyakutake, by Mumma and others [e.g., 12, 20, 21]. A large number of papers should appear in relation to the spectacular Hale-Bopp comet. So far, more than 33 chemical species have been detected in Hale-Bopp, most of which coincide with the organic and inorganic molecules present in the interstellar medium. Current knowledge of the organic composition of comets is reviewed by Mumma [20] who applied high resolution infrared spectroscopy for the first time to Hyakutake comet. He detected strong emissions from H_2O, HDO, CO, CH_4, C_2H_2, C_2H_6, CH_3OH, H_2CO, OCS, HCN, OH, and other chemical species. Of particular interest are the large amounts of methane and ethane in Hyakutake's nucleus. The question of comets and life has been discussed recently by Oró and Lazcano [21] and Oró and Cosmovici [12].

Comets can be seen as the bridge that connects life to the Universe at large. Perhaps this is one of the most important and ultimate consequences of the creative processes of cosmochemical evolution. The reality of the capture of cometary material by the Earth has been dramatically shown by the recent finding of Louis Frank, who has confirmed early speculations of the current fall of cometary material on the Earth. Frank had earlier found spots at a high altitude in the upper layers of the atmosphere that could correspond to water vapor. Now, with the help of the NASA Polar satellite, he has been able to photograph relatively large snowballs of several cubic meters falling continuously on the Earth. Even though the statistical numbers are not complete yet, the preliminary calculations indicate that the amount of water and other light compounds of supposed cometary origin that fall on the Earth in 1000 years would amount to 1 mm of water all around the planet. If we accept the above estimate literally, throughout the course of geological history this would total approximately 4.5 km of water all around the planet.

However, firstly it has to be realized that such incoming cometary material, especially water, is destroyed in the stratosphere by solar ionizing particles, with the hydrogen escaping into space. Secondly, that these cometary snowballs are not falling continuously but rather periodically, reflecting the passage of previous comets close to the orbit of the Earth. To sum up, the total amount of water accumulated by the Earth is probably quite smaller. At any rate this discovery confirms and extends the calculations presented in Table 2.

8.1. CARBONACEOUS CHONDRITES

In addition to comets, the Earth was and is bombarded by asteroids and meteorites. Thousands of meteorites have been recovered from Antartica and many parts of the world. Some of them are carbonaceous chondrites, which contain organic compounds. The Alais carbonaceous chondrite, which fell in France in 1806, was analyzed by Berzelius. In the past few years a large number of meteorites have been found in Antarctica. So far, one of the most interesting carbonaceous chondrites is the one that fell in Murchison, Australia, in 1969. Following the initial findings, in a very systematic research Cronin and his collaborators have analyzed this meteorite for organic compounds, and have recently reviewed this information [22]. The organic matter is largely macromolecular, and possibly related to the refractory organic mantle of interstellar dust grains of Greenberg's model [19].

The complex mixture of monomeric organic compounds in the Murchison meteorite includes carboxylic acids, amino acids, hydroxy acids, sulfonic acids, phosphonic acids, amides, amines, purines (adenine and guanine), and a pyrimidine (uracil), alcohols, carbonyl compounds, aliphatic, aromatic, and polar hydrocarbons. Some seventy five amino acids have been characterized. Of these, eight are common constituents of proteins, such as glycine, alanine, and aspartic acid. A few others are of metabolic interest, such as γ-aminobutyric (GABA), but many of them are not found in biological systems. With relation to the chirality of the α-carbon, all the amino acids are racemic (equal mixtures of D- and L-isomers). The analyses show the presence of many non-biological amino acids. This, together with other properties such as the higher D/H and $^{13}C/^{12}C$ ratios, suggests that these amino acids were synthesized from extraterrestrial precursors [22] predating the formation of the Solar System. Alkyl phosphonates, with the alkyl group (from C_1 to C_4) attached to the phosphorus atom of the phosphonate, are also present in this meteorite. This finding suggest a possible derivation of phosphonic acids from the interstellar molecule CP [22]. Phosphates are presumably derived from interstellar PO_2 as indicated by the IDPs of cometary origin [11].

9. Prebiotic chemical evolution

So far we have discussed a number of physical and chemical creative processes from the initial Big Bang to the formation of the Earth-Moon system. These creative processes include the formation of hydrogen and helium shortly after the Big Bang. Once (a) hydrogen was formed, it seems that the evolution of the rest of the Universe was followed by (b) the formation of galaxies and stars, (c) the nuclear synthesis of elements in stars and supernova explosions, (d) the formation of atoms and molecules in circumstellar and interstellar clouds, and eventually (e) the formation of planetary systems like our own. Once the Earth was formed and acquired water and the cometary precursors for the synthesis of biochemical compounds, there had to be the appropriate conditions that could allow the synthesis of biochemical monomers and polymers (polynucleotides, polypeptides, and lipids) that upon self-organization eventually gave rise to the appearance of the first self-reproductive living system. The processes of prebiological chemical evolution have been described in a number of papers from different laboratories, including our own. A review of this subject can be seen in Oro, Miller and Lazcano [23].

10. Early evolution of life

In addition to the chemical evolution approach, the study of the origin of life can be approached from an evolutionary biological point of view. The most ancient fossils are those found in sediments and rocks in western Australia (the Warrawoona formation). These fossils are 3,500 million years old and their morphology is similar to that of cyanobacteria, as shown by Awramik, Schopf, and colleagues [e.g. 24]. However, they cannot represent the oldest organisms on Earth because they are too complex, or too evolved. Of the three contemporary cell domains (Archaea, Bacteria, and Eucarya), the oldest are probably hyperthermophilic archaebacteria which live at high temperatures in the hot springs at the bottom of oceans. But those organisms are not widely believed to be the ancestral cells of all living beings either. It is to be hoped that by studying the sequences of some of the most primitive enzymes which are most preserved in ancient organisms, we shall be able to establish the root of the philogenetic tree of the three unicellular domains, as is being pursued by several investigators [e.g., 25]. Resolving this problem will not be easy, as the enzymes of any given unicellular organism existing today have had more than 3,500 million years to evolve and many changes may have occurred in the process, even in sequences that today seem very well preserved. However, recent progress in molecular biology raises hopes regarding the application of this method to the study of the origin of life.

A high concentration of oxygen (approximately 15% of present levels) appeared on Earth about 2,000 million years ago, or earlier. Presumably, this was the accumulation of molecular oxygen produced by the oxygenic cyanobacteria, once most of the exposed divalent iron had been oxidized to the trivalent form. This led to the evolutionary emergence of cells with a highly efficient oxidative phosphorylation system. These were the aerobic bacteria which, in the form of mitochondria and through the endosymbiotic evolution process studied by L. Margulis [26], gave rise to the emergence of eukaryotes at about the time of the appearance of significant concentrations of molecular oxygen. The newly emerged eukaryotic cells eventually evolved into multicellular organisms about 800 million years ago and branched into three great classes, fungi, plants, and animals.

10.1. THE K-T BOUNDARY EVENT: THE DISAPPEARANCE OF THE DINOSAURS AND OTHER COSMIC EVENTS

An important inflection in the evolutionary process took place about 65 million years ago when the collision of a comet, or asteroid, probably in the Yucatan Peninsula, caused a catastrophic darkening of our planet and left as a mark one of the largest craters on Earth, the Chicxulub crater. As is known this was first suggested by Alvarez and coworkers based on the presence of iridium as a cosmic marker in the Cretaceous-Tertiary (K-T) boundary sediments. Such a catastrophe is thought to have been responsible for the disappearance of dinosaurs, as well as, many other living species, and it must have opened a niche which allowed the evolutionary development of mammals and led to the eventual appearance of the ancestors of primates and man several million years ago. Had this not occurred, the intelligent beings that now populate Earth would probably have been descendants of the dinosaurs. Apparently, some little mammals were able to survive the K-T boundary explosion, presumably because they were omnivorous and living in caves. With the new vital space created by the disappearance of the dinosaurs, the little mammals were able to evolve freely giving rise to a multitude of species, some of which belonged to the suborder *Anthropoidea* and to the family *Hominidae* . The evolution of our ancestors, *Australophitecus* and hominid species leading to *Homo sapiens sapiens*, is briefly described elsewhere [e.g., 27].

A similar cosmic event occurred not long ago. In July 1994 we witnessed a relatively large collision of comet Shoemaker-Levy 9 with Jupiter. Had this collision occurred with the Earth, it may have wiped out most of the biosphere, including probably also humans, with the possible exception of microorganisms and their symbionts living at the bottom of the seas. It has recently been pointed out that the comet Swift-Tutle, responsible for the Perseid meteors, will pass again close to the orbit of the Earth. Brian Marsden indicated that that coincidence will only be 15 days apart, which is less than the margin of error in these calculations. Let's hope that we do not have a collision!

I believe the most important practical lesson of Solar System space science is to know and decide what can be done to avoid any such cosmic catastrophic impact with the Earth in the future. The technological means for detection and control of incoming cosmic bodies are already available to us. They only need to be organized into a global cooperative network of vigilance, action and reaction. There is no question that a unified policy should be established by all countries of the world through the United Nations Assembly. The more developed countries of the world should be sponsoring the educational efforts of UNESCO in this area of space science and other areas of global interest which are of major concern for the future of humankind.

11. Exploration of the Solar System: The Apollo mission

On July 20, 1969, Neil Armstrong became the first man to set foot on a heavenly body other than Earth, the Moon. Upon the return of the lunar samples, the moon dust was found to be blackish and the first reports to appear in Houston newspapers, when the crates containing the lunar samples were opened, suggested that it could well be graphite, rekindling hopes that life could have existed on the Moon some time in the past. After analysis of the lunar matter, we know today that life could never have existed on the Earth's satellite, and that the lunar dust dark color was mainly due to its constant irradiation by the solar wind. Analyses performed in our laboratory by combined gas chromatography and mass spectrometry (GC-MS) confirmed the absence of organic matter in the lunar samples, with the exception of a few parts per million of carbon monoxide, methane and other simple compounds. The traces of amino acids found were presumably the product of reactions of HCN of cometary origin, or implanted as carbon, nitrogen, and hydrogen atoms from the solar wind.

12. Life on Mars

Exactly seven years later, on July 20, 1976, the first Viking spacecraft landed on the surface of Mars. The dust of Mars was red, in keeping with the popular name by which this planet is known. In this connection, I was reminded by D. M. Anderson, a colleague on the Viking molecular analysis team, that the redness of the surface of Mars led me to forecast that we would not find life on this planet either. Molecular analysis for volatizable organic and inorganic compounds were carried out using an instrument (GC-MS) that was, in principle, similar to the one used in our laboratory for the lunar sample analysis. The new miniaturized apparatus was built according to the suggestions of K. Biemann, myself, and other members of the Viking project molecular analysis team. The major difference between the GC-MS instrument in our laboratory and that on the Viking Mars Lander, was that the latter, instead of weighing two tons, it only weighed 20 kilograms. What were our findings? We found no organic matter, not even in parts per million, in either of the two landing sites on Mars, the plains of Chryse and Utopia, where the Viking landers actually landed.

On the other hand, one of the three biological experiments pointed to the rapid formation of the radioactive carbon dioxide. Levin and Straat [28] suggested that this was the result of considerable microbial activity. I, however, countered that this was due to a relatively simple chemical reaction, in which the iron oxides and hydrogen peroxide in the analyzed Mars samples had oxidized to CO_2, the radioactive formic acid which Levin and Straat had included, among other metabolites, in the nutrient solution. This was 1976. In fact, 20 years earlier, while working on my doctoral thesis, I had studied the mechanism of formic acid oxidation in living beings. It is interesting to note that formic acid, when in the presence of hydrogen peroxide, is rapidly oxidized into carbon dioxide not only by enzymes such as catalase [29], which have an iron atom in their active center, but also by inorganic catalysts such as iron oxides (Fe^{2+}, Fe^{3+}), which are very abundant on Mars.

Furthermore, additional laboratory studies on the oxidation of organic substances by amounts of UV light which are comparable to those which fall on the surface of the red planet explained the total absence of organic matter on Mars. The half-life of residence of any meteoritic or cometary organic matter exposed on the surface of

Mars is barely a few months [30], an instant in the geological time scale. We can thus say that there is no evidence for life at the two Viking landing sites on Mars, although it could be a worthwhile proposition for future space missions to revisit the planet in order to look for fossils which would show if life has existed there in the past. It is unlikely, although not totally impossible, that life may exist in some unique sites of Mars, such as a deep crevice or underground thermal spot where some of the ice of the permafrost may be temporarily converted into liquid water. However, the speculation that the oxygen on Mars is biologically produced does not pass the rigorous test of Occam's razor.

12.1. RECENT DATA FROM MARS

The analysis of a Martian meteorite ALH 84.001 indicated the possible presence of relics of life on Mars about 3,600 million years ago. This assumption was based primarily on three facts: (i) the presence of polycyclic aromatic hydrocarbons (PAHs), (ii) the morphological presence of pyrite mineral grains, and (iii) the presence of what might be fossils of nanobacteria. In this respect I must say (i) that the same polycyclic aromatic hydrocarbons (phenantrene, pyrene, crysene and benzopyrene) were found in our laboratory in the Murchison meteorite in 1984 [14] They are not produced by microorganisms. On Mars the PAHs must have been infused in the rock by the collision that ejected the Martian meteorite [12]; (ii) that the pyrite mineral grains have been found to have a fibrous morphology that indicates high temperature formation, inconsistent with life; and (iii) that the supposed nanofossils are much smaller than any microfossil ever found on the Earth. They are similar to inorganic forms found in meteorites and terrestrial rocks. The evidence, therefore, is no unequivocal for the presence of life on Mars in the past.

12.2. EUROPA, SATELLITE OF JUPITER

Another interesting component of the Solar System, from the point of view of studying the possible existence of extraterrestrial microbial life is Europa, a satellite of Jupiter, which displays a fairly uniform frozen surface and which is traversed by very long crevices. According to various theoretical calculations, Europa has an 80 kilometer-deep ocean under a layer of 10-kilometers thick ice. The temperature of the liquid water under the surface is 4° C, and it is possible that hot springs lie at the bottom of the ocean, similar to those found in the oceans of the Earth. Given that the Jupiter primordial nebula must have contained many organic compounds, had the initial conditions on this satellite been favorable for the process of chemical evolution, it is possible that microorganisms similar to thermophilic archaebacteria that live at great depths in the hot springs of the Pacific Ocean and other seas, could exist at the bottom of Europa's great ocean. I have suggested several times to NASA that it would be interesting to conduct an unmanned space project for the exploration of Europa in order to determine the limits of the chemical evolutionary process in the Solar System.

13. Life beyond the Solar System

It is reasonable to think that life exists in other planetary systems. An example of this may well be the orbital system around the star Beta-Pictoris, which is some 54 light years away from Earth. This star is intriguing because it is surrounded by a great ring of comet-like matter which emits intense infra-red radiation. It was first detected by the infrared telescope of the IRAS astronomical satellite, and photographed from the Las Campanas Observatory in Chile by American astronomers Smith and Terrile. Studies conducted by French astronomers over the past five years suggest that more than 100 comets fall on the central star each year. More recent estimates indicate that a total of 1,000 comets per year, each a kilometer in size, is necessary to explain the observations of the disappearance of dusty cometary material into the central area of the disk where planets may be present [31]. If the orbital system had planets like Earth, life might well be emerging now on one of them around the star Beta-Pictoris. A recent report has been made on the finding of a planetary system, with two Earth-like planets around a very distant neutron star, or pulsar, in the Virgo constellation and there is additional indirect evidence of the existence of planetary companions to nearby stars, and direct evidence for protoplanetary disks in the Orion Nebula. The emergence of life and intelligence on a planet like the Earth is possible, but may not be common in the Universe [32, 33]. NASA is encouraging the difficult astronomical search for extrasolar planets.

13. 1. THE DISCOVERY OF EXTRASOLAR PLANETS

During the last two years one of the major discoveries of this century has taken place, that is the presence of planets beyond our Solar System at about 50 light years away. The first planet discovered was *51 Pegasi B*, by Michael Mayor and Didier Queloz of the Geneve Observatory, in Switzerland [34]. The planet has a mass equivalent to 0.6 Jupiter, and is very close to the star. There has been some debate about the nature of planets, so close to the central star. The following two planets were discovered by Geoff Marcy and Paul Butler, of the University of California at Berkeley, USA, from the Lick Observatory at Mount Hamilton [35].

Among the planets discovered by Marcy and Butler were *70 Virginis B*, with a very large mass of 8.1 in relation to that of Jupiter and an orbit close to that of Mercury in our Solar System. *70 Virginis B* it is an extremely giant gaseous planet where the temperature in the atmosphere must be very hot, about 1000°C. The following one was *47 Ursae Majoris*, with a mass also quite large, about 3.5 times that of Jupiter and located in an orbit a little further away than Mars in our Solar System. Therefore, it is possible that this giant planet has either frozen or liquid water, depending on the greenhouse effect of the gases of its atmosphere, the radiation received from the star and the heat developed at the interior of the planet by gravitational contraction.

In the past few years other planets and planetary systems have been discovered, but we will have to wait until the large infrared interferometry telescope developed by the NASA program ORIGINS is deployed in order to determine the possible presence of extrasolar planets with orbits and masses comparable to that of the Earth in the Solar System.

The spectral detection of water vapor, free molecular oxygen and ozone would be indicators of the possible existence of life in these extrasolar planetary systems. If such data are obtained in the future, the statistics for the calculation of the possibilities of extraterrestrial intelligence would be significantly improved. It is difficult to guess but we will probably have to wait for at least two decades before we obtain more specific information bearing on these extremely interesting and related discoveries.

14. Search for extraterrestrial intelligence (SETI)

It is still more difficult to say whether technologically advanced civilizations with the ability to transmit and receive intelligent signals exist in other planetary systems. Frank Drake [32] developed an equation containing a series of variables from which he calculates that there is an advanced planetary civilization for every 10 million stars in our galaxy. How can we resolve the great enigma of the possible existence of extraterrestrial civilizations? Several researchers, of whom Drake is the leading one, have suggested that it would be impossible using manned spaceships to explore our galaxy, since just one spacecraft would need all the energy produced by humans on this Earth to conduct an interstellar voyage to one of the nearer stars.

A group of American researchers, including Drake, Morrison and Tarter, as well as Oliver and Sagan –both of which passed away recently–, suggested tackling the problem passively, that is to say, by listening out for intelligent signals emitted by other civilizations. The radio-search for extraterrestrial intelligence, under the NASA SETI program started in 1992, at a fraction of the cost of any other space project. The project was initiated by means of radiotelescopes using two different approaches. For the first, specific target approach, about 1,000 stars similar to the Sun have been selected, which are studied in detail by the radiotelescopic antennae in Goldstone (California) and Australia. The selected microwave range corresponds to that of the emission of hydrogen (H) as well as that of the hydroxyl group (OH). In common terms, this frequency range of the spectral region is referred to as the water window (H_2O). The second approach of detecting "intelligent" signals is by conducting a general sidereal sweep or scan using radiotelescopes. The projects were managed jointly by NASA's Jet Propulsion Laboratory (JPL) in Pasadena, California and by NASA's Ames Research Center (ARC) in Moffett Field, California.

14.1. THE PHOENIX PROGRAM

The funding for NASA's High Resolution Microwave Survey was canceled in October of 1993, by the United States Congress. However, the SETI Institute of Palo Alto, California, an independent corporation headed by Dr. Frank Drake, which has been the primary private institution involved in the search for extraterrestrial intelligence, is actively obtaining support from private corporations and individuals under the name of a new project, "Phoenix," promoted initially by the late Bernard M. Oliver, in order to continue the efforts of such an important scientific endeavour. Project Phoenix will continue the targeted search for extraterrestrial intelligence in the vicinities of 1,000 nearby sun-like stars, using the Australian Parkes Radio Astronomy Observatory and then the Arecibo radio telescope in Puerto Rico and other observatories in the northern hemisphere. Perhaps the most interesting aspect of the overall project is the use of new supercomputers which can select the received spectral signals with more than one thousand million times greater efficiency than before. It is difficult to interpret the nature or meaning of some of the signals received so far. If we receive intelligent messages and establish contact with another terrestrial civilization in our galaxy, it will be mankind's greatest discovery and the major event in our entire history.

15. Epilogue: Peace from cosmic evolution

As we have seen, the Universe is essentially made of H_2 and He, but is rich in organic compounds and therefore conducive to the emergence of life, given the right conditions. But even more surprising, especially for those who study the nervous system, the brain, memory and other mental processes, is the realization that some of the simple molecules involved in the transmission of nervous impulses in living beings (neurotransmitters), such as glycine, glutamic acid and gamma amino butyric acid (GABA), have been found in the Murchison meteorite. One could therefore say that the Universe is not only prepared for the emergence of life, but also for the appearance of intelligence! This leads us to the intriguing corollary that the Universe might be populated by civilizations much more intelligent than those living on our small blue planet. Perhaps one day these advanced civilizations will be able to instruct us through interstellar communication as to the answers to human problems such as war, disease and old age. But until this utopian notion becomes a reality, we would do well to cherish our own small blue planet with all its varied and wonderful forms of life, the very ones that Darwin studied. After all, there is only one Earth in the Solar System.

Table 3. Ethical principles that may derive from a better knowledge of the cosmos

1. **Humility**: The life of all cells descends from simple molecules.
2. **Solidarity**: Our genes have a common origin (*Homo sapiens*).
3. **Cooperation**: We live in a resource-limited small blue planet.
4. **Hope**: Some day we may communicate with more advanced civilizations.
5. **Universality**: We come from stardust and in a stardust we shall return.
6. **Peace**: We should change our *culture of war* into a *culture of peace*.
7. **Golden rule**: Treat others as you would like them to treat you.

The landing on the Moon by the Apollo astronauts, allowed them to see the Earth as a small, distant body lost in the intensity of space. They did not see the borders that separate people into different nations nor the color of the people's skin. The astronauts developed a global collective consciousness that we were all citizens of the Earth. Furthermore, in harmony with our common genetic heritage, we are all brothers and sisters, and we would do well to share the limited resources of our little blue planet, and live in fraternal peace. After all, realistically, there is no other place better than the Earth. I believe this is the moral lesson of landing on the Moon, or as Neil A. Armstrong, said, "That's one small step for man. One giant leap for mankind." We could extend these cosmic reflections on evolution into the ethical principles shown in Table 3.

Acknowledgments

I thank Professors Julian Chela-Flores and Francois Raulin for having invited me to the Abdus Salam Memorial. This paper is dedicated to the memory of Abdus Salam. I would also like to recognize the memory of good friends and colleagues working in this general field who have passed away in 1996 and 1997. They include Melvin Calvin, Bernard M. Oliver, Carl Sagan, Eugene Shoemaker and Richard S. Young. I am pleased to thank A. Lazcano and G. V. Coyne, S.J. for having corrected the manuscript, also Merce Piqueras , L. Canas and V. Elizalde for helping in the preparation of this article.

References

1. Oro, J., (1963) Studies in experimental organic cosmochemistry, *Annals of the New York Academy of Sciences* **108**, 464-481.

2. Tryon, E.P., (1973) Is the universe a vaccum fluctuation? *Nature* **246**, 396-397.

3. Hubble, E., (1929) A Relation between distance and radial velocity among extra-galactic nebulae. *Proceedings of the National Academy of Sciences* **15**, 168-173.

4. Penzias, A.A. and Willson, R.W. (1965) A measurement of excess antenna temperature at 4080 Mc/s.*Astrophysical Journal* **142**, 414-419.

5. Mather, J.C. et al. (1990) A preliminary measurement of the cosmic microwave background spectrum by the Cosmic Background Explorer (COBE) satellite, *Astrophysical Journal* **354**. L37-40.

6. Guth, A.H., (1994) in Ekspong, G. (ed.), The Big Bang and cosmic inflation,*The Oskar Klein Memorial Lectures* **2**, World Scientific Publishing Co., Stockholm, pp. 27-70.

7. Linde, A., Linde, D. and Mezhlumian, A. (1994) From the Big Bang theory to the theory of a stationary universe, *Physical Review D* **49**, No. 4, 1783-1826.

8. Coyne, S.J., G.V. (1996) Cosmology: The universe in evolution, in Chela-Flores, J. and Raulin, F. (eds.), *Chemical Evolution: Physics of the Origin and Evolution of Life* , Kluwer Academic Publishers, Dordrecht, pp. 35-49.

9. Hawking, S.W. and Penrose, R. (1996) The nature of space and time, *Scientific American*, Vol. **275**, No. 1, pp. 60-65.

10. Senovilla, J.M.M. (1990) New class of inhomogenous cosmological perfect-fluid solutions without big-bang singularity. *Physical Review Letters* **64**, 2219-2221.

11. Macia, E. Hernandez and Oro, J. (1997) Primary sources of phosphorus and phosphates in chemical evolution, *Origins of Life and Evolution of the Biosphere* **27**, 459-480.

12. Oro, J. and Cosmovici, C.B. (1997) Comets and life on the primitive Earth, in Cosmovici, C.B., Bowyer, S. and Werthimer, P. (eds.), in *Astronomical and Biochemical Origins and the Search for Life in the Universe,* Editrice Compositori, Bologna, pp. 97-120.

13. Allamandola, L. J. , Tielens, A. G. G. M. and Barker, J. R. (1989) Interstellar polycyclic aromatic hydrocarbons: the infrared emission bands, the excitation-emission mechanism and the astrophysical implications, *Astrophys. J. Suppl.* **71**, 733-755.

14. Basile, B., Middleditch, B. S. and Oro, J. (1984) Polycyclic hydrocarbons in the Murchison meteorite, *Organic Geochemistry* **5**, 211-216.

15. Hoppe, et al. (1997) Type II supernova matter in a silicon carbide grain from the Murchison meteorite, *Science* **272**, 1314-1317.

16. Cameron, A. G.W. and Benz, W. (1991) The origin of the Moon and the single impact hypothesis IV, *Icarus* **92**, 204-216.

17. Oro, J., Mills, T. and Lazcano, A. (1992) Comets and the formation of biochemical compounds on the primitive Earth - A review, *Origins of Life* **21**, 267-277.

18. Delsemme, A. (1992) Cometary origin of carbon, nitrogen and water on the Earth, *Origins of Life* **21**, 279-298.

19. Greenberg, J.M. and Hage J.I. (1990) From interstellar dust to comets: A unification of observational constraints, *Ap. J.* **361**, 260-274.

20. Mumma, M. J. (1997) Organics in comets, in Cosmovici, C. B., Bowyer, S. and Werthimer, D. (eds.), *Astrononical and Biochemical Origins and the Search for Life in the Universe*, Editrice Compositori, Bologna, pp. 121-142.

21. Oro, J. and Lazcano, A. (1997) Comets and the origin and evolution of life, in Thomas, P. J. Chyba, C. F. and McKay, C. P. (eds.), *Comets and the Origin and Evolution of Life*, Springer, New York, pp. 3-27.

22. Cronin, J. R. and Chang, S. (1993) Organic matter in meteorites: Molecular and isotopic analyses of the Murchison meteorite, in Greenberg, J.M et al., (eds.),*The Chemistry of Life's Origins* , Kluwer Academic Publishers, Dordrecht pp. 209-258.

23. Oro, J., Miller, S.L. and Lazcano, A. (1990) The origin and early evolution of life on Earth, *Annu. Rev.Earth Planet Sci.* **18**, 317-256.

24. Schopf, J.W. (1983) *The Earth earliest biosphere: Its origin and evolution*, Princeton University Press, Princeton

25. Lazcano, A. (1994) The RNA world, its predecessors and descendants, in Bengston, S. (ed.), *Early Life on Earth*:: Nobel Symposium **84**, Columbia University Press, pp. 70-80.

26. Margulis, L. (1982) *Symbiosis in Cell Evolution*, Freeman, San Francisco, 419 pp.

27. Oro, J. (1996) Cosmic evolution, life and man, in J. Chela-Flores and F. Raulin (eds.), *Chemical Evolution: Physics of The Origin and Evolution of Life*, Kluwer Academic Publishers, Dordecht, pp. 3-19.

28. Levin, G.V. and Strat, P. A. (1976) Viking labeled release biology experiment: Interim results, *Science* **194**, 1322-1329.

29. Oro, J. and Rappoport D. A. (1959) Formate metabolism by animal tissues II. The mechanism of formate oxidation, *J. Biol. Chem.* **234**. No. 7, 1661-1665.

30. Oro, J. and Holzer, G. The photolytic degradation and oxidation or organic compounds under simulated Martian conditions, *J. Mol. Evol*. **14**, 153-160.

31. Lagage, O. O and Pantin, E. (1994) Dust depletion in the inner disk around Beta-Pictoris as a possible indicator of planets, *Nature* **369**, 628-630.

32. Drake, F. D. (1963) The radio search for intelligent extraterrestrial life, in :Mamikunian, G and Briggs, M.H. (eds.), *Current Aspects of Exobiology,* New York: Pergamon Press, pp. 323-345.

33. Oro, J. (1995) The chemical and biological basis of intelligent terrestrial life from an evolutionary perspective, in Shostak, G. Seth (ed.), *Progress in the Search for Extraterrestrial Life*, ASP Conference Series, Vol. 74, pp. 121-133.

34. Mayor, M. Queloz, D. Udry, S. and Halbachs, J.L. (1997) From brown dwarfs to planets, in Cosmovici, C. B., Bowyer, S. and Werthimer, D. (eds.), *Astronomical and Biochemical Origins and The Search for Life in the Universe*. Editrice Compositori, Bologna, pp. 313-330.

35. Butler, R. P. and Marcy G. W. (1997) The Lick observatory planet search, in Cosmovici, C. B, Bowyer, S and Werthimer, D (eds.), *Astronomical and Biochemical Origins and the Search for Life in the Universe*, Editrice Compositori, Bologna, pp. 331-342.

Opening Lecture

THE THEORY OF COMMON DESCENT

RICHARD D. KEYNES
Department of Physiology, University of Cambridge
Physiological Laboratory, Downing Street, Cambridge CB2 3EG, U.K.

1. Introduction

When Charles Darwin published *On the Origin of Species* in 1859, the Bishop of Worcester's wife was most distressed. "Let us hope it is not true", she is said to have remarked. "But if it is, let us pray that it does not become generally known!".

I am afraid that her prayer was not answered, nor is there much doubt today about the basic truth of Darwin's ideas. But during the past 130 years, Darwinism has not always been on top, and serious doubts remain about quite a number of the links in the argument. Can the Darwinian principle of Natural Selection really provide a complete explanation of the extraordinary facts on some of the modes of reproduction in the animal kingdom, and for the evolution of their patterns of migration and their highly specialized sensory organs? And how in any case were the initial steps taken between a world without living creatures or even the smallest of macromolecules, and the complexity of the most primitive of micro-organisms, let alone that of the innumerable species of plants and higher animals.

By way of introduction, I should explain that my presence at this conference is to a large extent the result of the fact that my middle name — which would be my last one in some parts of the world — is Darwin, for my mother was a grand-daughter of Charles. When in the summer of 1968 I was returning home from the pursuit in Chile of my main line of research, which is concerned with the biophysics of the conduction of nervous impulses, it so happened that my family connections led in Buenos Aires to my being taken to call on Dr Armando Braun Menendez in order to see his Darwin collection. This turned out to include two little portfolios of drawings and paintings made by Conrad Martens, the official artist on H.M.S. *Beagle* during 1833 and 1834. At that moment in time I have to confess that I actually knew and cared much less than I should have about my great-grandfather and his work. But the vividness and immediacy of the pictures in the portfolios suddenly brought the voyage of the *Beagle* to life for me, and introduced me to a new interest that I have since found extremely rewarding.

In due course, I have become a part-time but fully accredited member of what is often known nowadays as The Darwin Industry. First, I produced a book entitled *The Beagle Record* (Keynes 1979) in which I drew up a catalogue of Martens's *Beagle* pictures, and reproduced a number of them in colour, accompanied by selections from the correspondence and diaries of Darwin and Captain FitzRoy. Then I re-edited the Commonplace Journal kept by Darwin during the voyage,

J. Chela-Flores and F. Raulin (eds.),
Exobiology: Matter, Energy, and Information in the Origin and Evolution of Life in the Universe, 35–49.

generally known as *The Beagle Diary* (Keynes 1988) in order to distinguish it from his better known *Journal of Researches* (Darwin 1839}. And I am currently engaged on the much larger task of editing his previously unpublished *Beagle Zoology Notes and Specimen Lists*, which will be ready to go to the publishers very shortly.

Having thus become rather well acquainted with Darwin's earliest writings on natural history, my next step (Keynes 1997) was to go further back in order to understand better the initial course of development of his ideas. There were two basic principles that led to his arrival at *On the Origin of Species* (Darwin 1859). The first of these was what became known as The Theory of Common Descent, which amounted to the statement that evolution had indeed taken place, and that all living creatures were ultimately descended from a single ancestor. The second was the Theory of Natural Selection, which provided a mechanism to account for speciation, that is to say for the creation of new species. Since I am neither a professional historian of science, nor a scientist with any expertise in most of the fields represented at this conference, what I shall present here is a strictly personal view on the manner in which ideas first arose on the origin of life itself. I will follow the route that I myself took, starting from Charles Darwin and working backwards, and will consider first the Theory of Natural Selection and then the Theory of Common Descent.

2. The Theory of Natural Selection

Darwin had set out well versed as a theology student in Paley's *Evidences of Christianity* and *Natural Theology*, and was entirely orthodox in his religious beliefs. It is important to appreciate that until almost the end of the voyage he remained a believer in Special Creation. When, for example, he arrived at Charles Island in the Galapagos on 26 September 1835, he wrote in his pocketbook (see Barlow 1945 p. 247):

'The Thenca very tame and curious in these islands. I certainly recognise S. America in the ornithology — would a botanist. Three quarters of plants in flower.'

And in his daily journal he wrote (see Keynes 1988 p. 356):

'I industriously collected all the animals, plants, insects & reptiles from this Island. It will be very interesting to find from future comparison to what district or "centre of creation" the organized beings of this archipelago must be attached.'

So at that time he still believed that the creation of animals took place in particular parts of the world.

Four months later, when Darwin was in Australia, his faith in the existence of a Creator was apparently still unshaken, for while thinking about the kangaroos, duck-billed platypus and other curious creatures that he had seen, he wrote in his journal (see Keynes 1988 pp. 402-3):

'A little time before this, I had been lying on a sunny bank & was reflecting on the strange character of the Animals of this country as

compared with the rest of the World. An unbeliever in everything beyond his own reason might exclaim "Surely two distinct Creators must have been [at] work; their object however has been the same & certainly the end in each case is complete." Whilst thus thinking, I observed the conical pitfall of a Lion-Ant. A fly fell in & immediately disappeared; then came a large but unwary Ant; his struggles to escape being very violent, the little jets of sand described by Kirby were promptly directed against him. NB The pitfall was not above half the size of the one described by Kirby. His fate however was better than that of the poor fly's. Without a doubt this predacious Larva belongs to the same genus, but to a different species from the Europæan one. Now what would the Disbeliever say to this? Would any two workmen ever hit on so beautiful, so simple & yet so artificial a contrivance? It cannot be thought so. The one hand has surely worked throughout the universe. A Geologist perhaps would suggest that the periods of Creation have been distinct & remote the one from the other; that the Creator rested in his labor.'

It was near the end of the voyage, between mid-June and August 1836, when Darwin was re-organizing his Ornithology Notes (see Barlow 1963 p. 262) that his first admission of doubt about the immutability of species was expressed when he wrote of the three kinds of Galapagos mocking bird *Mimus*:

'These birds are closely allied in appearance to the Thenca of Chile or Callandra of La Plata. In their habits I cannot point out a single difference. They are lively, inquisitive, active, *run fast*, frequent houses to pick the meat of the Tortoise which is hung up, sing tolerably well; are said to build a simple open nest; are *very* tame, a character in common with the other birds. I *imagined* however its note or cry was rather different from the Thenca of Chile? Are very abundant, over the whole Island; are chiefly tempted up into the high & damp parts by the houses & cleared ground.

I have specimens from four of the larger Islands; the two above enumerated [from Charles and Chatham Islands], & female, Albermarle Isd. & male James Isd. The specimens from Chatham & Albermarle Isd. appear to be the same; but the other two are different. In each Isld. each kind is *exclusively* found: habits of all are indistinguishable. When I recollect the fact that from the form of the body, shape of scales & general size, the Spaniards can at once pronounce from which Island any Tortoise may have been brought. When I see these Islands in sight of each other, & possessed of but a scanty stock of animals, tenanted by these birds but slightly differing in structure & filling the same place in Nature, I must suspect they are only varieties. The only fact of a similar kind of which I am aware is the constant asserted difference between the wolf-like Fox of East & West Falkland Islds.

If there is the slightest foundation for these remarks, the zoology of Archipelagoes will be well worth examining; for such facts would undermine the stability of Species.'

When the *Beagle* returned to England on 2nd October 1836, Darwin proceeded to entrust his collections to the several specialists who were ready to look at them. Soon he received two highly significant reports. Richard Owen reported that the fossils that had been found in Patagonia confirmed for South America what was known as the "Law of Succession", concerning the close relationship between the past and present mammalia of large continents, which had previously been shown to apply to Australia. And John Gould not only confirmed that Darwin's three island forms of mocking birds were distinct species, but also found that almost all of his Galapagos land birds were new species, although some were closely related to those found on the American continent. Moreover, the finches were not, as Darwin had supposed, members of several widely different genera and families, but all belonged to a remarkable new family now known as the Geospizinae.

In his private journal for 1837 (see de Beer 1959 p. 7), Darwin made a famous entry:

'In July opened first note book on "Transmutation of Species" — Had been greatly struck from about Month of previous March on character of S. American fossils — & species on Galapagos Archipelago. These facts origin (especially latter) of all my views.'

The development of Darwin's biological, geological and metaphysical theorizing during the next seven years was duly recorded in a series of 12 notebooks and some separate sheets. Owing to his reprehensible habit of subsequently cutting out what he considered to be especially useful pages, and storing them elsewhere, the collection of manuscripts now preserved in the Cambridge University Library eventually became rather confused, to say the least, but a group of experts (Barrett *et al.* 1987) has in the end got together to produce an edition of the Transmutation Notebooks in what is agreed to be their original order.

The precise timing of Darwin's arrival at the Theory of Natural Selection is not in serious doubt, since he wrote in his *Autobiography* (see Barlow 1958 p. 120) that in October 1838:

'I happened to read for amusement Malthus on *Population*, and being well prepared to appreciate the struggle for existence which everywhere goes on, from long-continued observation of the habits of animals and plants, it at once struck me that under these circumstances favourable variations would tend to be preserved, and unfavourable ones to be destroyed. The result of this would be the formation of new species.'

It has sometimes puzzled historians of science to explain why a reading of Malthus was necessary in order to set Darwin thinking about natural selection, for the concept of the universal struggle for existence was already familiar to him from the volumes by Lyell (1830-3) that he had with him on board the *Beagle*. As I have explained elsewhere (Keynes 1997), my view is that the reading of Malthus served primarily to trigger the surfacing of an idea already dormant in his mind. However, I will not pursue the point any further here, because as far as this conference is concerned, the question of greatest interest is to consider how the Theory of Common Descent originated.

3. Erasmus Darwin

In the *Autobiography* written near the end of his life, Charles Darwin wrote somewhat disparagingly of his grandfather Erasmus, in an account of a conversation with Robert Grant at Edinburgh some 50 years earlier (see Barlow 1958 p. 49):

'He one day, when we were walking together, burst forth in high admiration of Lamarck and his views on evolution. I listened in silent astonishment, and as far as I can judge, without any effect on my mind. I had previously read the *Zoönomia* of my grandfather, in which similar views are maintained, but without producing any effect on me. Nevertheless it is probable that the hearing rather early in life such views maintained and praised may have favoured my upholding them under a different form in my *Origin of Species*. At this time I admired greatly the *Zoönomia*; but on reading it a second time after an interval of ten or fifteen years, I was much disappointed, the proportion of speculation being so large to the facts given.'

Since Erasmus Darwin is a less familiar figure than his grandson, and is perhaps better known as a poet than as a scientist, a brief biographical note may be in order; for a fuller biography, King-Hele (1977) should be consulted. He was born in 1731 and died in 1802, went up to St John's College in Cambridge, and took his Cambridge M.B. in 1755 after completing his medical training at Edinburgh. He set up his practice first for some years in Lichfield, and later in Derby, and his fame as a physician became prodigious, matching his energy and, it would seem, his appetite. In 1757 he married Mary Howard, by whom he had five children, followed by two illegitimate daughters after she died, for whom he set up a school and published a little book on *A Plan for the Conduct of Female Education*. His eldest son, Charles, was a doctor of brilliant promise who died tragically young from a dissecting wound. The second son, Erasmus, became a lawyer; and his third son Robert was a successful doctor at Shrewsbury, and father of the second Charles. At the age of 50, a corpulent, stammering widower, he fell in love with one of his patients, a rich and witty young widow, Elizabeth Pole. The marriage was again a happy one with seven further children.

His intellectual interests were all embracing, and in 1766 he and the leading scientists and technologists of the area — Matthew Boulton, Josiah Wedgwood, James Watt, John Baskerville, William Small, James Keir and Joseph Priestley among them — founded the Lunar Society, which met monthly for discussions, and outshining the Royal Society in its distinction and vigour, was an important driving force behind the Industrial Revolution in England.

He was a prolific and technically competent poet, and in his lifetime was best known to the general public as the author of *The Botanic Garden*, two long poems entitled *The Loves of the Plants* (1789) and *The Economy of Vegetation* (1792), though the overpraise that greeted their initial appearance was succeeded by a period of ridicule. His aim was both to amuse and to impart serious instruction, and *The Botanic Garden* deserves respect as the first and probably the last attempt to popularize science in verse.

Erasmus Darwin is best remembered nowadays for the two huge volumes that he published in 1794 and 1796 entitled *Zoönomia, or the Laws of Organic Life*, which was in effect a comprehensive medical treatise, based on 40 years of practical experience, that covered not only medicine and psychology, but much of animal biology, with physics, philosophy and religion thrown in as well. Finally, his best poem, *The Temple of Nature*, appeared posthumously in 1803.

4. The Theory of Common Descent

Despite the unenthusiastic views about his grandfather expressed years afterwards by Charles, the second of his Transmutation Notebooks, opened in June 1837, was headed *Zoonomia*, and begins with a discussion of Erasmus's views on generation. On pages 21 to 24 of this notebook (see Kohn 1987 pp. 177-80), Charles proposed that:

'organized beings represent a tree *irregularly branched*, some branches far more branched — Hence Genera.— as many terminal buds dying, as new ones generated. There is nothing stranger in death of species, than individuals. If we suppose monad definite existence, as we may suppose is the case, their creation being dependent on definite laws, then those which have changed most. owing to the accident of positions must in each state of existence have shortest life; Hence shortness of life of Mammalia.—

Would there not be a triple branching in the tree of life owing to three elements air, land & water, & the endeavour of each one typical class to extend his domain into the other domains, & subdivision three more, double arrangement.— if each Main stem of the tree is adapted for these three elements, there will be certainly points of affinity in each branch'

This was illustrated on page 36 of Notebook B by the first of his branching diagrams, showing the descent from a common ancestor labelled 1 to species A, B, C and D, the gap between A and B being the largest, between C and B the smallest, and between B and D there being a rather greater distinction:

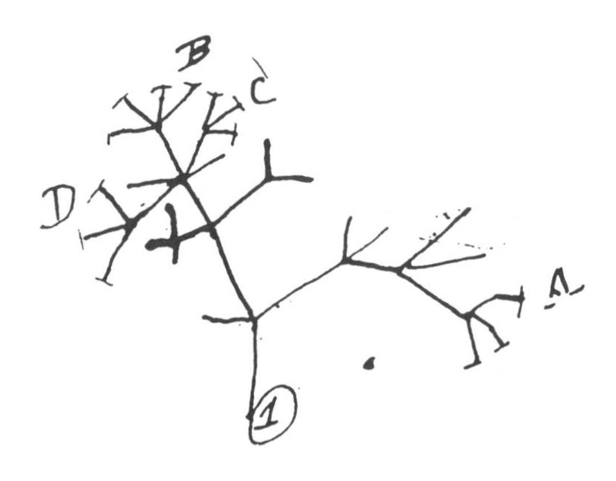

A few months later, after an exploration of this line of thought in fascinating detail, Charles was pondering further about the implications of the existence of a single germ of life, and wrote in Notebook B (see Kohn 1987 pp. 227-8):

'My theory would give zest to recent and Fossil Comparative Anatomy, & it would lead to study of instincts heredetary & mind heredetary, whole metaphysics.— it would lead to closest examination of hybridity, to what circumstances favour crossing & what prevents it — and generation, causes of change in order to know what we have come from & to what we tend. This & direct examination of direct passages of structure in species, might lead to laws of change, which would then be main object of study, to guide our speculations with respect to past & future. The Grand Question, which every naturalist ought to have before him, when dissecting a whale, or classifying a mite, a fungus, or an infusorian, is "What are the laws of life"?'

However, on page 58 of Notebook C, written during the first months of 1838, he was beginning to see complications in his theory (see Kohn 1987 pp. 237-328):

'As we have birds impressions in Red Sandstone, great lizards in do.— Coniferous wood in Coal Measure.— highest fish in Old Red Sandstone.— Nautili in [this he left blank]. it is useless to speculate not only about beginning of animal life.: generally, but even about great division, our question is not, how there come to be fishes & quadrupeds, but how many there come to be, many genera of fish &c &c at present day.—'

On page 102 he recorded some ideas on the possible origin of life:

'The intimate relation of Life with laws of Chemical combination, & the universality of the latter render spontaneous generation not improbable.'

And on page 104:

'It is very remarkable as shown by Carus how intermediate plants are between animal life & "*inorganic life*".— Animals only live on matter already organized.—'

However, he was left with some doubts about the concept of descent from a single ancestor, and he made no direct reference anywhere in his notes to precisely where he had found it.

In his first brief summary of his ideas, the Essay of 1842 (see Francis Darwin 1909), although he said initially that **'Be it remembered I have nothing to say about life and mind and *all* forms descending from one common type'**, he did refer for the first time to the "theory of descent" near the end of the Essay, and to "The Theory of Common Descent" at some length in the considerably expanded Essay of 1844. Finally he came clean in their final paragraphs with the words:

'There is a simple grandeur in the view of life with its powers of growth, assimilation and reproduction, being originally breathed into matter under

one or a few forms, and that whilst this our planet has gone circling on according to fixed laws, and land and water, in a cycle of change, have gone on replacing each other, that from so simple an origin, through the process of gradual selection of infinitesimal changes, forms most beautiful and most wonderful have been evolved.'

This passage remained almost unchanged in *On the Origin of Species*. However, except in his notes and some private correspondence with Joseph Hooker, Charles preferred never to speculate on the manner in which life might have been breathed into matter. Thus in a letter written to J.D.Hooker on 29 March 1863 (Burkhardt and Porter 1998) he said:

'Who would have ever thought of the old stupid Athenæum taking to Oken-like transcendental philosophy written in Owenian style! It will be some time before we see "slime, snot or protoplasm" (what an elegant writer) generating a new animal. But I have long regretted that I truckled to public opinion, and used the Pentateuchal term of creation, by which I really meant "appeared" by some wholly unknown process. It is mere rubbish, thinking at present of [the] origin of life; one might as well think of [the] origin of matter.'

On 1 February 1871, again writing to Hooker, he said:

'It will be a curious discovery if Mr Lowne's observation that boiling does not kill certain moulds is proved true; but then how on earth is the absence of all living things in Pasteur's experiment to be accounted for? I am always delighted to see a word in favour of Pangenesis, which some day, I believe, will have a resurrection. Mr Dyer's paper strikes me as a very able Spencerian production.

It is often said that all the conditions for the first production of a living organism are now present, which could ever have been present. But if (and oh what a big if) we could conceive in some warm little pond, with all sorts of ammonia and phosphoric salts, light, heat, electricity, etc., present, that a protein compound was chemically formed, ready to undergo still more complex changes, at the present day such matter would be instantly devoured or absorbed, which would not have been the case before living creatures were formed.'

So his views on the origin of life always remained rather negative.

Erasmus Darwin had suffered from no such inhibitions, and writing about generation in Section 39 of the first volume of *Zoonomia* (Darwin, 1794) he suggested on page 505 that:

'From thus meditating on the great similarity of the structure of the warm-blooded animals, and at the same time of the great changes they undergo both before and after their nativity; and by considering in how minute a portion of time many of the changes of animals above described have been produced; would it be too bold to imagine, that in the great length of time, since the earth began to exist, perhaps millions of ages before the commencement of the history of mankind, would it be too bold to imagine, that all warm-blooded animals have arisen from one living filament, which THE FIRST GREAT CAUSE endued with animality, with the power of acquiring new parts, attended with new propensities, directed by irritations, sensations, volitions, and associations; and thus possessing the faculty of continuing to improve by its own inherent activity, and of delivering down those improvements by generation to its posterity, world without end!

He continued in the same exuberant style to consider the coming into being of cold-blooded animals and of plants, and said on page 507:

'Linnaeus supposes, in the Introduction to his Natural Orders, that very few vegetables were at first created, and that their numbers were increased by their intermarriages. . . . . Many other changes seem to have arisen in them by their perpetual contest for light and air above ground, and for food or moisture beneath the soil. . . . . From these one might be led to imagine, that each plant at first consisted of a single bulb or flower to each root, as the gentianella and daisy; and that in the contest for air and light new buds grew on the old decaying flower stem, shooting down their elongated roots to the ground, and that in the process of ages tall trees were thus formed, and an individual bulb became a swarm of vegetables. Other plants, which in this contest for light and air were too slender to rise by their own strength, learned by degrees to adhere to their neighbours, either by putting forth roots like the ivy, or by tendrils like the vine, or by spiral contortions like the honeysuckle; or by growing on them like the mistleto, and taking nourishment from their barks; or only by lodging or adhering on them, and deriving nourishment from the air. as tillandsia.'

Erasmus's style is very different from that adopted by his grandson 50 years later, whose dismissal of *Zoonomia* as excessively speculative was by no means unfair. Moreover he could be accused of taking a somewhat Lamarckian view of evolution. Nevertheless, Erasmus redeemed himself by continuing:

'Shall we then say that the vegetable living filament was originally different from that of each tribe of animals above described? And that the productive living filament of each of those tribes was different originally from the other? Or, as the earth and ocean were probably peopled with vegetable productions long before the existence of animals; and many families of these animals long before other families of them, shall we conjecture, that one and the same kind of living filaments is and has been the cause of all organic life?'

The Theory of Common Descent could hardly be presented more unequivocally than that. It seems clear that Charles must already have read this passage in 1837, but it is less clear when he would have read Erasmus's later and more eloquent pronouncement about the origin of Life and Common Descent, which was published posthumously in 1803 in his last and undoubtedly best poem entitled *The Temple of Nature.*

Canto I of the poem deals with the *Production of Life*, and after discussing the growth of the earliest civilizations, continues on line 227:

"Ere Time began, from flaming Chaos hurl'd
Rose the bright spheres, which form the circling world;
Earths from each sun with quick explosions burst,
And second planets issued from the first. 230
Then, whilst the sea at their coeval birth,
Surge over surge, involv'd the shoreless earth;
Nurs'd by warm sun-beams in primeval caves
Organic Life began beneath the waves.

"First HEAT from chemic dissolution springs,
And gives to matter its eccentric wings;
With strong REPULSION parts the exploding mass,
Melts into lymph, or kindles into gas.
ATTRACTION next, as earth or air subsides,
The ponderous atoms from the light divides, 240
Approaching parts with quick embrace combines,
Swells into spheres, and lengthens into lines
Last, as fine goads the gluten-threads excite,
Cords grapple cords, and webs with webs unite;
And quick CONTRACTION with ethereal flame
Lights into life the fibre-woven frame.—
Hence without parent by spontaneous birth
Rise the first specks of animated earth;
From Nature's womb the plant or insect swims,
And buds or breathes, with microscopic limbs." 250

V. "ORGANIC LIFE beneath the shoreless waves
Was born and nurs'd in Ocean's pearly caves;
First forms minute, unseen by spheric glass,
Move on the mud, or pierce the watery mass;
These, as successive generations bloom,
New powers acquire, and larger limbs assume; 300
Whence countless groups of vegetation spring,
And breathing realms of fin, and feet, and wing.

"Thus the tall Oak, the giant of the wood,
Which bears Brittania's thunders on the flood;
The Whale, unmeasured monster of the main,
The lordly Lion, monarch of the plain,
The Eagle soaring in the realms of air,
Whose eye undazzled drinks the solar glare,
Imperial man, who rules the bestial crowd,
Of language, reason, and reflection proud, 310
With brow erect who scorns this earthy sod,
And styles himself the image of his God;
Arose from rudiments of form and sense,
An embryon point, or microscopic ens!

"In countless swarms an insect-myriad moves
From sea-fan gardens, and from coral groves;
Leaves the cold caverns of the deep, and creeps

On shelving shores, or climbs on rocky steeps. 330
As in dry air the sea-born stranger roves,
Each muscle quickens, and each sense improves;
Cold gills aquatic form respiring lungs,
And sounds aerial flow from slimy tongues."

Both in his account of the initial origin of life, and of the subsequent steps in the evolution of higher animals, it must be agreed that Erasmus Darwin was painting a picture in language that might almost have been used today. It also seems that in that superb concluding paragraph from the Essays of 1842 and 1844, and from the *Origin*, there are definite echoes of *The Temple of Nature* with its circling world.

Accepting, then, that Erasmus Darwin has a good claim to have originated the Theory of Common Descent at around the end of the 18th century, from whom did he derive his ideas?

One important source he did acknowledge in *The Botanic Garden* and in a footnote to line 381 of Canto IV of *The Temple of Nature*, when he wrote:

'*Which buds or breathes*, l. 381. Organic bodies, besides the carbon, hydrogen, azote, and the oxygen and heat which are combined with them, require to be also immersed in loose heat and loose oxygen to preserve their mutable existence; and hence life only exists on or near the surface of the earth; see Botan. Garden, Vol. 1. Canto IV, l. 419'

For he continued by noting that his great contemporary in France, Antoine Laurent Lavoisier, born in 1743 and guillotined as a tax-gatherer in 1794, had written:

'L'organization, le sentiment, le movement spontané, la vie, n'existent qu'à la surface de la terre, et dans les lieux exposés à la lumiére. Traité de Chimie par M. Lavoisier, Tom. I. p. 202.'

Lavoisier is indeed a pivotal figure in the argument, because it was he who had demolished the phlogiston theory and shown that although matter may alter its state in a series of chemical reactions, it does not change in amount; and that the constituents of water were the gases hydrogen, the water-forming element, and oxygen, the acid-forming element. He had also introduced the nomenclature used today in chemistry, and had for the first time given a precise meaning to the word 'molecule'.

Three of the greatest naturalists of the eighteenth century, Linnaeus (1707-1778), Buffon (1707-1788) and Lamarck (1744-1829) must not be forgotten. The system of classification introduced by Linnaeus with a hierarchy of categories: class, order, genus and species, of course has substantial implications for evolution. But although Linnaeus liberalized his belief in the fixity of species towards the end of his life, he never deserted it. So he was certainly not a believer in the Theory of Common Descent.

Buffon, author of 44 volumes of the *Histoire Naturelle*, whose publication began in 1749, believed that living matter, which he termed *molécules organiques*, although organic chemistry as we understand it today had yet to be discovered, was continuously formed as the result of spontaneous chemical combination. These molecules in turn combined spontaneously to become the first individual of a species, with a *moule intérieure*, an internal mould that guaranteed permanence to its descendants.

Many lower organisms were continuously produced in this fashion, and there were as many kinds of animals and plants as there were viable combinations of organic molecules. Buffon performed experiments which he considered to prove that under the right conditions, invisible atomic and molecular particles formed minute primordial organisms that could be seen under his microscope to be capable of spontaneous motion, but a complex controversy involving his claims arose with Spallanzani and others, moving later into the debate over the interpretation of Brownian motion.

Buffon was not an evolutionist, because he firmly regarded species as fixed, but he may be credited with having formulated the theory of common descent for the first time in 1766, though applying only to a single family of vertebrates such as the horses, and not extensible to a wider grouping. However, in his 44 volumes he contradicted himself a good many times on a good many issues, and did not make a great many contributions of lasting significance.

When in 1793 Lamarck (1744-1829) was appointed as Professor of the inferior animals in France, which he renamed in a less uncomplimentary fashion as the invertebrates, he came up with new and valid reasons for believing in evolution and common descent, but then spoiled his case by endowing all animals with a power to interact with the environment and acquire ever greater complexity or perfection. Such a 'soft' inheritance of acquired characters has certainly to be rejected, and has given Lamarckism a dirty name, though in other respects Lamarck deserves to be remembered as a scientist of considerable distinction.

It may thus be concluded that shares in originating the Theory of Common Descent towards the end of the 17th century should go to Buffon, Lamarck and Erasmus Darwin. But Erasmus surely deserves the prize for expressing the theory in verse!

5. The contribution of the ancient Greeks

Writing to Lyell about Lamarck in 1863 (see Burkhardt & Porter 1998), Charles Darwin said:

'Plato, Buffon, my grandfather, before Lamarck and others, propounded the *obvious* view that if species were not created separately they must have been descended from other species.'

It must therefore be asked what contribution was made by the ancient Greeks in this field.

According to a book entitled *From the Greeks to Darwin* by Henry Fairfield Osborn, published in 1894, the presocratic philosopher Empedocles got the whole story right in the 5th century BC, holding that life began spontaneously and gradually in primordial slime, that plants preceded animals, and that those creatures that had by chance been suitably formed, survived, while those that were unsuitable, perished.

So there on the face of it you have both common descent, and survival of the fittest.

The extant writings of Empedocles are, however, mainly disordered fragments of his works that can be pieced together to suit the fancies of the editor, and in modern studies his theories have been firmly dismissed as preposterous. Moreover they were also rejected by Aristotle, the greatest by far of early naturalists as an observer, who regarded the natural order as eternal and unchanging, and was not remotely an evolutionist.

I turned finally to the Socratic dialogues of Plato, and in particular to his work entitled *Timaeus* (Cornford, 1937). This makes interesting reading, and quickly explains why Plato has been called the great antihero of evolutionism, for everything was created by the gods according to mathematical and geometrical principles very remote from biology. I will not attempt to explain his theories in detail, but will simply quote from the concluding chapter of *Timaeus*, translated by Professor Francis Cornford, husband of Charles Darwin's grand-daughter Frances.

'And now, it would seem, we have fairly accomplished the task laid upon us at the outset: to tell the story of the universe so far as to the generation of man. For the manner in which the other living creatures have come into being, brief mention shall be enough, where there is no need to speak at length; so shall we, in our own judgement, rather preserve due measure in our account of them.

Let this matter, then, be set forth as follows. Of those who were born as men, all that were cowardly and spent their life in wrongdoing were, according to the probable account, transformed at the second birth into women; for this reason it was at that time that the gods constructed the desire of sexual intercourse, fashioning one creature instinct with life in us, and another in women. The two were made by them in this way. From the conduit of our drink, where it receives liquid that has passed through the lungs by the kidneys into the bladder and ejects it with the air that presses upon it, they pierced an opening communicating with the compact marrow which runs from the head down the neck and along the spine and has, indeed, in our earlier discourse been called 'seed'. This marrow, being instinct with life and finding an outlet, implanted in the part where this outlet was a lively appetite for egress and so brought it to completion as an Eros of begetting. Hence it is that in men the privy member is disobedient and self-willed, like a creature that will not listen to reason, and because of frenzied appetite bent upon carrying all before it. In women again, for the same reason, what is called the matrix or womb, a living creature within them with a desire for child-bearing, if it be left long unfruitful beyond the due season, is vexed and aggrieved, and wandering throughout the body and blocking the channels of the breath, by forbidding respiration brings the sufferer to extreme distress and causes all manner of disorders; until at last the Eros of the one and the Desire of the other bring the pair together, pluck as it were the fruit from the tree and sow the ploughland of the womb with living creatures still unformed and too small to be seen, and again differentiating their parts nourish them till they grow large within, and thereafter by bringing them to the light of day accomplish the birth of

the living creature. Such is the origin of women and of all that is female.

Birds were made by transformation: growing feathers instead of hair, they came from harmless but light-witted men, who studied the heavens but imagined in their simplicity that the surest evidence in these matters comes through the eye.

Land animals came from men who had no use for philosophy and paids no heed to the heavens because they had lost the use of the circuits in the head and followed the guidance of those parts of the soul that are in the breast. By reason of these practices they let their fore limbs and heads be drawn down to earth by natural affinity and there supported, and their heads were lengthened out and took any sort of shape into which their circles were crushed together through inactivity. On this account their kind was born with four feet or with many, heaven giving to the more witless the greater number of points of support, that they might be all the more drawn earthwards. The most senseless, whose whole bodies were stretched at length upon the earth, since they had no further need of feet, the gods made footless, crawling over the ground.

The fourth sort, that live in water, came from the most foolish and stupid of all. The gods who remoulded their form thought these unworthy any more to breathe the pure air, because their souls were polluted with every sort of transgression; and in place of breathing the fine and clean air, they thrust them down to inhale the muddy water of the depths. Hence came fishes and shell-fish and all that lives in the water; in penalty for the last extreme of folly they are assigned the last and lowest habitation. These are the principles on which, now as then, all living creatures change one into another, shifting their place with the loss or gain of understanding or of folly.

I am afraid that Charles Darwin had for once not done his homework properly, and had not gone back to sources as all historians should, in including Plato among his predecessors in evolutionary theory.

One of the ironies of the history of science is that Darwin was destined to remain until the end of his life innocent of all knowledge of the actual mechanism of heredity, and he never knew that the truth of the Theory of Common Descent would finally be proved nearly 100 years later by the demonstration that eukaryotes and prokaryotes share a genetic code with their common ancestor.

6. References

Barlow, N. (ed.) (1945) *Charles Darwin and the Voyage of the Beagle.* Pilot Press, London.

Barlow, N. (ed.) (1958) *The Autobiography of Charles Darwin* 1809-1882. Collins, London.

Barlow, N. (ed.) (1963) Darwin's Ornithological Notes. *Bull. Br. Mus. nat. Hist.* (hist. Ser.) 2, 203-278.

Barrett, P.H., Gautrey, P.J., Herbert, S., Kohn, D. and Smith, S. (eds.) (1987) *Charles Darwin's Notebooks, 1836-1844. Geology, Transmutation of Species, Metaphysical Enquiries.* British Museum (Natural History) and Cambridge University Press.

de Beer, Sir Gavin (1959) Darwin's Journal. *Bull. Br. Mus. nat. Hist.*

(hist Ser.) 2, 3-21.
Burkhardt, F. and Porter, D.M. (eds.) (1998) *The Correspondence of Charles Darwin*, Vol. 11 (in the press). Cambridge University Press.
Cornford, F.M. (1937) *Plato's Cosmology. The Timaeus of Plato translated with a running commentary.* Kegan Paul, Trench, Trubner & Co. Ltd., London.
Darwin, C. (1839) *Journal of researches into the geology and natural history of the various countries visited by H.M.S. Beagle, under the command of Captain FitzRoy, R.N. from 1832 to 1836.* Henry Colburn, London.
Darwin, C. (1859) *On the Origin of Species.* John Murray, London.
Darwin, E. (1789/91) *The Botanic Garden: a Poem, in Two Parts.* Part I, *The Economy of Vegetation.* Part II, *The Loves of the Plants.* J.Johnson, London.
Darwin, E. (1794) *Zoönomia; or, The Laws of Organic Life*, Vol. 1. J.Johnson, London.
Darwin, E. (1803) *The Temple of Nature; or The Origin of Society: a Poem, with philosophical notes.* J.Johnson, London.
Darwin, F. (ed.) (1887) Letter of 1 February 1871 from Darwin to Hooker quoted on p. 18 of Vol. III. In *The Life and Letters of Charles Darwin.* John Murray, London.
Darwin, F. (ed.) (1909) *The Foundations of the Origin of Species. Two Essays written in 1842 and 1844 by Charles Darwin.* Cambridge University Press.
Keynes, R.D. (ed.) (1979) *The Beagle Record.* Cambridge University Press.
Keynes, R.D. (ed.) (1988) *Charles Darwin's Beagle Diary.* Cambridge University Press.
Keynes, R.D. (1997) Steps on the Path to the Origin of Species. *J. theor. Biol.* 187, 461-471.
King-Hele, D. (1977) *Doctor of Revolution. The life and genius of Erasmus Darwin.* Faber & Faber, London.
Kohn, D. (1987) (ed.) Notebooks B and C. In *Charles Darwin's Notebooks, 1836-1844,* edited by P.H.Barrett and others. British Museum (Natural History) and Cambridge University Press.
Osborn, H.F. (1894) *From the Greeks to Darwin. An outline of the development of the evolution idea.* Macmillan and Co., New York.

The Cyril Ponnamperuma Lecture

THE SEARCH FOR INTELLIGENT LIFE IN THE UNIVERSE

FRANK DRAKE
SETI Institute
Mountain View, California

It is an honor and a pleasure to be chosen to give this lecture in memory of Cyril Ponnamperuma. Cyril was a good friend and colleague for more than thirty years, during which time we shared in the excitement of many events and discoveries in the growing scientific activity associated with the origins of life, and the possibilities of extraterrestrial life. I have the fondest memories of a long visit to Sri Lanka which I and some colleagues made, with Cyril as our generous host and tour guide. It was an eye-opener, once again revealing Cyril's scientific expertise, and the energy he brought to developing science programs all over the world. He was a unique individual of great talents who will not be duplicated.

Cyril's primary interest was, of course, in the chemistry of the origins of life. He was, nevertheless, a polymath, and was not just interested in discovering extraterrestrial life, but took actions to enhance support and enthusiasm for searches for such life. Whenever I would meet him, he would always first smile, and then ask "Have you found anything yet?". Just as we in the SETI field are not discouraged by a lack of discoveries, despite diligent searching, so he was not discouraged when the answer to his question was always, so far "no". He clearly recognized formidable obstacles to success. So many stars, so many radio frequencies to be searched. He knew that a successful search would take much effort and, perhaps, a very long time. But he knew that, given sufficient effort by talented people, there was a good chance of success. In this, as usual, he was quite right.

Indeed, over those thirty years when Cyril watched over our shoulders, SETI, the Search for Extraterrestrial Intelligence has made enormous strides. Our largest radio telescopes are a hundred times larger than they were. Our radio receivers are as much as ten times more sensitive. Through this pair of developments, we can now detect from the most distant parts of the galaxy signals of the same intensity as the strongest we are able to radiate. To be sure, such signals will be hard to find, since they may well be tightly beamed, and the beam may only rarely point towards us. We may have to look at the "right" star on the "right" frequency a large number of times before we will be observing when the extraterrestrials are sending a signal in our direction.

J. Chela-Flores and F. Raulin (eds.),
Exobiology: Matter, Energy, and Information in the Origin and Evolution of Life in the Universe, 53–57.

The greatest development in those thirty years was unanticipated. This was the development of exceptionally powerful multi-channel radio receivers whose cost was affordable. This development is an outcome of the fantastic progress in computer technology. Because of this progress, the cost per channel of multi-channel radio receivers has dropped by a factor of more than 100,000. This advance is exploited in modern systems such as those of Project Phoenix of the SETI Institute, Project Serendip of the University of California, Berkeley, and Project Beta of Harvard University. In each case multi-channel radio receivers with approximately 100 million channels have been developed. Not only do they operate very effectively, but they are reliable and easy to use. Bandwidths of as small as a fraction of a hertz are easily achieved.

Given these many channels of information being received at any given time, the flood of data from the systems becomes challenging in two major ways: 1) One must search through the flood of data for signals which might of be extraterrestrial intelligent origin; and 2) One must quickly and easily reject the many received signals which are from our own civilization. The solution to both of these challenges again depends on the exploitation of computer technology. The first challenge has been dealt with most thoroughly in Project Phoenix, where a combination of special computer hardware and software has been used to create a signal detection instrument which can, in a sufficiently short time so as not to delay observations, detect continuous wave and pulsed signals of constant frequency, or those drifting in frequency. Any signals so detected, and there are always many in any given observation, are then presented to the system for further automatic analysis.

The first step in this further automatic analysis is to deal with the second challenge, the identification and elimination of human-generated signals, or "radio frequency interference" or "RFI". Again, methods to do this are most highly developed in Project Phoenix. In this project, signals already observed and identified as RFI are recorded in an RFI library. Newly found signals are compared with the contents of this library. When there is a match of a new signal with an entry in the library, the signal is rejected.

The next step in this process, in Project Phoenix, is a costly but very effective one. Here the approach is to utilize a second telescope located at a substantial distance, say several hundred kilometers, from the main search telescope. At the second telescope there is a much smaller multi-channel receiver with far fewer channels, perhaps a few thousand. Now, when a promising signal is detected at the main telescope, and is not immediately rejected as RFI, the information about the radio frequency of this signal is sent to the second telescope. The receiver of the second telescope is tuned to signal frequency and a search is made for the signal. If no signal is seen, this is clear evidence that the signal is RFI and originated in the vicinity of the main telescope. If the signal is seen, there is now strong evidence that the signal source is far above the Earth, possibly from an airplane, a spacecraft, or - the civilization of another star.

How does one then determine which of these possibilities is the correct one? By making use of the Doppler effect. Each of the two telescopes is moving at a speed of the order of 1000 km/hour due to the rotation of the earth. Of crucial importance, each telescope will have a specific and different velocity component in the direction of the radio source. These velocity components will be changing in time in an exactly predictable way due to the rotation of the earth. The computer computes the "proper" difference in radio frequency to be expected for the two telescopes, due to their different locations, and also computes the expected change in radio frequency at the two telescopes due to the changing direction of the velocity vector. These predicted frequency differences and drift rates are compared with what is actually observed at the telescopes. If there is not a match, the signal is RFI. If there is a match, then the signal is a very good candidate to be a true extraterrestrial signal. A further test is to move the telescopes so that they are no longer looking at the target star. If the signal persists, it is being captured by the "sidelobes" of the telescope and is RFI. If the signal disappears when the telescope is not looking at the star of interest, the signal could well be coming from that star. The observations are repeated then several times to increase the probability that the signal is of interstellar intelligent origin. So far, in Project Phoenix, no signal has passed these tests.

Other approaches are used in other projects. In Project Serendip, where the project has no control of the telescope pointing, signal detection depends on the occurrence, by chance, of repeated observations of a location in the sky where the telescope was pointed when a candidate signal is detected. This may take many months. In Project Beta two feed horns are used on the telescope, aligned east-west. The telescope does not move, so it is, in fact, scanning the sky at the speed of rotation of the Earth. With this geometry, a signal should appear in one feed horn and then the other with a precisely predictable time delay. In this way promising candidate signals can be identified, and the location in the sky from which they came can be reobserved later.

At the present time the portion of the radio frequency spectrum which is searched is limited to the range 1 to 3 Gigahertz, with most of this coverage being provided by Project Phoenix. There are some preliminary attempts being made to do SETI searches at optical wavelengths, but so far no comprehensive searching has been done at wavelengths other than microwave wavelengths.

The improvement in SETI instrumentation over the years has been truly striking. The overall search capability, taking into account improved sensitivity, increased number of channels, increased variety of detectable signal types, but not counting the advantages of the automatic signal identification procedures mentioned above, has improved by a factor of some 100 million million times over the last 35 years. This improvement has grown exponentially with a doubling time of about 280 days. This is considerably faster growth than the similar exponential grown in computer capability, expressed by "Moore's Law", where the doubling time is 1 1/2 years. Perhaps just as remarkable is that most of this growth has occurred without financial support from governments. Although NASA sponsored a major SETI project in the United States, the High Resolution Microwave Survey, and this project eventually evolved into Project Phoenix, the primary source of financial support for these projects, world-wide, has been gifts from private donors and foundations. Project Phoenix has benefited particularly through the generous support of a number of very successful leaders of the computer and electronic industries in the United States.

It is hoped that governmental support for SETI projects may resume soon. Surely the interest in this field, and the scientific basis for it has grown enormously in recent years. Indeed, in NASA the prime project these days is "Origins", in which the origin and distribution of life, including intelligent life, are the key subjects. Bolstering the excitement of these fields has been the discovery now of a number of other planetary systems, as well as clear evidence of planetary systems in formation. Further excitement has been generated by the beautiful pictures from the Galileo spacecraft showing compelling evidence that there is an ocean beneath the ice of Europa. The discovery of a number of kinds of evidence for ancient life on Mars, in the meteorite ALH 84001, although inconclusive, has sparked enthusiasm for the idea that there is a great deal of life to be found in space. We have been reminded, as never before, that space is an awesome place in which striking phenomena, seemingly incredible, are legion. Foremost of these is the life of Earth, and now, it is believed, extraterrestrial life.

SETI scientists are already planning the next steps in producing ever more powerful instrumentation. Already they are talking about receiving systems with a thousand million channels. As the cost of computers has plummeted, such systems are easily within our grasp. The only decisions to be made concern the most cost-effective way to achieve such systems. There is no doubt that such systems are affordable, even when support is only in the form of gifts.

Of fast growing interest is the possible construction of a large radio telescope dedicated to SETI. At the present time, SETI observations are carried out either on small telescopes, or through guest use of large telescopes, always utilizing only a small percentage of the available telescope time, or not having control of where the telescope is pointed. This causes the programs to proceed very slowly, and greatly reduces their cost-effectiveness. It is also inefficient because equipment may often have to be moved, reinstalled, recalibrated and adjusted, meaning in effect that the equipment is rarely used while in optimum working condition. Personal hardships for staff members are created by this mode of operation.

It is hoped that the resources will be found to allow the construction of a telescope or telescopes dedicated to SETI. Indeed, SETI scientists are cooperating with those who are planning the "One Square Kilometer Telescope" for general radio astronomy use. It is hoped that a version of this, which could be smaller, more limited in capability, and thus less expensive, might be built. A favored approach at the present time is to use an array of a very large number of small steerable parabolas, similar to those used to receive commercial direct-broadcasts from satellites. Using several hundred up to several thousand such antennas in an array would provide a receiving capability equal to that of the largest existing radio telescopes. Knowing how resourceful SETI scientists have been in the past, such an instrument could well come into being in the near future.

I wish that Cyril were still with us to witness the relentless march of progress which has characterized this field in recent years. He would be thrilled. He would be delighted to see that, if anything, the rate of progress is actually increasing. He was always an optimist, and the improvements in SETI activities would have been just what he expected. Especially, I wish he were still with us so that some day I could answer his standard question, "Have you heard anything yet?", with a smile and a "yes!".

Public Lecture

ARE WE ALONE?

PAUL DAVIES
Centre for Basic Research
P.O. Box 389
Burnside
South Australia 5066

1. Biological determinism

Is the universe teeming with life, or are we alone? This is surely among the biggest of the big questions of existence. Since there is good reason to suppose that there are many earthlike planets in the universe, the answer to the question hinges on one of the great imponderables of modern science: the likelihood that life will form under earthlike conditions.

The idea that, given half a chance, "life will out" is widespread in both the scientific community and the media. When the Galileo spacecraft beamed back its first pictures of Europa in April 1997, the word on everybody's lips was "life!". The excitement focused on the discovery of ice covering an ocean of liquid water. The assumption made by most commentators was that water implied life, or at least a good possibility of life. The reasoning seemed to be that as life is found on Earth wherever liquid water exists, therefore if there is liquid water on Europa there should be life there too. It doesn't require an Einstein to spot the logical fallacy in that reasoning. Water may be a necessary ingredient for life to form, but it is far from sufficient.

The belief that earthlike conditions are all that is necessary for life to emerge, given enough time, is known as biological determinism. This viewpoint proceeds from the assumption that life is somehow written into the laws of nature, and is therefore inevitable given the right conditions. Christian de Duve is a well-known biological determinist, who describes life as a "cosmic imperative". In his book *Vital Dust*, he writes: "Life is almost bound to arise, in a form not very different from its form on Earth, wherever physical conditions are similar...This conclusion seems to me inescapable [1]."

Since even the simplest known life form is incredibly complex, and could not conceivably spring into existence fully-formed by chance alone, biological determinism supposes that there is a prebiotic stage where chemistry favours the formation of those molecules that are biologically relevant. In other words, the laws of physics and chemistry are slanted in favour of life. Biochemists often cite as evidence for such bio-friendly laws laboratory experiments of the sort pioneered in the 1950s by Stanley Miller

J. Chela-Flores and F. Raulin (eds.),
Exobiology: Matter, Energy, and Information in the Origin and Evolution of Life in the Universe, 61-68.

and Harold Urey. These scientists sought to recreate the conditions thought to have prevailed on the Earth several billion years ago, to test the hypothesis that life emerged from a type of primordial soup of chemicals. Miller and Urey added methane, ammonia and hydrogen to water in a flask, and sparked electricity through the mixture for several days. The result was the production of amino acids, the building blocks of proteins. The experiment was hailed as the first step towards the production of life "in a test tube".

In spite of this early success, I shall argue that it is a serious mistake to regard Miller-Urey type experiments as the beginning of a long road along which chemical mixtures are inexorably conveyed towards the destination "life". It may well be the case that life does form readily given the right circumstances, but the known principles of physics and chemistry do not support this conclusion. The basic laws of physics are simple, mathematical and general. Life is complex, non-mathematical and specific. Clearly the laws of physics and chemistry alone will not explain life. That is not to say that life violates the laws of physics and chemistry, only that one requires more than these laws to explain biogenesis. Therefore, if life is widespread in the universe, it suggests that there are organising principles at work in addition to basic physical laws that can fast-track matter along the road to life. The philosophical consequences of this are profound.

2. Ancient ideas of extraterrestrial life

The hypothesis that we are not alone in the universe dates back to ancient Greece, and the deliberations of the Atomist philosophers in particular. Atomism held that the universe is nothing but identical sorts of particles moving in a void. It then followed that if those particles combined to make living organisms on Earth then they might equally do so on other worlds too.

This hypothesis rested on the assumption of three principles. The first was the principle of uniformity - that the laws of nature are the same everywhere. Few scientists would deny this principle today. The second was the principle of mediocrity - that Earth is not special or privileged. Astronomical observations do indeed suggest that Earth is a typical planet around a typical star in a typical galaxy. Might Earth's biology be typical too? Unfortunately, life is the one exception to the rule that what we observe on Earth is unexceptional, for the simple reason that we, the observers, are a product of that life. If there was only one inhabited planet in the universe we would have to be living on it! So we cannot use the existence of life on Earth to argue for the existence of life on other earthlike planets.

The third assumption is known as the principle of plenitude - that nature tends to realise all available opportunities open to it. If true, this principle implies that life will tend to form wherever and whenever conditions permit. However, the principle of plenitude is not a law of nature, and few scientists would regard it as a reliable guide. Moreover, living organisms represent but one very special sort of complex system. If the principle of plenitude were generally true, then life would be swamped by the vastly greater number of alternative, non-living, complex arrangements of matter.

Traditionally there have been three hypotheses for the origin of life. The first is that it was a miracle. Quite properly scientists reject this. The second is that it was a chemical fluke, a stupendously improbable molecular accident, so unlikely that it will not have happened twice in the observable universe. Many biologists, steeped as they are in the tradition of Darwinism, subscribe to this view. Thus Jacques Monod famously wrote, "man at last knows that he is alone in the unfeeling immensity of the universe, out of which he has emerged only by chance [2]." It is the essence of Darwinism that evolution is blind, and that nature has no foresight. No species is preordained, it is merely the result of random accidents of fate. It is therefore offensive to the spirit, though not the letter of Darwinism to suppose that the formation of life is preordained, or predetermined, by the laws of nature.

The third hypothesis is that life is the natural and inevitable product of the outworking of bio-friendly laws. This is the position adopted by, for example, de Duve, and the great majority of scientists working on biogenesis. They point out that, since Darwinism applies only after life (of some sort) gets going, it cannot be used to draw conclusions about the emergence of the first living thing. Just because the course of biological evolution is blind and directionless does not mean that prebiotic chemical evolution must also be blind and directionless. Maybe chemistry channels matter towards complexity with unexpected efficiency, coaxing life into existence against the raw odds.

It is possible to compute the probability against life forming as a result of the accidental shuffling of molecules. For example, the chances of making a short protein by chance, given a supply of appropriate amino acids as building blocks, is about 10^{130} to one. If this was the way that proteins initially came to exist, it is clear that they will have synthesised themselves only once, and that we will be alone in the universe. People often say that the universe is so vast it is unreasonable to believe we are alone, in spite of the unlikelihood of life. This argument is nonsense. The number of earthlike planets within the observable universe is probably no larger than 10^{20}. This number is trivially small compared to the odds against making even a single protein by random molecular combination, let alone all the other complex molecules needed for life. Most scientists conclude from this arithmetic that chance alone isn't the correct explanation for biogenesis. The enormous odds against life forming by random molecular shuffling may be taken as a measure of one's justification to disbelieve the chance hypothesis.

3. Bricks alone don't make a house

A frequently-used argument against the chance hypothesis is that the building blocks of life - organic molecules such as amino acids - form readily under a wide range of circumstances. Comets, asteroids and icy planetesimals are rich in organics, and such objects are likely to be abundant throughout the universe. Moreover, the fossil record suggests that life on Earth got going almost as soon as conditions permitted. Until about 3.8 billion years ago, bombardment by comets and asteroids would have effectively sterilised the surface of the planet, yet there is good evidence that life was flourishing on the surface 3.5, and perhaps as early as 3.8, billion years ago.

Unfortunately this argument is shaky. First, we cannot conclude very much from a sample of only one. More seriously, the fact that amino acids are easy to make is no guarantee that life is easy to make. It is a long and difficult road from amino acids to life. An investigation of the thermodynamics of protein synthesis, for example, reveals the essential problem. Synthesising amino acids from simple organic materials is thermodynamically favoured, that is, it is a "downhill" process in which energy is released - rather like the formation of crystals from a solute. It is then no surprise that amino acids are found to be so widespread and easy to make. By contrast, linking amino acids together in long chains to form peptides, the stuff of proteins, is thermodynamically "uphill". This means it will not happen spontaneously any more than water will run uphill spontaneously. To make peptide bonds means driving against the natural thermodynamic gradient, which requires an input of energy.

There was no lack of energy available on the early Earth to achieve molecular synthesis, but the problem cannot be solved just by throwing energy at it. The energy has to be coupled into the right degrees of freedom to make the right molecules. The situation can be compared to finding a pile of bricks and concluding that a house must be round the next corner. To turn a pile of bricks into a house is also an uphill process; they bricks must be literally raised up. This needs energy. You could supply enough energy by exploding a stick of dynamite under the bricks, but the chances of making a house that way are remote. The energy and materials need to be organised in a very specific way to turn bricks into a house. In the same way, turning amino acids into proteins demands a high degree of organisation.

Sidney Fox and his colleagues reported at this conference that heat energy applied to amino acids can achieve the synthesis of some proteinoid molecules [3]. The key question is, are these molecules biologically relevant? Fox claimed they are, and went on to say that the experiments are reproducible and that the proteinoids are not random associations of amino acids, but biased towards those particular combinations that are needed by life. If that is true, then it suggests a type of biological determinism of the most dramatic sort. It suggests that dynamite can build houses. Let us look further at what this claim entails.

4. Life as an information processing system

The problem of biogenesis is often presented as a problem of the origin of complexity. However, nature is full of complexity, from the pattern of raindrops on the ground to the rings of Saturn. What makes life special is not complexity *per se*, but the fact that its complexity is organised or specified in a very particular way.

The logical architecture of life is the same as that of a digital computer. Genetic information is stored in bits as a sequence of nucleotides on DNA or RNA, and this information acts as the input data for protein synthesis in little factories called ribosomes. There is thus an information flow from nucleic acids to proteins, and from one generation to the next via replication of nucleic acid. Life as we know it turns on the deal struck between nucleic acids and proteins - two very different classes of molecules. There is a well-known chicken-and-egg puzzle concerning proteins and nucleic acids. Each needs the other to function correctly. Here I want to present what I think is a more basic difficulty of biogenesis, a problem of a conceptual nature. Indeed, I believe it is *the* fundamental problem of biogenesis. Cooperation between these "chalk-and-cheese" molecules, nucleic acids and proteins, is possible only because of a sophisticated code, the so-called genetic code, that turns nucleotide sequence information into amino acid sequence information, so that the right proteins get made according to the DNA or RNA instructions. The chalk thus talks to the cheese indirectly, via a coded information channel.

There is a nice analogy to this arrangement. Ask: why is a radio-controlled airplane so much easier to fly than a kite? A kite is (quite literally) hard-wired to the controller (the person on the ground). By contrast, a radio-controlled plane uses a software or informational channel to deliver commands. The radio waves do not themselves push and pull the plane about as do the kite wires; they merely encode the information that harnesses other forces to do the necessary work. Information channels, or digital control, can couple chalk and cheese, and free the cheese from the forces and restrictions that apply to the chalk. Life is like the radio-controlled plane. The problem is that no amount of sophisticated evolution of a kite will turn it into a radio-controlled airplane, because *radio control is a completely different concept* from kite wires. To understand how an information channel developed in living organisms means discovering how software emerged spontaneously from hardware.

The problem is actually much tougher than I have stated it because biological information is not just any old information. It is information of a very special sort. To see what sort, consider the sequence of nucleotides in a typical DNA molecule. It can be written as a string of letters A,C,G and T, these being abbreviations for the four different nucleotides that life uses. The question then arises of whether a given string of such letters has any pattern to it, or whether it is random. There is a branch of mathematics called algorithmic information theory that enables precise definitions to be given for randomness, order and organisation, concepts that are often used very sloppily in discussions of living organisms [4]. It is ideally suited to the analysis of genomes, because a genome is in fact an algorithm for the construction of an organism.

According to algorithmic information theory, a random sequence is one that has no pattern, and so cannot be abbreviated, or compressed, into a simpler formula. That is, there is no compact algorithm that will reproduce, or generate, the sequence. The information content of a nonrandom sequence, such as a periodic sequence, is rather low, because it can be encoded in a rather short algorithm, whereas random sequences are information-rich. Biochemists believe that genome sequences are mostly random (discounting gross exceptions, such as where whole genes are repeated or transposed). This is as you might expect given that the role of the genome is to store information; the enormous complexity and variability of life suggests that genomes should be information-rich. However, you can't tell just by looking that a sequence is random. Some apparently random sequences of digits, such as the decimal expansion of pi, turn out not to be random at all. (Pi can be generated by a very compact algorithm.) Similarly, geometrical forms known as fractals (e.g. the Mandelbrot set) appear to have immense complexity and information content, but in fact they can be generated by very simple algorithms.

Could it be that life is really simplicity masquerading as complexity - like a fractal? Is there a hidden code (I am not referring to the genetic code) or pattern lurking in your genome that means all your genetic information could be expressed by a simple mathematical formula? The idea seems preposterous, yet it is essentially what is entailed in the claim of biological determinism. (To be fair, that claim concerns the first living thing, not you or I.) According to biological determinism, life's organised complexity emerges in a lawlike manner from an incoherent soup of chemicals. Thus the laws of nature, which are nothing but algorithms for converting input states of matter into output states, are somehow supposed to generate biological information. Yet this information cannot come from the laws themselves, so long as we are dealing with the known laws of physics, because these laws have very low algorithmic information content.

Accepting that the genome is random and information-rich, then appealing to non-random chemistry to make life is a clear contradiction. Non-randomness is the exact *opposite* of what is needed to produce a random macromolecule. The whole point of the genetic code, for example, is to *free* life from the strictures of non-random chemical bonding. A genome can choose which amino acid sequence it wants, independently of the chemical preferences of molecules. It achieves this by deploying special enzymes designed precisely to override the non-random biases of chemistry. That is why life goes to all the trouble of using coded information and software-mediated assembly. Life works not by submitting to the directionality of chemistry, but by *circumventing* what is chemically and thermodynamically "natural", and, through the exercise of software control, harnessing energy when needed to build what it requires.

Once this essential point is grasped, the real problem of the origin of life is clear. Since the heady successes of molecular biology, most investigators have sought the secret of life in the physics and chemistry of molecules. But they will look in vain for conventional physics and chemistry to explain life. The secret of life lies in the logical and informational rules it exploits, not in its chemical basis. Life succeeds precisely by *escaping* from chemical imperatives.

As I have stated, biological organisms comply with the laws of physics and chemistry, but these laws are only incidental to their operation. Their main role is to permit an appropriate logical and informational system to come into being. The origin of life lies in the transition from a state where molecules slavishly follow mundane chemical pathways, to one in which they organise themselves to create their own pathways. The chalk-and-cheese mixing ability of software control, as exemplified in the use of a genetic code, is the clearest example of this transcendence.

5. The origin of biological information

I have stressed that the laws of physics as we currently know them are algorithmically simple so cannot of themselves generate biological complexity, i.e. a large amount of biological information. So long as we restrict our considerations to familiar physical laws, therefore, biological information can come only from the environment. But how?

You might suppose the answer is obvious. The world is full of random and chaotic processes. Could they not generate a random object like a genome? Unfortunately this won't do. The genome, although random, is not arbitrary. The sequence of nucleotides along the genome spell out the message of how to make proteins. If these "letters" A,G,C,T of the genome are scrambled, the chances are very high that the message will be biological nonsense. Only an exceedingly small and special subset of all random sequences code for functional proteins. Thus the genome has two properties that seem almost contradictory: it is both *random* and *specific* (i.e. narrowly specified).

To make this point clear, let me use another analogy. Imagine tipping a jar of coffee beans onto the floor. The resulting pattern of beans will be more or less random: randomness is not hard to achieve. But suppose we wanted to attain a *specific* random pattern from among the countless irrelevant random patterns available? That would seem to be almost impossible. We can appeal to chance to give us randomness and we can appeal to law (or necessity as Monod called it) to give us a specified end result. But what peculiar combination of chance and law gives *specified randomness*?

There *is* a known combination of chance and law that generates random, specified molecular sequences: it is Darwinian evolution. Chance in the form of random mutations and law in the guise of natural selection certainly elaborates existing genomes. But Darwinian evolution is a long and arduous process, in which biological information is painstakingly accumulated after an enormous amount of information processing. The problem is to explain whether the prebiotic soup swiftly and effortlessly produced the first genome, as biological determinism attests.

One possibility is "Darwinism all the way down". If Darwinian principles of variation and selection can be applied to the prebiotic soup too, then the construction of an otherwise "impossible object" - a genome that is both random and specific - might be explained. No doubt this is part of the story. Since the random synthesis of RNA or DNA from scratch is exceedingly improbable, most investigators assume that some simpler precursory replicator molecules must have started the thing going. However, few biochemists believe that molecular evolution by Darwinian means is the whole story. Appeal is also made to chemical self-organisation. According to this theory, complex informational molecules arise spontaneously through nonlinear, far-from-equilibrium chemical feedback loops. Unfortunately, self-organisation suffers from the same problem as any known lawlike process: it can only shift information around, it cannot generate it.

My conclusion is that life will *not* follow from soup with lawlike inevitability, unless we broaden the notion of law. Can we imagine a law of nature that generates information? Most scientists regard the spontaneous appearance of information as absurd, although they accept the spontaneous appearance of matter and even spacetime (in the big bang). It is possible that the research into complexity theory and artificial life will discover organising principles that in effect create information from nothing, or at least harvest it from the environment with unexpected efficiency. It is also possible that quantum mechanics holds the key. Molecular mechanics is, after all, quantum mechanics. It has been proved that quantum computation can process information far quicker than classical computation [5]. As the problem of biogenesis is essentially an informational problem, there is a hint of a link here.

These ideas are pure conjecture of course. I develop them further in my forthcoming book [6]. Here I merely wish to say that I do believe we live in a bio-friendly universe in which life is widespread. But I do not think the known laws of physics and chemistry *alone* are as life-encouraging as some scientists make out. The fact that we may need to go beyond these known laws to explain life gives added urgency to the search for life elsewhere in the universe. I am a passionate supporter both of the space programme, which I hope will find independently-arising life on Mars, Europa or in comets, and of SETI. These searches are important because they test not only our theories of life, but also the very scientific principles on which these theories are based. If life is found elsewhere, and panspermia contamination can be positively ruled out, then the philosophical ramifications will be immense [6]. It would in effect support Freeman Dyson's vision of a bio-friendly universe, so poetically captured in his inspiring words, "I do not feel like an alien in this universe. The more I examine the universe and study the details of its architecture, the more evidence I find that the universe in some sense must have known we were coming [7]."

References

1. De Duve, C.: *Vital Dust*, Basic Books, New York (1995), 292.
2. Monod, J.: *Chance and Necessity*, trans. A. Wainhouse, Collins, London (1972) , 167.
3. See the contribution by S. Fox in this volume.
4. See, for example, Kuppers, B.: *Information and the Origin of Life*, MIT Press, Cambridge, Mass., 1990; Yockey, H.: *Information Theory and Molecular Biology*, Cambridge University Press, Cambridge, 1992.
5. Milburn, G.: *Feynman's Processor*, Allen & Unwin, Sydney, 1998.
6. Davies, P.: *The Fifth Miracle*, Simon & Schuster, New York, 1998.
7. Dyson, F.: *Disturbing the Universe*, Harper & Row, New York (1979), 250.

Section 2: Matter in the Origin and Evolution of Life in the Universe

THE CONCEPT OF MATTER AND MATERIALISM IN THE ORIGIN AND EVOLUTION OF LIFE

GEORGE V. COYNE, S.J.
Vatican Observatory
V-00120 Vatican City

1. The Beginnings in Ionian Greece

When the Greeks of fifth century Ionia sought to understand the world and themselves in it they had at their disposition neither the natural sciences nor philosophy, as we know these disciplines. The virtue of this is that they were free to speculate as they thought best and thus they gave birth, more slowly in the case of the natural sciences, to these disciplines. They started with three quite obvious observations: things are of many kinds and most of them can be classified, things have a beginning and many of them an end, things change. They sought to get a comprehensive explanation of these three facts. Comprehensive is the important word. How to explain everything? What they did is to assert that there is a "stuff", matter, that is shared by everything they experienced. Matter is what endured through change; matter is what was universally shared but then became specified for different types of things and different individuals of the same types. Did matter have a beginning? Did it have an end? How could this matter come to be so many different things? Were the human being, dust and water of the same matter? This notion of matter was very empirical and sensory; it was "stuff". As time went on it created more questions than answers. But the questions were probing ones.

Aristotle abstracts from this empirical "stuff" and then talks of matter in an ambiguous way which creates great problems for the future. He first talks of matter as the subject of change(s): the leaf which is green today and brown tomorrow; the man who has black hair today and white ten years from now. The leaf and the man are the "matter" of change. But then he also talks of matter when there is a change from one kind of being to another: a seed to a tree, an egg to a chicken. We can speak no more now of a change but rather of a coming to be. Is their a continuity in this coming to be? Is there a "stuff" which endures and, if so, can we detect it? Surely the chicken does not differ from the egg simply by having more "stuff". It is not the quantity of matter which makes a new being. What does? For Aristotle there appears to be a dynamic content to matter. New types of beings seem to come from it.

J. Chela-Flores and F. Raulin (eds.),
Exobiology: Matter, Energy, and Information in the Origin and Evolution of Life in the Universe, 71–80.

Neither the empirical "stuff" of the ancient Greeks nor the abstract "matter" of Aristotle, even dynamic matter, can stand up reasonably to this barrage of questions. The notion of matter is in difficulty. And the difficulties increase. The rarefaction and condensation of matter become important because they explain the differences in the four fundamental types of matter: earth, air, fire and water. But how does one explain the density of matter? The atomists, of course, have the easiest approach. Density is a measure of the space between atoms. But then the difficulties mount: what is space? is there a void? We are nonetheless at a crucial point; we are beginning to speak of the amount of matter in a given space or of the "quantity of matter". We will soon see that this notion, at least in the physics of motion, will put an end to the notion of matter altogether and give rise to that of mass. And then the notion of mass will become closely allied to that of space.

In addition to the three observations of the ancient Greeks mentioned at the beginning there is the further experience to explain: matter may move. It is in the analysis of motion that further problems about matter come to light. The ancient Greeks discussed falling bodies and suggested that the speed of the fall depended upon the weight of a body and air resistance. They also noted that it is more difficult to move a heavy body and to stop its motion than to do so for a lighter body. But what did it mean for matter to have weight, to be heavy or light? It is interesting that in Aristotle's analysis of both natural and forced motions he makes no use of his concept of matter. It appears that once Aristotle had abstracted from the empirical "stuff" of the Ionians, he would not return to it and consequently would never approach the notion of a quantity of matter. For Aristotle matter became a principle of being.

This principle is referred to as "primary matter" and it is that which underlies the changes from one type of being to another: the seed to the tree, the egg to the chicken. Further analysis of this principle shows that it is completely indeterminate, cannot be named or described but has the potentiality to be "informed", i.e., to be made definite, intelligible, actual, predictable, spiritual, etc. It is no exaggeration, I think, to say that it is in this Aristotelian notion of "primary matter" that we have the origin of the sharp, and sometimes misleading, distinctions that are to arise in Western thought between material, spiritual, living, non-living, etc. It is a notion that invites a challenge. If "primary matter" is indeterminate, pure potentiality and cannot be named or described, how does it explain anything? Does it exist? What is meant by calling it a "principle"? It was the defining of such principles which became the main task of Greek physics. At a minimum we can assert that this made Greek physics into philosophy and that the birth of modern science in the 17th century was to initiate a completely new, and even contrasting, way of doing physics, separating it altogether from philosophy.

2. The Birth of Modern Science. Newton and His Contemporaries

Having established the movement in the notion of matter from the "stuff" of the Ionians to the philosophical principle of the Aristotelians, allow me to leap over the intervening centuries, which were dominated by Aristotelianism, to the 17th century. For present purposes I will concentrate on the thought of Isaac Newton, while contrasting him with other scientists of his time. Ernan McMullin introduces his study of *Newton on Matter and Activity*[1] by noting that, if one considers only his major work, the *Principia*, one could assert that Newton virtually eliminated the concept of matter from physics by replacing it with the notion, "quantity of matter", which will soon become "mass". However, notes McMullin, by examining Newton's other writings and correspondence one sees that he struggled for sixty years with the notion of "matter".

Newton was primarily interested in the notions of force, of body and of the existence or not of the void. But basic to all of these was the concept of matter and he was always drawn to the neo-Platonic emphasis on matter as inert and totally passive. Newton's preoccupation here was a theological one. He would not risk having an autonomous world which did not depend on God. But if there was no seat of action in matter, where was it? How and where did change originate? Although in the *Principia* Newton claims that he is seeking only for a mathematical explanation of motion, he is actually searching, if one judges by his other writings, for the real, physical source of changes in motion. His approach is a very inductive one. Matter for him is not at all the co-principle of substance as in Aristotle. Typical of his approach is his claim that

> The laws and properties of all bodies on which experiments can be made, are the laws and properties of bodies universally.[2]

In Rule III of the *Principia* Newton states that "to be material is to be extended, solid, mobile, to possess inertia, to attract and to be attracted by all other bodies". For the first time attraction is added to the classical list of the qualities associated with materiality and it is, of course, in analyzing attraction that Newton begins a totally new discourse on the nature of matter.

Contrary to Descartes, Newton states that while all matter is extended not all extension is matter. He is here undoubtedly alluding to the nature of space and of the void and this will lead him to a lengthy treatment of action at a distance. Contrary to his own tendencies Newton appears to make matter active since all material bodies attract and are attracted. He seeks to avoid this by seizing upon force as the principle of activity. Here again he is in contrast to Descartes who immediately went to God as the source of activity. Furthermore, Descartes delayed the development of the concept of mass by his exclusively geometrical notion of matter. For him matter and extension were equivalent and volume substitutes for mass.[3]

Because of his desire to maintain matter as inactive there is a tension in Newton's thinking on the relationship of matter and force. He had the insight to realize that force was only required to explain a change of motion, including going from rest to motion. But for him matter offered a resistance to motion and resistance to him was activity. This leads him to the concept of inertial force proportional to the "quantity of matter". But where is this inertial force? Leibniz with his monads clearly puts this force in matter. Newton cannot do that. His hesitation to make matter active and his struggle to deal with the notion of action at a distance are, in my opinion, correct instincts which will eventually lead to the development in physics of force fields. It is of some interest to note that in the cases of both Leibniz, who made matter active, and of Newton, who refused to do so, the motivation was theological. Leibniz thought it would be demeaning of God to require that he always be acting in matter and Newton felt that God would easily be excluded altogether if he were not the immediate source of activity in matter.

In light of these diverse views as to activity in matter we might ask: Was gravity essential to matter? For Newton the answer was clearly negative because he could not accept the notion of essential. To him essence was an ontological principle of being inherited from the Aristotelians and his was an inductive and empirical notion of matter. For him gravity was universal, but it was not essential. For Newton matter continued to be a substratum for all empirical properties, including attraction.

As strange as it may seem, it was from his interest in alchemy that Newton became convinced of the intrinsic dynamism of natural processes.[4] Since he could not find that dynamism in bare matter and motion, he was first led to consider the arcane forces of the alchemists whereby all changes in nature were explained by animating principles disguised under code names. He went on, however, to pursue chemical experiments, becoming more and more reliant upon them than upon his vast knowledge of alchemy, and he became convinced that nature is permeated with intrinsic activity of many kinds. This passage of Newton from what we call today physics to biology and chemistry is an important one in the historical development of the notion of matter and we shall return to discuss it.

We now summarize the most important elements of Newton's physics with respect to matter. He rejected the Aristotelian notion of prime matter as an ontological principle of all created being in favor of a return to the Ionian notion of matter as a substratum for the empirical qualities which he studied by induction. He struggled to eliminate the notion of matter being active but, in so doing, he could not ultimately explain his notion of the "force of inertia" and the attractive force of all matter. In proposing the notion of "quantity of matter" he leads the way to the concept of mass. From Newton on matter is eliminated from the discourse of physics and is replaced by mass. As has been noted by others[5] it is paradoxical that the rise of materialism as a philosophy in the 17th and 18th centuries is attributed to birth of modern science, when in reality matter as a workable concept had been eliminated from scientific discourse. Matter, in the new physics, is not measurable.

3. Science on the Move after Newton

It has been my purpose to use Newton as an example of the struggles with the notion of matter in physics at the time of the birth of modern science. To complete this picture somewhat, let us now examine a bit of the aftermath of Newton. In his *Theory of Natural Philosophy* (1763) Roger Boscovich introduces the notion of point centers of force, instead of extended solid corpuscles. These point centers have inertia and a single force acting at them is supposed to explain all of the qualities of matter. There is thus a tendency in Boscovich to the reification of force, to making it the "stuff" of the Ionians and the substratum of Newton. This tendency is furthered, although not intentionally, by Euler in his mathematics on the mechanics of fluid media. Euler opposed the reification of forces and he believed in an aether. But in his *Principia motus fluidorum* (1761) he presented an idealized mathematical model for the transmission of action in a continuous medium. This mathematization soon led others to develop the notion of force fields whereby the relationship of bodies to one another is characterized by the specification of forces in a space-time coordinate system. Was a medium required for the exercise of these forces? What was a "field"? Faraday specified these questions by formulating empirical criteria for claiming that a particular field involved real processes in intervening space.[6] By his criteria optical, electrical and magnetic fields are real, whereas gravity is action at a distance. Except for gravity this seemed to require an aether but all attempts to find such an aether failed. Maxwell's unification of light, electricity and magnetism in a single mathematical formulation still left the question of an aether unresolved. Hertz led the movement towards the inevitable defeat of aether theories with his statement: "Maxwell's theory is Maxwell's system of equations" and with his reproach of those who tried to cloth the mathematical equations with the "gay garment" of a physical counterpart.[7] The reality is the mathematical formulation of the reality. The advent of relativity theory, of quantum field theory, of quantum cosmology, etc. will only further complicate the discussion of the nature of matter as to action at a distance.

The mathematization of physics that we have just been addressing will continue through the classical revolution in physics of the 17th and 18th centuries and will become, in a different way as we shall see, an essential ingredient of the new physics of the 20th century. As usual in scientific revolutions, what was happening only came to full realization after it had happened. A three-layered conception of the universe, only partially inherited from the Platonic-Pythagorean tradition, came to be accepted implicitly, and only slowly did it come to consciousness. There was the layer of the true mathematics, the mathematical structures of which the world is truly made. Then there was the second layer, the mathematics of we humans, structures which were in a Platonic sense only the shadows of the first layer. Finally there were at the third layer the images in concrete reality of the true mathematical structures which we humans attempted to understand with a our shadow mathematics. However, there is a subtle development, described well by Michael Heller,[8] in which at the second layer mathematics is not only the language or the interpretative tool of physics, but it becomes also the "stuff" of the ideal world of physics. For the present this "stuff" remained under the control of empirical verification i.e., the third layer. The images in concrete reality, remained the test of how true the human mathematical structures were.

The rise of quantum mechanics and of relativity theory at the beginning of this century soon weakened the connection between the second and third layers described above and, in fact, reemphasized the connection between the second and first layers. The images in concrete reality made very little, if any, sense as a test of mathematical "stuff" of the ideal world of physics. There are no natural images or representations which correspond to Hilbert spaces, the mathematical "stuff" of quantum theory. And while general relativity has passed all of the experiments yet made to test its empirical predictions there are no adequate images or representations which correspond to motions at relativistic velocities or under very large gravitational forces. In its "purest" form the physics of both the sub-quantum world and the world "beyond-relativity" is strictly mathematical in the tradition of Plato and Pythagoras and has little to do with any sensory component.

There is another area in which the new physics has advanced our understanding of the nature of matter. The studies of the dynamics of non-linear systems has given birth to the fields of chaos theory and complexity. This represents, in some sense, a return from quantum physics to the world of macroscopic physics and it is, in another limited sense, a revindication of Aristotle's view that the world of the senses is too rich to be limited to or comprehended by mathematics. There are really two parts of this: deterministic chaos arising from classical mechanics and non-linear systems in thermodynamics. The immense variety of forms, shapes and structures which we find in both the inorganic and organic world challenges any theory that they could have come forth from some deterministic set of laws of physics. And yet, using the mathematical analysis of non-linear systems and the laws of physics, we can come to understand the structural design for changes, but we cannot predict the result because we cannot know what result small perturbations, accumulated in a non-linear way, will produce. Thus while we can analyze mathematically and, in that sense, understand the structure of such dynamical systems, we cannot predict the outcome because of the accumulative effect of non-linear perturbations. In the end the world of the senses has a richness which defies ultimate mathematical analysis.

4. From Physics to the Biosciences

This leads us rather naturally from the world of physics to that of biology and chemistry, of biophysics and biochemistry. The very fact that we have such developed fields of dual denomination is an indication of the direction in which the discourse is now directed. The well established scientific evidence of the complexification of matter in the evolutionary process leads me to suggest that we have returned once again to the notion of matter as a substratum, but now that notion is much enriched both by the mathematics of non-linear systems and by our knowledge of biochemistry. What now dominates our thinking, as it did for physics in the case of the historical development towards field theories, is the concept of relationship.

No part of the universe can be understood except in its dynamical and evolutionary relationship to all other parts of the universe. The specification of matter (an electron, a quark, DNA, the human brain, etc.) is attained by its relationship to and interaction with all other parts and with the whole. The best of scientific knowledge tells us that all of the diverse objects in nature have had a common origin and have shared in and come from a common evolutionary process.

An initial eruption of energy soon gave birth to the first matter in the universe in the form of quarks which in turn formed the first sub-atomic particles until finally the simplest of all atoms, hydrogen, was formed. As the universe continued to expand and cool, matter continued to organize itself in ever more complex structures:

molecules, dust, galaxies, stars, proto-organic substances, vegetation, mammals. This evolution in an ever expanding, universe evolving towards ever more complex organization of material required also a diversification of the original energy of the universe into various forces: nuclear, electromagnetic and gravitational.

Over the centuries the debate has raged over the relative place of chance and determinism in this evolutionary process. While the laws, for example, of physics as we know them are quite deterministic, - given a cause the effect follows inevitably - we know today of many systems which are non-deterministic, or in the language of mathematics, non-linear systems. These are systems, whether physical, chemical or biological, where, although the causes or concatenation of causes are all defined and known, we cannot predict the final effect, because an undetermined and undeterminable series of fluctuations intervenes between cause and effect. This non-linearity becomes more dominant as evolution proceeds to ever more complex systems.

Pure chance is not in itself a satisfactory explanation for this increasing complexification. But we might still ask to what extent chance played a role. The philosophical inclinations of Albert Einstein are not well understood. In his physics, however, he was clearly deterministic. In the debate over the meaning of quantum mechanical indeterminism he claimed that God does not play with dice. Recently an eminent biochemist, Christian de Duve, has replied: Yes He does, because He is sure to win.[9] In attempting to frame his conclusion in the context of Einstein's statement, what de Duve is actually claiming is that intrinsic to the universe there is an interplay of determinism, chance and opportunity. His response to Einstein was to state that it is in the very nature of the universe that intelligent life inevitably come to be, although a long and complicated process involving laws, chance happenings and propitious opportunities was required.[10]

5. Dualism Challenged by Continuity in Evolution

Throughout the historical development of the notion of matter there has been a dominant tendency towards dualism in the origins and evolution of the universe. I have suggested above that this tendency was strongly supported by the Aristotelian notion of prime matter as a principle of indeterminacy and potentiality for change, even to new kinds of beings. A further contribution to this dualistic tendency comes from religious considerations. We have seen an example of this in Newton and Descartes and in their contemporaries. Descartes in particular required two levels of being: matter and spirit, and matter was completely inactive, motion having been communicated to it by God at creation. This tendency endures to our day. In order to preserve the primacy of God, of the spiritual, of the supernatural some have found it necessary to insert discrete moments in the continuous evolutionary process which we have described above. According to this position, organic could not arise from inorganic, life could not come from non-life, the human intelligence and spirit could not come from matter. God must have intervened at these critical phases in evolution. Such positions appear to contradict the most recent scientific evidence available which sees a continuity in the natural processes which lead to the complexification of matter in the universe.

This dualistic tendency is usually resolved in one of two extreme ways: materialism or divine intervention. Put in its most simple expression crass materialism will not allow that complexification in the evolutionary process can lead to new kinds of beings. All beings, however, complex, are reducible to their material parts. At the other extreme, is the position that essentially new beings, and especially the human being, require a direct intervention by God in the evolutionary process. Materialism, as I have described it, is, I believe, essentially refuted as an inadequate view of what is meant today by material, that all beings are related to all other beings in the universe in their common origin and common evolution towards more complex systems. The need for an interventionist God is essentially refuted by the scientifically well established continuity in the evolutionary process and its explicability in a scientific analysis which need not be a threat to informed religious thought.

I would like to give one explicit example of what I mean. Recently on 22 October 1996 John Paul II issued a brief message on evolution[11] to the members of the Pontifical Academy of Sciences during their Plenary Session. He introduced his message by asking: "How do the conclusions reached by the various scientific disciplines coincide with those contained in the message of revelation?". While the encyclical of Pope Pius XII in 1950, *Humani Generis*, considered the doctrine of evolution a serious hypothesis, worthy of investigation and in-depth study equal to that of the opposing hypothesis, John Paul II states in his message:

> Today almost half a century after the publication of the encyclical [*Humani Generis*], new knowledge has led to the recognition that the theory of evolution is no longer a mere hypothesis.[12]

The sentences which follow this statement indicate that the "new knowledge" which the Pope refers to is for the most part scientific knowledge. He had, in fact, just stated that "the exegete and the theologian must keep informed about the results achieved by the natural sciences". The context in which the message occurs strongly supports this. As the specific theme for its plenary session the Pontifical Academy of Sciences had chosen: *The Origin and Evolution of Life*, and it had assembled some of the most active researchers in the life sciences to discuss topics which ranged from "Molecular Phylogeny as a Key to Understanding the Origin of Cellular Life" to "The Search for Intelligent Life in the Universe" and "Life as a Cosmic Imperative"; from, that is, detailed molecular chemistry to careful analyses of life in the context of the evolving universe. Only months before the plenary session of the Academy the renowned journal, *Science*, published a research paper announcing the discovery that in the past there may have existed primitive life forms on the planet Mars. Furthermore within the previous two years a number of publications had appeared announcing the discovery of extra-solar planets.

This ferment in scientific research not only made the plenary session theme very timely, but it also set the concrete scene for the Papal message. Most of the scientific results cited were very tentative and very much disputed (as is true of almost all research at its beginning), but they were very exciting and provocative. Only three months after the plenary session the Pope would receive in private audience a group of scientists from Germany, Italy and the United States who were responsible for the high-resolution observations being made by the satellite, *Galileo*, of the Jovian planets and their satellites. Within a few months of that audience NASA would announce the discovery of a huge ocean on Europa, a satellite of Jupiter.

These are the circumstances surrounding the Papal message on evolution. Did they influence it? A careful reading of the message is consistent with the suggestion that they did. The crux of the message is the discussion of the opposing theories of evolutionism and creationism as to the origins of the human person. In the traditional manner of Papal statements the main content of the teaching of previous Popes on the matter at hand is reevaluated. And so the teaching of Pius XII in *Humani Generis* that, although it may be true that the human body takes its origins from pre-existent living matter, the spiritual soul is immediately created by God. And so, is everything resolved by embracing evolutionism as to the body and creationism as to the soul? Note that the word "soul" does not reappear in the remainder of the dialogue. Rather the message moves to speak of "spirit" and "the spiritual".

If we consider the revealed, religious truth about the human being, then we have an "ontological leap", an "ontological discontinuity" in the evolutionary chain at the emergence of the human being. Is this not irreconcilable, wonders the Pope, with the continuity in the evolutionary chain seen by science? An attempt to resolve this critical issue is given by stating that:

> The moment of transition to the spiritual cannot be the object of this kind of [scientific] observation, which nevertheless can discover at the experimental level a series of very valuable signs indicating what is specific to the human being.

The suggestion is being made, it appears, that the "ontological discontinuity" may be explained by an epistemological discontinuity. Is this adequate or must the search continue? Is a creationist theory required to explain the origins of the spiritual dimension of the human being. Are we forced by revealed, religious truth to accept a dualistic view of the origins of the human person, evolutionist with respect to the material dimension, creationist with respect to the spiritual dimension. The message, I believe, when it speaks in the last paragraphs about the God of life, gives strong indications that the dialogue is still open with respect to these questions.

I would like to use the inspiration of those closing paragraphs to suggest that reflections upon the God's continuous creation may help to advance the dialogue with respect to the dualistic dilemma mentioned above. We might say that God creates through the process of evolution and that creation is continuous. Since there can ultimately be no contradiction between true science and revealed, religious truths, this continuous creation is best understood in terms of the best scientific understanding of the emergence of the human being which I think is given in the following summary statement by the eminent evolutionary chemist, Christian de Duve:

> . . . evolution, though dependent on chance events, proceeds under a number of inner and outer constraints that compel it to move in the direction of greater complexity if circumstances permit. Had these circumstances been different, evolution might have followed a different course in time. It might have produced organisms different from those we know, perhaps even thinking beings different than humans.[13]

But does the contingency involved in the emergence of the human being contradict religious truth? Not, it appears to me, if theologians can develop a more profound understanding of God's continuous creation. God in his infinite freedom continuously creates a world which reflects that freedom at all levels of the evolutionary process to greater and greater complexity. God lets the world be what it will be in its continuous evolution. He does not intervene, but rather allows, participates, loves. Is such thinking adequate to preserve the special character attributed by religious thought to the emergence of spirit, while avoiding a crude creationism? Only a protracted dialogue will tell.

6. References

1. McMullin, E.: *Newton on Matter and Activity* , University of Notre Dame Press, Notre Dame, 1978.
2. Cohen, I.: Hypotheses in Newton's Philosophy, *Physics* 8 (1966) 163-184.
3. Jammer, M.: *Concepts of Mass in Classical and Modern Physics,* Harvard University Press, Cambridge, Massachusetts, 1961, pp. 58-61.
4. Westfall, R.: Alchemy in Newton's Career, *Reason, Experiment and Mysticism*, ed.: Righini, Bonelli and Shea, 189-232.
5. McMullin, E..: op. cit in Note 1, page 1 and Heller, M., Adventures of the Concept of Mass and Matter, *Philosophy in Science* 3 (1988) 15-35.
6. Faraday, *M.: Experimental Researches in Chemistry and Physics*, London, 1859.
7. Hertz states this in the Introduction to his *Untersuchungen über die Ausbreitung der elektrischen Kraft,* Leipzig, 1892; translated by Jones, D.E. as *Electric Waves,* London, 1893.
8. Heller, *M.: The New Physics and a New Theology,* Vatican Observatory Publications, Vatican City, 1996; distributed by the University of Notre Dame Press, Notre Dame, Indiana, pp. 36-39.
9. de Duve, C..: Life as a Cosmic Imperative, *The Origin and Early Evolution of Life,* Pontificia Academia Scientiarum, Vatican City, 1997; Commentarii, Vol. IV, No.3, p. 320.
10. de Duve, *C.: Vital Dust: Life as a Cosmic Imperative,* Basic Books, New York, 1995.
11. The original message in French was published in *L'Osservatore Romano* for 23 October 1996 and an English translation in the Weekly English Edition of *L'Osservatore Romano* for 30 October 1996. The French text is now also published in *The Origin and Evolution of Life,* Pontificia Academia Scientiarum, Vatican City, 1997; Commentarii, Vol. IV, No. 3, pp. 15-20.
12. The English translation of this sentence, published in the Weekly Edition of *L'Osservatore Romano* for 30 October 1996, is incorrect when it says: " ... the recognition of more than one hypothesis ...".
13. de Duve, C..: op. cit. in Note 9, page 320.

RE-APPRAISAL OF THE AGE OF THE OLDEST WATER-LAIN SEDIMENTS, WEST GREENLAND:

Significance for the existence of life on the early Archaean Earth

STEPHEN MOORBATH and BALZ SAMUEL KAMBER
Department of Earth Sciences, Oxford University,
Parks Road, Oxford, OX1 3PR, U.K.

Abstract. Two recent papers [1, 2] report geological, geochemical, and geochronological data from the early Archaean of West Greenland to suggest that life existed on Earth by ≥3850 Myr ago. The crucial evidence is based on zircon U-Pb age measurements of magmatic rocks (metamorphosed to orthogneisses) regarded to be younger than adjacent metasedimentary rocks containing components with a carbon isotope signature characteristic of a biological origin [1]. This new lower age limit for the existence of life on Earth overlaps the period at which massive, destructive impacts were occurring on the moon's surface (and presumably on Earth as well), probably terminating close to 3.8 Gyr ago [3].

A critical re-examination and re-interpretation of the relevant geochronological evidence in West Greenland suggests that the rocks of possibly biological significance are some 120-190 Myr younger than claimed in the cited papers. By this time, major cataclysmic impacts would long have been over, and more congenial conditions for earliest life established.

1. Introduction

On Akilia Island, in southern West Greenland, a layered enclave consisting of amphibolite, banded iron-formation (BIF) and ultramafic rock, which forms part of the so-called Akilia association, is cut by a quartz-diorite sheet from which ion-probe (SHRIMP) zircon U-Pb weighted mean ages of 3865±11 Myr and 3840±8 Myr have been reported [2]. This was taken as evidence that by ~3850 Myr there was a hydrosphere permitting chemical sedimentation of BIF. Furthermore, C-isotope data on graphite inclusions in apatite from this highly metamorphosed BIF on Akilia Island are consistent with a biogenic origin [1], extending the record of life on Earth back to ≥3850 Myr. Nutman et al. [2] state that "life and surface water by ~3850 Myr provide constraints on either the energetics or termination of the late meteoritic bombardment event (suggested from the lunar cratering record) on Earth".

In the Isua greenstone belt, some 160 km north-east of Akilia Island, a similar association of BIF apatite with possibly biogenic graphite (i.e. low $\delta^{13}C$) has been reported [1]. Nutman et al. [2] report that cherts and BIF are interlayered with amphibolites cut by a tonalite sheet yielding a zircon U-Pb date of 3791±4 Myr, which would thus be a minimum age for at least part of the Isua greenstone belt. They also report zircon U-Pb dates of 3806±4 Myr and 3708±3 Myr for two felsic (silica-rich) units

J. Chela-Flores and F. Raulin (eds.),
Exobiology: Matter, Energy, and Information in the Origin and Evolution of Life in the Universe, 81–86.

of probable volcanic origin from other parts of the Isua greenstone belt. Nutman et al. [2] consider that the Isua BIF could either be as old as those from Akilia Island or somewhat younger, namely between ~3865 and ~3790 Myr.

Here we critically examine the published evidence quoted above, and conclude that the rocks of possible relevance for early life may be significantly younger than ~3850 Myr.

2. The age of the Akilia association on Akilia Island and neighbouring islands

The term "Akilia association" was introduced by McGregor and Mason [4] for all enclaves of rock intruded by the regional Amîtsoq gneisses, except for the largest "enclave", the Isua greenstone belt.

The crucial zircon U-Pb dates for two quartz-diorite sheets from Akilia Island, regarded as younger than the local Akilia association, are 3865±11 Myr and 3872±10 Myr (Figure 1 in [2]). In detail, zircons from these rocks exhibit a much more complex U-Pb age pattern with dates in the range of ~3870 Myr to ~3550 Myr (Figure 4 in [2]). Nutman et al. [2] interpret the oldest peak on their age histograms as representing the true age of intrusion of the quartz-diorite sheets, thus yielding a minimum age for the Akilia association and its low $\delta^{13}C$ graphite inclusions.

In an earlier paper, Bennett et al. [5] report samarium-neodymium (Sm-Nd) data for a suite of Akilia association rocks from a wide region, including Akilia Island and Innersuartuut Island, some 10 km to the south, where rock types and field relationships are virtually identical (Figure 2 in [2]). The Akilia association rocks used for their Sm-Nd analyses are gabbros and leucogabbros which occur as enclaves in Amîtsoq gneisses, as do the BIF enclaves on both islands. Surprisingly, Bennett et al. [5] do not plot their Sm-Nd data for the Akilia association on a simple isochron plot of the measured parameters $^{147}Sm/^{144}Nd$ versus $^{143}Nd/^{144}Nd$, perhaps the most fundamental and widely used graphical representation in geochronology. In fact, their data [5] for five separate localities (seven data points) of Akilia association gabbros and leucogabbros yield an isochron age of 3675±48 Myr. Four of these samples come from Akilia Island (one sample) and from Innersuartuut Island (three samples) and are directly relevant to the present paper. These four samples yield a perfect isochron (MSWD <1) with an age of 3677±37 Myr, as shown in Figure 1. We regard it as probable that this represents a close estimate for the age of not only the gabbro/leucogabbro enclaves but also for the closely associated BIF enclaves which (on Akilia Island at least) contain apatite with graphite inclusions with probably biogenic C-isotope ratios [1]. On Innersuartuut Island, BIF enclaves also contain apatite grains with graphite inclusions and coatings (Mojzsis pers. comm., and Oxford unpublished work), but no C-isotope ratios are yet available. Remember that on both Akilia and Innersuartuut islands, the Akilia association enclaves are cut by gneissic granitoid sheets with some zircon U-Pb dates which are significantly older than 3677±37 Myr.

The only SHRIMP zircon U-Pb dates so far measured directly on an Akilia association rock were published by Schiøtte and Compston [6]. This is a biotite-schist regarded as of volcano-sedimentary origin from Innersuartuut Island, which was referred to above. The zircon dates ranged from ~3.75 Gyr to ~2.7 Gyr, of which the two oldest groups gave mean dates of 3756±22 Myr and 3685±8 Myr. After due evaluation of the geochronological evidence and morphology of the zircon grains, Schiøtte and Compston

[6] favour 3685 Myr as representing the original age of this part of the Akilia association, with the 3756 Myr zircons being xenocrysts in the volcanic rock, i.e. derived from older rocks. Note that the preferred age of 3685±8 Myr is within error of the Sm-Nd age of 3677±37 Myr obtained from our isochron plot (Figure 1) based on Bennett and Nutman's Sm-Nd data for Akilia and Innersuartuut islands [5]. There is no sign of any zircon approaching the value of 3865±11 Myr obtained by Nutman et al. [2] for a cross-cutting sheet of dioritic gneiss on Akilia Island.

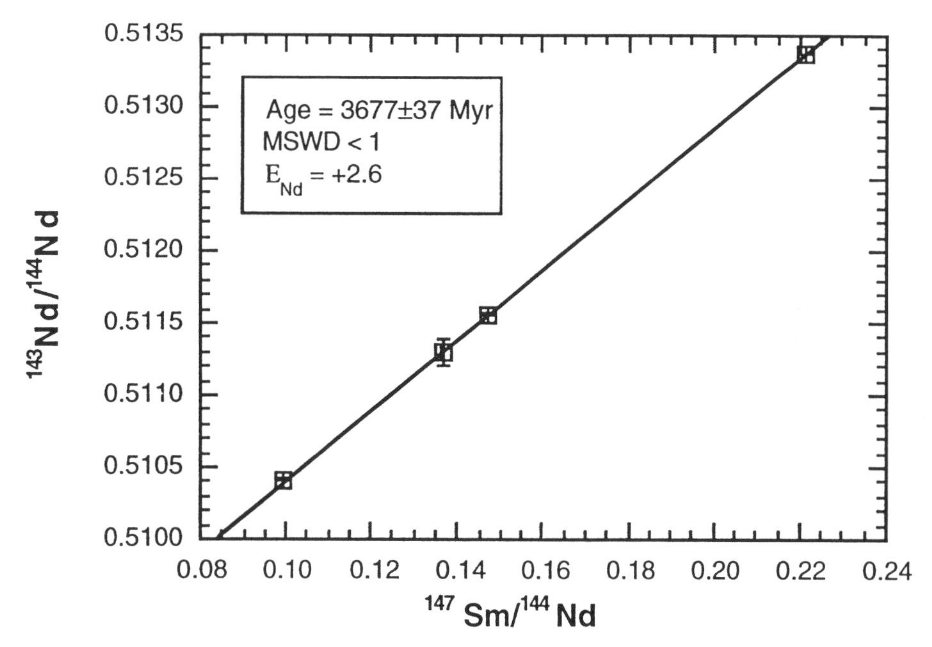

Figure 1. Sm-Nd isochron plot for data of Bennett et al. [5] on Akilia association enclaves from Akilia Island (one sample - the highest point), and nearby Innersuartuut Island (three samples). The mean square weighted deviate (MSWD) of <1 shows that this is a statistically perfect isochron. E_{Nd} is a measure of the initial $^{143}Nd/^{144}Nd$ ratio, which is of great importance for petrogenetic and geochemical studies [e.g., 7] as well as for modelling mantle evolution. The quoted error on the age is 2 sigma (95% confidence level). General background information on geochronology and related isotope geochemistry is given in [17] and references therein.

The evidence presented here strongly suggests that the age of the Akilia association on Akilia and Innersuartuut islands (and perhaps elsewhere in this region) may be as young as ~3680 Myr [7], nearly two hundred million years younger than the age claimed by Nutman et al. [2]. In that case, the much older zircon U-Pb dates of ~3865 Myr on cross-cutting dioritic gneiss sheets would represent the age of "inherited' zircon grains derived from a source region no longer exposed, and with no bearing on the true age of the Akilia association rocks containing biogenic remnants. For further discussion, see Section 4.

3. The age of the Isua greenstone belt

A review of previous work on this topic was presented in an earlier Volume of this series [8].

As pointed out in Section 1, the most recent geochronological work [2] is somewhat contradictory. On the one hand a minimum age for the Isua greenstone belt of 3791±4 Myr is claimed from the dating of a cross-cutting tonalite sheet whilst, on the other hand, several zircon U-Pb dates from Isua greenstone belt rocks themselves are significantly younger than this. Thus Nutman et al. [2] state that "...... units of impure BIF and chert contain rare ~3710 Myr zircons and are associated with felsic volcano-sedimentary rocks with an age of 3708±3 Myr. The largest cherts and BIFs in the eastern end of the Isua supracrustal belt have yielded rare, small, euhedral prismatic zircons with ages of 3690-3700 Myr (Nutman, unpublished data). If these zircons represent low-level volcanic contamination of these BIFs then they were deposited at only ~3700 Myr.".

The results of a detailed Sm/Nd age review on Isua greenstone belt rocks are closer to the younger age than to the older age. The Sm-Nd data are taken from the literature [7, 9, 10, 11]. Twenty-four samples of mica-schists (biotite+plagioclase+quartz+garnet +chlorite+hornblende) define a regression age of 3742±49 Myr (Figure 2). These rocks are closely associated in the field with felsic rocks of probably volcanic origin which yield zircon U-Pb dates of ~3.71 Gyr [2, 12], as well as with the BIF. Sm-Nd age measurements on the felsic rocks do not allow calculation of meaningful age regressions because of the very restricted range of Sm-Nd ratios, so that they are not included in Figure 2.

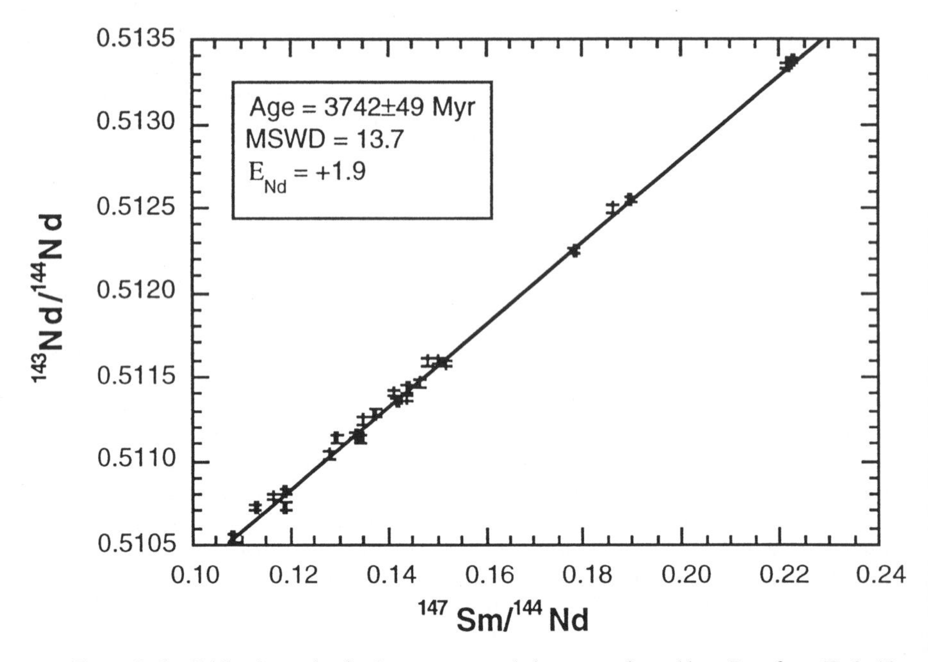

Figure 2. Sm-Nd isochron plot for Isua greenstone belt garnet-mica schists. Data from [7, 9, 10, 11]. For further discussion see text and caption to Figure 1.

The evidence presented here suggests that deposition of that part of the Isua greenstone belt relevant to the possibly biogenic C-isotope ratios in BIF [1] maybe somewhere in the range of 3.75 to 3.70 Gyr. In that case the older zircon U-Pb date of 3791±4 Myr reported for cross-cutting tonalite may represent the age of zircon grains inherited from an older source, just as in the case of the Akilia association, described in Section 2.

4. Significance of zircon U-Pb dates

Numerous studies over the past 25 years, partly reviewed by Mezger and Krogstad [13], have shown that zircons can survive partial melting, entrainment in hot magmas and regional metamorphism without completely loosing their age information. Indeed these authors conclude that "..... the only way for zircon to become completely reset under crustal condition is through dissolution and subsequent reprecipitation. Under all other conditions in the continental crust zircons are forever.". It is becoming recognised that all meaningful zircon U-Pb chronology must be accompanied by detailed structural and morphological studies of individual zircon grains by various methods (e.g. cathodoluminescence, electron microscopy etc.), in order to study crystal zoning patterns which may reflect the entire geological history of each grain. This is an essential aid in geochronological interpretation. It is encouraging to note that such combined studies are increasingly being reported. In the case of the West Greenland rocks discussed above, it seems probable that, subject to further work, the oldest zircon U-Pb dates from discordant, younger granitoid gneisses actually represent zircons from an older source which is no longer exposed and which therefore do not yield a true minimum age for those rocks of biological interest cut by the granitoids. We suggest that the true minimum age of deposition is given by the Sm-Nd age data on the supracrustal (i.e. sedimentary and/or volcanic) rocks themselves, despite their greater inherent analytical errors as compared to individual zircon U-Pb dates.

5. Conclusions

From the above discussions, we regard the Sm-Nd dates of 3677±37 Myr and 3742±49 Myr from the Akilia association of Akilia and Innersuartuut islands and from the Isua greenstone belt respectively, as the currently most reliable ages of deposition for these supracrustal rock units. Although the two dates are, strictly speaking, within error of each other, it is also possible that the Isua greenstone belt is slightly older than the Akilia association. As pointed out earlier, a few zircon U-Pb dates have been reported from within the Akilia association and the Isua greenstone belt which agree quite closely with the Sm-Nd dates [2, 6]. The revised dates presented here are some 120-190 million years younger than the minimum dates of 3865±11 Myr and 3791±4 Myr reported by Nutman et al. [2] for the Akilia association and the Isua greenstone belt respectively, both of which contain BIFs in which apatite has low $\delta^{13}C$ graphite inclusions of possible biogenic origin [1]. If our re-interpretation of these dates from West Greenland is correct, the question of overlap of earliest life with a bolide impact scenario terminating at ≥3800 Myr (as on the moon), as suggested by Nutman et al. [2], becomes irrelevant. Even if such cataclysmic impacts had occurred on Earth, demonstrably quiescent conditions by ~3750-3650 Myr ago appear much more compatible with existence of the early biosphere [14, 15]. At any rate, the sedimentary and volcanic rocks of the Akilia association and the Isua greenstone belt show no evidence whatever in the field for deposition under turbulent, impact-influenced conditions [16].

Acknowledgements

This is a contribution to the Isua Multidisciplinary Research Project, supported by the Danish National Science Research Council, the Geological Survey of Denmark and Greenland, the Commission for Scientific Research in Greenland, and the Minerals Office of the Greenland Government. Kamber is funded by a Swiss BBW grant (95.0833).

References

[1] Mojzsis, S.J., Arrhenius, G., McKeegan, K.D., Harrison, T.M., Nutman, A.P. and Friend, C.R.L.: Evidence for life on Earth before 3800 million years ago. *Nature*, 384 (1996) 55-59.

[2] Nutman, A.P., Mojzsis, S.J. and Friend, C.R.L.: Recognition of ≥3850 Ma water-lain sediments in West Greenland and their significance for the early Archaean Earth. *Geochim. Cosmochim. Acta*, 61 (1997) 2475-2484.

[3] Wilhelms, D.E.: The geologic history of the Moon. *U.S. Geol. Surv. Prof. Paper*, 1348 (1987) 127p.

[4] McGregor, V.R. and Mason, B.: Petrogenesis and geochemistry of metabasaltic and metasedimentary enclaves in the Amîtoq gneisses, West Greenland. *Amer. Mineral.*, 62 (1977) 887-904.

[5] Bennett, V.C., Nutman, A.P. and McCulloch, M.T.: Nd isotopic evidence for transient, highly depleted mantle reservoirs in the early history of the Earth. *Earth Planet. Sci. Lett.*, 119 (1993) 299-317.

[6] Schiøtte, L. and Compston, W.: U-Pb age pattern for single zircon from the early Archaean Akilia association south of Ameralik fjord, southern West Greenland. *Chemical Geology*, 80 (1990) 147-157.

[7] Moorbath, S., Whitehouse, M.J. and Kamber, B.S.: Extreme Nd-isotope heterogeneity in the early Archaean - fact or fiction? Case histories from northern Canada and West Greenland. *Chem. Geol.*, 135 (1997) 213-231.

[8] Moorbath, S. and Whitehouse, M.J.: Age of the Isua supracrustal sequence of West Greenland: a plausible repository for early life, In: J. Chela-Flores and F. Raulin (ed.), *Chemical Evolution: Physics of the Origin and Evolution of Life*. Kluwer Academic Publishers, Dordrecht, 1996, pp. 87-95.

[9] Hamilton, P.J., O'Nions, R.K., Bridgwater, D. and Nutman, A.: Sm-Nd studies of Archaean metasediments and metavolcanics from West Greenland and their implications for the Earth's early history. *Earth Planet. Sci. Lett.*, 62 (1983) 263-272.

[10] Moorbath, S., Taylor, P.N. and Jones, N.W.: Dating the oldest terrestrial rocks - fact and fiction. *Chem. Geol.*, 57 (1986) 63-86.

[11] Jacobsen, S.B. and Dymek, R.F.: Nd and Sr isotope systematics of clastic metasediments from Isua, West Greenland: identification of pre-3.8 Ga differentiated crustal components. *J. Geophys. Res.*, 93 (1988) 338-354.

[12] Nutman, A.P., McGregor, V.R., Friend, C.R.L., Bennett, V.C. and Kinny, P.D.: The Itsaq Gneiss Complex of southern West Greenland; the world's most extensive record of early crustal evolution (3,900-3,600 Ma). *Precamb. Res.*, 78 (1996) 1-39.

[13] Mezger, K. and Krogstad, E.J.: Interpretation of discordant U-Pb zircon ages: An evaluation. *J. Metamorphic Geol.*, 15 (1997) 127-140.

[14] Maher, K.A. and Stevenson, D.J.: Impact frustration of the origin of life. *Nature*, 331 (1988) 612-614.

[15] Sleep, N.H., Zahnle, K.J., Kasting, J.F. and Morowitz, H.J.: Annihilation of ecosystems by large asteroid impacts on the early Earth. *Nature*, 342 (1989) 139-142.

[16] Appel, P.W.U., Fedo, C.M., Moorbath, S. and Myers, J.S.: Well-preserved volcanic and sedimentary features from a low-strain domain in the ~3.7-3.8 Ga Isua greenstone belt, West Greenland. *Geology*, (submitted for publication).

[17] Dickin, A.P.: *Radiogenic Isotope Geology*, Cambridge University Press, Cambridge, 1995.

THE ORIGIN OF MIND AND LIFE

Sidney W. Fox[1], Aristotel Pappelis[2], and Randall Grubbs[1], [1]University of South Alabama, Mobile, AL, USA; [2]Southern Illinois University, Carbondale, IL, USA

Abstract

Successful retracement in the laboratory under terrestrial conditions of synthesis of the first cell from progenitor thermal protein has been recorded (Fox et al., 1995; Enger and Ross, 1997). This retracement has benefited from recognition of some reversibility in molecular evolution (Atkinson and Fox, 1951; Fox et al., 1995a; Fox et al., 1995b). A number of other new advances were also required. Sequence analysis of amino acid residues was a method needed early (Fox, 1945). (While the 1945 prospectus preceded analysis of insulin and other related developments by other workers), the evolutionary applications led to recognition of self-sequencing of amino acids into polymers when heated, self-assembly of the resultant thermal proteins to first living cells (Fox, 1971; Fox et al., 1995; Enger and Ross, 1997) and a number of other advances.

1. Introduction

The experimental results (Fox et al., 1995), which others have begun to recognize as a scientific answer to the problem of the origins of living systems (Enger and Ross, 1997; World Wide Web, 1997), agree in one major aspect with the view of Stanley Miller. The first steps had to be simple. This is in contrast to the multipage theoretical anticipation of Oro (1996) in this meeting and by others. Yet others, such as Behe (1996), represent those who seem not to have read what they cite, or contrast their views with some perspectives of unbased supposition. What stands out in contrasting our theory with those of dissidents is that the experimental results of our group are repeatable. Jukes (1997) represents another source of misinterpretation when he refers to microspheres merely as polymers instead of as cells. He fails to recognize the emergent properties of proteinoid microspheres(protocells) which have been catalogued and that processes of molecular biology were operative in the cellular world only after cells arose.

The positive results have been repeated by thousands of high school students from available manuals. Following this support, Pope John Paul II and his Vatican scientists financed three trips to Rome for the senior author in which Fox and associates' new results and interpretations were presented. The presentations preceded the 1996 pronouncement by the Pope that evolution is more than a hypothesis. Those presentations are again explained here.

J. Chela-Flores and F. Raulin (eds.),
Exobiology: Matter, Energy, and Information in the Origin and Evolution of Life in the Universe, 87–92.

In experiments that have retraced the initial construction of emergent polybiofunctional cells, we observe that the progenitor processes are simple phenomena at the same time that the mechanisms are very complex. In their essence, the simple phenomena require ammonia, formaldehyde, water, and heat.

2. Necessary Compounds and Energy

Heat is the only type of energy that has been needed to reconstruct the evolutionary sequence of

a) HCHO + NH_3 + heat -----> amino acids

b) amino acids + heat -----> thermal protein

c) thermal protein + H_20 -----> cells

much as Herrera (Negron-Mendoza, 1995) used. These were ammonia, formaldehyde, heat, and water all of which more than thirty years after Herrera's experiments (Fox and Dose, 1972) were found to be prominent components of interstellar matter. This is the kind of stuff that Urey (1952) proposed condensed into the first planets.

Thermal protein, proteinoid indexed as such by Chemical Abstracts since 1972, is unique in its propensity to form cells(when moistened) as we have observed in hundreds of tests of a simple type replicated by thousands of high school students. Also unique is the finding that the characteristics of these cells, in their role as protocells, include so many attributes of a neuronal type as to indicate that the first cell was already a neuron.

The molecules most often invoked as progenitors of living systems have been RNA and DNA. Recently these have been repudiated as progenitors by leaders in the research on them such as Ferris (1993) and Crick (1994). This left open two possibilities: proteins, as we know them from biochemical studies; and, thermal proteins, as we know them from synthesis as described above.

The answer to the question of how protein as we know it came into existence now appears to follow on how thermal protein came into existence. Recognition of the phenomena of the origins of thermal protein was a blend of fortuity and serendipity. Thermal protein appeared during a search for how some kind of precellular protein arose simply (Fox and Middlebrook,1954; Fox and Harada, 1959). The origins of thermal protein were suggested by the finding that amino acids could be heated together without decomposition if the mixture contained minimal amounts of aspartic acid, glutamic acid or both.The discovery of cellular origins was an almost inevitable early consequence of the synthesis of thermal proteins, since cells were first seen as a "milky layer" while washing out thermal proteins from test tubes in which they were made.The processes involved in the formation of protocells are molecular but are now being visualized by oil immersion-light microscopy at 800-1200 magnifications

3. The Cell

According to almost all biologists, the sine qua non of life is the cell, with its vital characteristics. This point has been missed or ignored, until recently, by chemists who have dominated the activity under the origin-of-life research banner. Since the first laboratory production of a cell that meets many of the definitions of life (Fox, 1971; Fox et al., 1996) the characteristics have included growth, metabolism, reproduction, and response to stimuli. Response to stimuli, however, did not seem as meaningful as it now does until it included electric potential (Ishima et al., 1981).

Since 1981 the electrical responses can be said to be dominant among the characteristics displayed. By 1995 (Fox et al., 1995) the electrical activity had been studied sufficiently that the retraced protocell was seen also as a retracement of origin of the first neuron. Since the cell is the symbol of life and the neuron is the symbol of the brain (mind) and CNS, the emergence of either cell or neuron signified the emergence of life and mind, not just life.

When the origin and evolution of the protoneuron is retraced we can assert that the first (abstract) mind emerged at much the same time as the physical structure of the cell. We can then ask whether the roots of life or the roots of mind emerged first. From our experiments we regard the discriminating interactions of molecules as representing the first roots of judgment (mind) while the cooperative properties of molecular self-assembly and resultant emergent attributes constituted the immediate precursors of life. For this reason we chose as title The Origin of Mind and Life. Due to the almost instantaneous rapidity of the self-assembly of thermal protein to protocells displaying emergent vital characteristics defining life, we now think of the emergence of the evolutionary protocell as the Big Bang of Biology.

What seems to be most apparent is the simplicity of the origin of mind and life. Such comments as those of De Duve (1995) and Oro (1996) are dominant views in biology which connote maximal evolutionary complexity. This arises from reductionistic analyses of modern organisms. What our experiments have shown is of maximal simplicity of reaction (not mechanism) as seems appropriate for terrestrial protobiogenesis (Fox, 1995).

Reversal in evolution does not require DNA nor RNA as progenitors within the process (Nakashima and Fox, 1980, 1981), a fact which relates to Prusiner's self-reproduction of proteins in Mad Cow Disease without participation of DNA nor RNA. An experimental report on a self-replicating peptide (Lee et al., 1996) also seeks to explain a stage in protein self-replication. However, the experimentally based evolution of protein replication reported herein depends upon (1) self-sequencing of amino acids when heated, (2)the fact that the resultant thermal proteins uniquely generate CELLS, (3)the cataloguing of vital emergent characteristics of the cell, (4) the complementary functions of precellular thermal proteins and cellular proteins, (5) the function of the demonstrated cell membrane in gathering and transmitting environmental energy, and (6) the contribution of other answers such as the origin of the neuron. This last provides the basis for the Origin of Mind and Life.

Outstanding of the new findings is that the protocell was already a protoneuron. From that series in molecular and cellular evolution is inferred the Origin of Mind and Life. The larger overview of thermal protein as the progenitor of mind and life is best understood from the contribution of Dose and coworkers (Heinz et al., 1980) as due to the synthesis of photoactive flavins.

4. The Neuron

In conventional thinking, the mind arose after the body. As the result of the experiments that indicate that the neuron was a compatriot of the cell that arose approximately at the same time, the question has been reopened. One way to rephrase the question is to ask if the roots of mind followed or preceded the roots of life. By this criterion, roots of the mind appeared before roots of the living body, in our view. The roots of mind, which we identify as molecular discrimination, were here before the precursor of the cell, i.e. thermal protein.

Most significant is that both the origin of thermal protein and assembly thereof to the cell occur spontaneously (the later almost instantly at 100°C and within minutes at 4°C); Grubbs et al.,(1998). Experiments indicate that so much happened so fast in the production of thermal protein and assembly thereof, that this may well be referred to as the Big Bang of Biology.

The neuronal properties include electrical activity, relationship of the pattern of activity to the combination of amino acids brought together to make the thermal protein, junctional propensity of these units and other secondary factors.

In none of the many microspheres from thermal protein in which a microelectrode has been implanted and tested has the microsphere failed to respond electrically. In some of those cases, lecithin has been included; the lecithin appears to associate with the double layer membrane and to modify the activity.

5. Updated Overview

The first cell was already a neuron.

The resultant new review of the evolutionary sequence yields the irreversible sequence: amino acids + heat --> thermal protein + water --> protoneuronal cells (in which DNA --> RNA --> protein). This sequence describes essentially the evolution of matter to mind and life.

Cellular protein carries further evolution of the characteristics of thermal protein both intrinsically and in emerging systems.

Acknowledgments

The authors thank Drs. Timothy D. Sherman and John A. Freeman of the Dept. of Biology, and Dustin H. Grubbs for discussions and numerous acts of assistance.

References

Atkinson, D.E. and Fox S.W. (1951). Synthesis of aromatic amino acids by mutant strains of Lactobacillus arabinosus. Arch. Biochem. Biophys 31, 212-223.

Behe, M.J. (1996). Darwin's Black Box, Simon and Schuster.

Chela-Flores, J. and Raulin, F. (eds.) (1996). Chemical Evolution: Physics of the Origin and Evolution of Life. Kluwer Academic Publishers, Dordrecht.

Crick, F. (1994). Interview with Jane Clark. J. Consciousness Studies 1, 15.

De Duve, C. (1995). Vital Dust/Life as a Cosmic Imperative. Basic Books, New York, p. 93.

Enger, E.D. and Ross, F.E. (1997). Concepts in Biology, Eighth Edition. Wm. C. Brown Publishers, Dubuque, IA, p. 362.

Ferris, J. (1993). Catalysis and prebiotic RNA synthesis. Origins of Life/Evolution of Biosphere 23, 307-315.

Fox, S. W. (1945). Terminal amino acids in peptides and proteins. Advances Protein Chem., 2 pp.155-177.

Fox, S.W. (1966). Experiments with a precellular model, in Conference on Theoretical Biology, NASA, De novo Cell Synthesis, G. J. Jacobs (ed). Kluwer Academic Publishers, p. 1-6.

Fox, S.W. (Dec. 6, 1971). Chemical origins of cells-2. Chem. Eng. News 49, 46-53.

Fox, S.W. (1985). Protobiological Selforganization. in Structure and motion: Membranes, Nucleic Acids, and Proteins, E. Clementi, C.Corongiu, M.H. Sarma, and R.H. Sarma (eds). Adenine Press,Guilderland, N.Y., pp. 101-114.

Fox, S.W. (1995). To cellular life and neurocellular assemblies, in Evolution, Biochemistry, and Related Areas of Physicochemical Biology, B. Poglazov et al (eds). Bach Institute of Biochemistry and ANKO, Moscow, pp. 161-175.

Fox, S.W., Bahn P.R. et al. (1995). Experimental retracement of the origins of a protocell; it was also a protoneuron. in Chemical Evolution: Structure and Model ofthe First Cell, C. Ponnamperuma and J. Chela-Flores (eds). Kluwer Academic Publishers, Dordrecht, pp. 17-36.

Fox, S.W., Bahn P.R., Pappelis A., and Yu, Bi. (1996). Experimental retracement of terrestrial origin of an excitable cell. in Chemical Evolution: Physics of the Origin and Evolution of Life. J. Chela-Flores and F. Raulin (eds). Kluwer Academic Publishers, Dordrecht, pp. 21-32.

Fox, S.W. and Dose, K. (1972) Molecular Evolution and the Origin of Life. W.H. Freeman and Co., San Francisco.

Fox, S. W. and Middlebrook, M. (1954) Anhydrocopolymerization of amino acids under the influence of hypothetically primitive terrestrial conditions. *Federation Proc.* *13*, 703.

Heinz, B., Ried, W., and Dose, K. (1980). Thermal production of pteridines and flavins from amino acid mixtures. Angew. Chem. 91, 510-511.

Ishima, Y., Przybylski, A.T., and Fox, S.W. (1981). Electrical membrane phenomena in spherules from proteinoid and lecithin. BioSystems 13, 243-251.

Jukes, T.H. (1997). Oparin and Lysenko. J. Mol. Evol. 45, 339-346.

Lee, D.H., Granja, J.R., Martinez, J.A., Severin, K., and Ghadri, R. (1996). A self-replicating peptide. Nature 382 525-528.

Nakashima, T. and Fox, S.W. (1980). Synthesis of peptides from amino acids and ATP with lysine-rich proteinoid. J. Mol. Evol. 15, 161-168.

Nakashima, T. and Fox, S.W. (1981). Formation of peptides from amino acids by single or multiple additions of ATP to suspensions of nueleoproteinoid microparticles. BioSystems 14, 151-161.

Negron-Mendoza, A. (1995). Alfonso L. Herrera, a Mexican pioneer in the studies of the origin of life. In Chemical Evolution; Structure and Model of the First Cell. C.Ponnamperuma and J. Chela-Flores (eds.) Kluwer Academic Publishers, pp. 11-13.

Oro, J. (1996). Cosmic evolution, life and man. In Chemical Evolution: Physics of the Origin and Evolution of Life. J. Chela-Flores and F. Raulin (eds.) Kluwer Academic Publishers, pp. 3-19, especially pp. 6-11.

Urey, H. C. (1952) The Planets, Yale University Press, p.150

World Wide Web Address-
(http://entropy.me.usouthal.edu/harbinger/articles/rel_sci/fox.html)

Section 3:
Energy from Inert to Living Matter

ENERGY FOR THE ORIGIN OF LIFE

HERRICK BALTSCHEFFSKY, ANDERS SCHULTZ AND
MARGARETA BALTSCHEFFSKY
Dept. of Biochemistry, Arrhenius Laboratories
Stockholm University, S-106 91 Stockholm, Sweden

Abstract

In our universe, life may exist on many planets, more or less similar to Tellus, the Earth. As the only life we know is life on Earth, our ideas about life in the universe are naturally based on what we know about this life and its various aspects, such as its origin from an energetic point of view.

Life has been characterized in terms of a specific flow of energy, matter and information. The transition from general molecular information to specific genetic information, resulting in an energy requiring process of replication of a kind allowing for mutations and natural selection, may be considered as a most essential anastrophe (H. Baltscheffsky, 1997) in the chains of events resulting in the origin and subsequent evolution of life. Before, during and after its origin, particularly significant energy sources may have been light energy and chemical energy. The metabolism of living cells is known to be based on their capacity for capture and conversion of energy, with energy coupling allowing the energy requiring reactions to be driven by those reactions, which under prevailing conditions can supply the necessary energy.

Among the many still open questions about the energetics of the origin of life are: 1) was the source of energy for the origin of life photic or chemical or both, and were there other significant energy sources, such as electrical discharges etc.?; 2) were the first energy-rich chemical compounds of late chemical or early biological evolution inorganic or organic or of both kinds?; and 3) were there several systems for energy conversion at the prenucleotide level, such as, for example, those which have been suggested to be based on inorganic pyrophosphate ("PPi world"), organic thioesters ("thioester world") and iron-sulphur compounds ("iron-sulphur world"), and were there close chemical links between such "worlds"?

Some major anastrophic transitions in the energetics of the origin and early evolution of life will be presented. Our search for very early or "primitive" enzymes will be discussed, with examples taken from recent results with both soluble and membrane-bound inorganic pyrophosphatases (PPases).

1. Introduction

Living organisms of this day and age essentially depend on sunlight, directly or indirectly, as the driving force for all their different energy requiring reactions. Light of

J. Chela-Flores and F. Raulin (eds.),
Exobiology: Matter, Energy, and Information in the Origin and Evolution of Life in the Universe, 95–102.

suitable wavelengths is transformed in the photosynthetic organisms, green plants, algae and various photosynthetic bacteria, to electric, protic or other ionic potential differences between reactive molecules or over membranes, thus giving rise in coupled reactions also to the formation of energy-rich chemical compounds.

In this photosynthetic energy conservation, photophosphorylation, the energy-rich compound formed was found to be ATP (adenosine triphosphate) (Arnon et al., 1954), the major "energy currency" in all known living cells. The first and still only known alternative to ATP in photophosphorylation was obtained in photobacterial chromatophores and found to be PPi (inorganic pyrophosphate) (H. Baltscheffsky et al., 1966). The discovery of this alternative light-induced conversion and conservation of energy in the structure of PPi was made possible by simply adding to the reaction mixture only Pi (inorganic orthophosphate) and leaving out ADP, which was required for ATP formation.

The photosynthetic apparatus of known organisms is far too complex to allow very detailed assumptions about the sequences of steps involved in its very early evolution. On the other hand, excellent attempts have recently been made to obtain reasonable overviews on the evolution of different kinds of reaction center and photosystem in prokaryotes, algae and plants (Rutherford and Nitschke, 1996; Nitschke et al. 1996). Owing to the great complexity of all currently known photosynthesis in living cells, there seems to exist today an overwhelming opinion in support for the assumption that energy conversion for the origin of life on Earth was not driven by light. This may be correct, but it cannot be excluded that traces of an existing evolutionary path from unknown early photosystems to prokaryotic photosynthesis have existed but disappeared, perhaps already long ago, or have not yet been discovered.

A more optimistic view, that still existing traces are only awaiting discovery, may also be entertained. After all, the finding of the photosynthetic PPi formation alternative to the ATP formation, the former showing less complexity both at the substrate level and the enzyme level (the PPi synthase has one kind of protein subunit whereas the ATP synthase has at least eight kinds, which may be considered to be a most striking example) shows that plausible primitive variations in the present metabolic panorama may still exist within the field of bioenergy conversion.

After this general introduction we shall briefly discuss three different prebiotic energy conversion systems or "worlds", which have been suggested as important for the origin of life. Thereafter one of these "worlds", the "PPi world" will be treated more fully. We will focus attention on 1) recent theoretical results indicating the existence of a "primitive" soluble PPase in the archaeon *Methanococcus jannaschii* (H. Baltscheffsky et al., 1997) and 2) experimental data with the gene of the so far unique membrane-bound and proton-pumping PPi synthase from *Rhodospirillum rubrum* (M. Baltscheffsky et al., 1997). In this connection attention will be given to some plausible, major anastrophic transitions of energy conversion.

2. Energy "Worlds" for the Origin of Life

In this last decade of the 20th century, three different systems or "worlds" for prebiotic energy conversion leading to the origin of life have been proposed: the "thioester world" (de Duve, 1991), the "iron-sulphur world" (Wächtershäuser, 1992), and the "PPi world" (H. Baltscheffsky, 1993). The possible occurrence and prebiotic significance of more than one of these suggested worlds, as well as molecular interactions between them, have been discussed (H. Baltscheffsky, 1996).

Good arguments have been presented in support of each of the three prebiotic energy conversion worlds, and one may speculate at length about which of the three may have been the oldest or the most important. Considering our present knowledge about cellular metabolism and the central role of ATP as energy donor to its energy requiring reactions one finds that the logical chemical candidate to the role as immediate predecessor to ATP is PPi. Also from the energetic point of view PPi is structurally very similar to ATP.

In this respect, and extrapolating backwards from ATP metabolism (Lipmann, 1965), PPi metabolism and the PPi world stand out as being particularly pertinent. The earlier given views (H. Baltscheffsky and M. Baltscheffsky, 1995) that PPi may have been important already in chemical evolution leading to the origin of life and that traces may still be found in the structures of proteins of a transformation from PPi metabolism to ATP metabolism will be discussed. Some new results concerning PPi and ATP metabolizing proteins and their apparent significance will be specially considered.

3. On the "PPi world"

Biological, chemical, and geological support has been given for our hypothesis concerning a central, prebiotic and early biotic role for PPi on the Earth at the time when the evolutionary process led to the origin of life and biological evolution, *i. e.* for the suggested "PPi world" (H. Baltscheffsky and M. Baltscheffsky, 1995; H. Baltscheffsky, 1997). As has been mentioned above, a subsequent and possibly still traceable switch from PPi metabolism to ATP metabolism may have occurred thereafter.

Two new developments appear to be of significance in this context. Both are concerned with PPases, one with what appears to be a most "primitive" one (H. Baltscheffsky et al., 1997), the other with the most advanced one, which physiologically is capable also of functioning as a PPi synthase, at the expense of electron transport driven by light or suitable substrates in isolated chromatophores from the photosynthetic bacterium *R. rubrum* (M. Baltscheffsky et al., 1997).

When the first complete genome of an archaeon, the hyperthermophilic *Methanococcus jannaschii*, was published (Bult et al., 1996), the list of derived and identified proteins did not contain any inorganic pyrophosphatase. Assuming that the presence of a functioning PPase is necessary for the survival of any known free-living organism, we decided to search for it with somewhat reduced identity and similarity demands when compared to known soluble PPases and using a consensus sequence found in soluble PPases.

One of the originally deduced putative proteins, MJ0882_hyp, appeared to qualify much better than other possible alternatives as a plausible PPase, with most of the known active site characteristics present. Of the seven charged amino acids known to be of particular importance in the active sites of soluble PPases (Heikinheimo et al., 1996), six were found in corresponding sequence positions in MJ0882-hyp. The seventh, an arginine, in the known soluble PPases, was not present in MJ0882_hyp, but a lysine residue was found in the corresponding position (H. Baltscheffsky et al., 1998). Thus, the positive charge was present and apparently required also here. Furthermore, the size of the molecule, the three-dimensional modelling of it with the energy minimization method and some apparent capability for oligomerization, all seemed to be in agreement with what could be expected for a "primitive", soluble prokaryotic PPase.

In addition, a sequence motif, similar to that found in vacuolar membrane-bound and most of the soluble PPases (D/EX$_7$KXE, where D or E may occupy the first position and where X stands for any of the 20 amino acids found in proteins) (Rea et al., 1992), was found to be present in appropriate position, in the sequence DILRGKKLKof MJ0882-hyp (H. Baltscheffsky *et al.*, 1998), giving the general sequence

D/EX_7K

which has been found to be a metal and substrate binding stretch in the active-site model for PPase catalysis of PPi hydrolysis (Heikinheimo et al., 1996), where D/E (Asp/Glu) binds to metal (Mg) and K (Lys) to substrate. The unusually high content of charged amino acids in the sequence motif of MJ0882_hyp may be assumed to be due to high stability requirement for proteins at the high temperatures for growth of this hyperthermophilic archaeon.

Of special interest for us in this connection was the finding that the derived primary structure for the protein of PPi synthase from *R. rubrum* contained the motif discussed above, in the sequence situated between the fourth and the fifth of its 15 deduced transmembrane segments (M Baltscheffsky et al., 1998), namely

DVGADLVGK

When this motif is compared with the motif A (Walker et al., 1982) or P-loop (phosphate-binding loop) (Saraste et al., 1990) of ATP synthase subunit β, namely

$GX_2GVGKTV$

it is seen that the sequence VGK is shared by all three. An additional G in equivalent position is found in both the *R. rubrum* PPi synthase and the ATP synthase subunit β. These four identities are also found in the known vacuolar PPases and may well be more of functional than of evolutionary significance. Additionally, it may be more than a coincidence that, of the soluble-type PPases, only the one obtained from *Saccharomyces cerevisiae* mitochondria (Lundin et al., 1991) has a G preceding the K (and also a T preceding the G, as in the ATP synthase subunit α, which is mentioned in view of the fact that the α and β subunits of the ATP synthase are known to be homologous and thus have a common origin).

Summarizing the results of these sequence motif comparisons, the following facts have emerged:

1) a flexibility providing glycin residue immediately precedes a lysine residue, which appears to be involved in substrate binding in the membrane-bound, proton-pumping PPases, PPi synthase and ATP synthases;

2) among the soluble PPases only the "soluble-type" yeast mitochondrial PPase has a glycine in the position immediately preceding lysine.

In view of these described similarities, one may speculate that a suitably located glycyl-lysyl unit is required for the biological coupling between the pumping of protons and phosphorylation. One may also wonder whether the existence of this unit, uniquely in the "soluble-type" yeast mitochondrial PPase among known solube PPases, indicates a capability of this protein to participate in chemiosmotic phosphorylation.

These observations may, at least in current contexts, be more of dynamic than of evolutionary significance. One major evolutionary, and anastrophic, transition from PPi metabolism to ATP metabolism has, however, recently been discovered (Michels *et al.*, 1998). Clear evidence has been obtained, in the structure of the proteins involved, for the molecular evolution from an ancestral PPi-dependent phosphofructokinase to the glycosomal ATP-dependent enzyme in *Trypanosoma brucei*. Results of this kind give additional support for the idea that current but still unknown metabolic patterns may be a rich source of information for increased understanding of the origin and early evolution of biological energy conversion.

4. Comments about PPases on different levels of sophistication

This presentation has to a large extent been focussed on different levels of sophistication in the biological structure and dynamics of one of the two central participants in the PPi world evolutionary perspective, namely the PPase protein. The present picture is one of very rapid development at the two extremes of PPase molecular structure as well as at the relevant, but in most of its aspects still only dimly perceived, presumed evolutionary level.

4a. A hypothetical PPase at low level of sophistication

Our new results concerning the hypothetical protein MJ0882_hyp, derived from the genome of the hyperthermophilic archaeon *M. jannaschii*, indicating that it is a PPase (H. Baltscheffsky et al., 1998), an enzyme which was earlier missing from the list of putative enzymes derived from the complete genome, point toward the existence of a less structurally and functionally advanced soluble PPase than those which have so far been sequenced. In a number of homologous sequences of hypothetical proteins, the identities with amino acids of MJ0882_hyp are above 30 - 40 %. However, most of the functionally important amino acids in the acive site of known PPases are missing, in contrast to the case with MJ0882_hyp.

We assume that at least some members of this earlier unrecognized family of hypothetical proteins may earlier have been of PPase nature (H. Baltscheffsky *et al.*, 1998). The bacteria with these putative, "earlier" PPases are endowed with the normal variety of soluble PPase. Thus such a less advanced and less efficient PPase, presumably of a more ancient kind, may not have been useful anymore but gradually evolved away from PPase nature, and perhaps in some cases to a novel metabolic function.

This view would appear to take us closer to the origin of PPase function and its biological evolution. At the enzyme level the takeover of soluble PPase function from a less to a more efficiently functioning structure may be considered a catastrophe for the less and an anastrophe for the more advanced PPase protein, in those organisms where both were present. For biological evolution as a whole, such significant mutational steps in the direction of increasing perfection may be regarded as typical metabolic anastrophes.

4b. A highly sophisticated PPase, functioning physiologically also as a PPi synthase

The sequencing of the whole gene of a PPi synthase (M. Baltscheffsky et al., 1998) has given new impetus to the evolutionary perspective concerning biological energy conversion in two directions, both backwards, toward the other PPases and forward, toward the ATP synthases. The new knowledge about the derived amino acid sequence of the PPi synthase has already provided novel information about the position of the active site and the probable area of proton-pumping (M. Baltscheffsky *et al.*, 1998). It may be emphasized, that they are both on the same polypeptide, which is so far unique for a membrane-bound synthase protein involved in biological electron transport phosphorylation.

Since the derived amino acid sequence of the PPi synthase is homologous with but not entirely similar to those of the known vacuolar, membrane bound and proton-pumping PPases, the possibility has now arisen to ask detailed questions about the differences between them with respect to the active sites, uniquely allowing the PPi synthase to function physiologically as a synthase. Such information may be expected to be of value also for detailed molecular comparison between the PPi synthase and ATP synthases.

4c. Some common properties

The above discussed existence of a similar metal and substrate binding motif in all known different types of PPase now includes also the PPi synthase. Some similarity was found also between this motif and the phosphate binding motif (P-loop) of ATP synthase β-subunits. Thus these now clearly discerned common properties include all known and putative PPases, and the PPi and ATP synthases. This structural link extending from the most "primitive", putative soluble PPase over known soluble and membrane-bound PPases and PPi synthase, to the β-subunit of the ATP synthase has today no distinct evolutionary counterpart in these systems. And any detailed speculation about what the current wealth of new and incoming information may bring in this respect would be premature.

Nevertheless, it seems valuable to point out that studies on the evolution of sequence motifs and enzyme acivity have contributed to our knowledge about homologies and thus evolutionary links even between enzymes with functions as different as reductase and sulfurylase (Bork and Koonin, 1994).

4d. A look ahead

Macromolecular sequence and three-dimensional structure comparisons as well as the new field of bacterial genomics would appear to show great promise for further solutions of problems also in the area of molecular energetics in the origin and evolution of life.

The evolutionary steps leading to those energy conversion reactions which drove the earliest life processes may well become amenable to more detailed hypotheses with the help of further advances in our understanding of "primitive" versions of obligatory enzymes. Extrapolation backward from them, in addition to experimentation with prebiotic models and molecules should contribute to throw more light on the major anastrophic transitions which led to the necessary continuity in the supply and conversion of energy for the processes of life to begin and to go on.

This point of view can be elucidated with an apparent potential link between prebiotic, chemical evolution and an amino acid sequence of assumed functional significance, which has been found in all membrane-bound and proton-pumping PPases, including the PPi synthase. It was mentioned above, that they contain a putative substrate and metal binding motif, which in the *R. rubrum* PPi synthase is located in the loop between the fourth and fifth of its 15 putative transmembrane segments. This loop in *R. rubrum* consists of 57 amino acid residues, of which those in the mentioned motif are in positions 11 to 19. The loop may be presumed to contain the entire or the most essential part of the active site for the actual phosphorylation/dephosphorylation, as coupled to the proton-pumping through the membrane. Another similarly common sequence motif is located in positions 37 to 45 (DNVGDNVGD) of the loop. Its three aspartyl residues could well be of importance for cation binding and/or hydrogen bonding in the active site. It may have emerged from an early tetrapeptide repeat:

DNVG or Asp-Asn-Val-Gly

One is struck by the fact that this tetrapeptide structure contains three (Gly, Val and Asp) of the four amino acids claimed to have been coded for by the first four triplets of the early, evolving genetic code (Eigen and Schuster, 1978) and of the six amino acids found in the Murchison meteorite (Ponnamperuma, 1971). And the fourth, Asn, also an "early" amino acid, can be formed from Asp by single mutation. So, early chemically formed amino acids, and peptides from them, for example tetrapeptides, may have become useful in connection with the origin of life and incorporated into essential parts of enzymes still operating in pyrophosphate bioenergetics. This development would seem to have involved anastrophic molecular events in both chemical and biological evolution.

5. Acknowledgements

Supported by Carl Tryggers Stiftelse för Vetenskaplig Forskning, Magnus Bergvalls Stiftelse and Wenner-Grenska Samfundet.

6. References

Arnon, D.I., Allen, M.B., and Whatley, F.R. (1954) Photosynthesis by isolated chloroplasts, *Nature* **174**, 394-396.

Baltscheffsky, H. (1993) Chemical origin and early evolution of biological energy conversion, in C. Ponnamperuma and J. Chela-Flores (eds.), *Chemical Evolution: Origin of Life*, A. Deepak, Hampton, pp. 13-23.

Baltscheffsky, H. (1996) Energy conversion leading to the origin of life: did inorganic pyrophosphate precede adenosine triphosphate?, in H. Baltscheffsky (ed.), *Origin and Evolution of Biological Energy Conversion*, VCH Publishers, New York, pp. 1-9.

Baltscheffsky, H. (1997) Major "anastrophes" in the origin and early evolution of biological energy conversion, *J. Theor. Biol.* **187**, 495-501.

Baltscheffsky, H. and Baltscheffsky, M. (1995) Energy, matter and self-organization in the early molecular evolution of bioenergetic systems, in J. Chela-Flores, M. Chadha, A. Negron-Mendoza, and T. Oshima, (eds.), *Chemical Evolution: Self-Organization of the Macromolecules of Life,* A. Deepak, Hampton, pp. 83-89.

Baltscheffsky, H, Baltscheffsky, M., Nadanaciva, S., Persson, B., and Schultz, A. (1997) Possible origin and evolution of inorganic pyrophosphatases, Abstract, in R. Lahti (ed.), *Proc. 1st Int. Meeting on Inorganic Pyrophosphatases*, Turku, pp. 1-3, (ISBN 951-29-0990-1).

Baltscheffsky, H. Schultz, A., and Persson, B. (1998) MJ0882_hyp: An archaeal "early" pyrophosphatase related to methyltransferases and protoporphyrinogen oxidases?, (submitted).

Baltscheffsky, H., von Stedingk, L.-V., Heldt, H.W., and Klingenberg, M. (1966) Inorganic pyrophosphate: formation in bacterialphotophosphorylation, *Science* ***153***, 1120-1122.

Baltscheffsky, M., Schultz, A., and Nadanaciva, S. (1997) The inorganic pyrophosphate synthase from *Rhodospirillum rubrum* and its gene, Abstract, in R. Lahti, (ed.), *Proc. 1st Int. Meeting on Inorganic Pyrophosphatases*, Turku, pp. 16-18, (ISBN 951-29-0990-1).

Baltscheffsky, M., Nadanaciva, S., and Schultz, A. (1998) A pyrophosphate synthase gene: Molecular cloning and sequencing of the cDNA encoding the PPi synthase from *Rhodospirillum rubrum*, (submitted).

Bork, P. and Koonin, E.V. (1994) A P-loop-like motif in a widespread ATP pyrophosphatase domain: Implications for the evolution of sequence motifs and enzyme activity, *Proteins* **20**, 347-355.

Bult, C.J. et al. (1996) Complete genome sequence of the methanogenic archaeon, *Methanococcus jannaschii*, Science **273**, 1058-1072.

de Duve, C. (1991) *Blueprint for a Cell. The Nature and Origin of Life*, Patterson, New York.

Eigen, M. and Schuster, P. (1978) The hypercycle, *Naturwiss.* **65**, 341-369.

Heikinheimo, P. et al. (1996) A site-directed mutagenesis study of *Saccharomyces cerevisiae* pyrophosphatase. Functional conservation of the active site of soluble inorganic pyrophosphatases, *Eur. J. Biochem.* **239**, 138-143.

Lipmann, F. (1965) Projecting backward from the present stage of evolution of biosynthesis, in Fox, S.W. (ed.) *The Origins of Prebiological Systems and their Molecular Matrices*, Academic Press, New York, pp. 212-226.

Lundin, M., Baltscheffsky, H., and Ronne, H. (1991) Yeast *PPA2* gene encodes a mitochondrial inorganic pyrophosphatase that is essential for mitochondrial function, *J. Biol. Chem.* **266**, 12168-12172.

Michels, P.A.M., Chevalier, N., Opperdoes, F.R., Rider, M.H., and Rigden, D.J. (1998) The glycosomal ATP-dependent phosphofructokinase of *Trypanosoma brucei* must have evolved from an ancestral pyrophosphate-dependent enzyme, *Eur. J. Biochem.* (in press).

Nitschke, W., Mattioli, T., and Rutherford, A.W. (1996) The FeS-type photosystems and the evolution of photosynthetic reaction centers, in H. Baltscheffsky (ed.), *Origin and Evolution of Biological Energy Conversion*, VCH Publishers, New York, pp. 177-203.

Ponnamperuma, C. (1971) Primordial organic chemistry and the origin of life, *Quart. Revs. Biophys.* **4**, 77-106.

Rea, P.A. et al. (1992) Vacuolar H^+-translocating pyrophosphatases: a new category of ion translocase, *TIBS* **17**, 348-353.

Rutherford, A.W. and Nitschke, W. (1996) Photosystem II and the quinone-iron-containing reaction centers: comparisons and evolutionary perspectives, in H. Baltscheffsky (ed.), *Origin and Evolution of Biological Energy Conversion*, VCH Publishers, New York, pp. 143-175.

Saraste, M., Sibbald, P.R., and Wittinghofer, A. (1990) The P-loop - a common motif in ATP- and GTP-binding proteins., *TIBS* **15**, 430-434.

Walker, J.E., Saraste, M., Runswick, M.J., and Gay, N.J. (1982) Distantly related sequences in the α- and β-subunits of ATP synthase, myosin, kinases and other ATP-requiring enzymes and a common nucleotide binding fold, *EMBO J.* **1**, 945-951.

Wächtershäuser, G. (1992) Groundworks for an evolutionary biochemistry: the iron-sulphur world, *Prog. Biophys. molec. Biol.* **58**, 85-201.

VISUALLY RETRACING THE EMERGENCE OF THE EVOLVABLE PROTOCELL

RANDALL GRUBBS[1], SIDNEY W. FOX[1], ARISTOTEL PAPPELIS[2], JOHN BOZZOLA[2], AND PETER R. BAHN[3]
1. Department of Marine Sciences, University of South Alabama, Mobile, AL 36688 USA 2. Department of Plant Biology, MC 6509, Southern Illinois University at Carbondale, IL 62901 USA 3. Bahn Biotechnology, RR 2, Box 239A, Mount Vernon, IL 62864 USA

Abstract. Seven thermal proteins and two of their dialyzed products were observed (oil-immersion light microscopy 1,000x @ 60° C) as they were moistened with 1.0% aqueous solutions of each of seven salts to yield protocells by self-assembly. The spherical protocells ranged in diameters from 0.1 to 10 μm, but formed two populations most commonly ~ 0.5 and ~ 3.0 μm. The small protocellular population was first visible within 30 sec and the larger protocells formed about 30 sec to 4 min. later. Both populations increased slightly in diameter within the following 15-30 min.

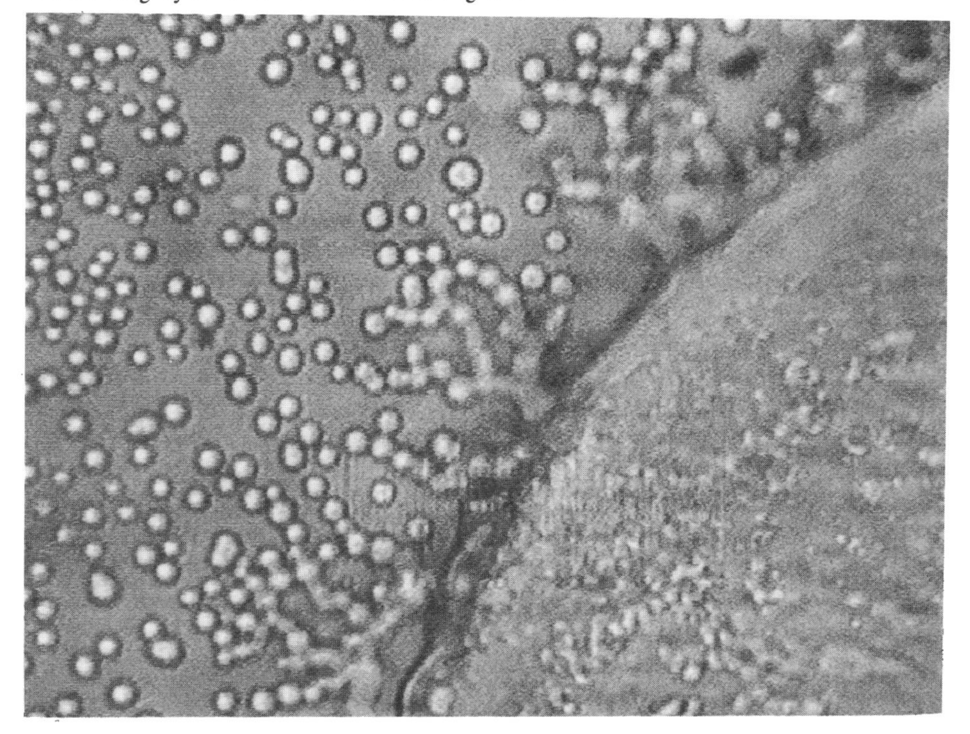

Figure 1.Protocells (~ 3.0 μm in diameter) resulting from the contact of aqueous 1.0% NaCl solution with a granule of thermal protein Poly (asp, glu, pro : 1:1:1: by weight) at 60 °C.

J. Chela-Flores and F. Raulin (eds.),
Exobiology: Matter, Energy, and Information in the Origin and Evolution of Life in the Universe, 103–106.

1. Introduction

In 1954, amino acids were known to copolymerize with only simple heat as an energy source to create peptide bonds (6). Four years later, in 1958, it was published that amino acids will order themselves into Proteinoid (4) (listed as Protein; thermal by Chemical Abstracts since 1972). It has been shown through repeated experiments that Thermal Proteins convert to protocells (5) upon contact with water or aqueous solutions. Protocells produced by the self-organization of thermal protein have only recently become recognized by many as such due to over forty years of experimentation and characteristic cataloguing which have revealed that they exhibite attributes(or the roots thereof) used to describe prokaryotic and eukaryotic life(7). Though the emergent vital attributes are present in protocells, the mechanisms by which they function in modern cells appear to have been specialized through evolution. The one attribute that protocells do not share with prokaryotic nor eukaryotic cells is their immediate genesis from inanimate thermal protein. This event has been filmed by use of a new technique which allows the visualization of the protocell formation not possible using the traditional method of (1-3 mg thermal protein/1ml boiling water). When moistened with water or (1.0%) aqueous solutions of salts common in sea water, protocells (~ 0.1-10 μm in diameter) formed from the thermal protein by self-assembly.

2. Materials and Methods

Seven thermal protein complexes were synthesized at 180° C for 6 h (1-3). The cooled products were powdered and aliquots dialyzed (Spectra/por tubing of 1,000 MW cutoff for 24 h at 4° C with four changes of deionized water; products dried at 37° C). Hydrolyzed samples and paper chromatography indicated recovery of the original amino acids. The thermal protein complexes synthesized for this study were; TP 9 = Poly (asp, glu, trp; equimolar); TP 64 = Poly (asp, glu, lys: equimolar); Poly (glu, gly: 1.47:1.50 by weight); Poly (asp, glu, pro: 1:1:1 by weight); leu-rich thermal protein Poly 18 equimolar amino acids (2.0g 18 equimolar amino acids: 2.0g leu by weight); Poly (glu, leu, tyr: 1.5:1.3:1.8 by weight); Poly (asp, glu, DL-dopa; 1:1:1 by weight); and their dialysis products. No protocells, to our search or knowledge, have ever been formed from proteins obtained from prokaryotic or eukaryotic cells.

Granules of thermal protein complexes were heated on alcohol-cleaned microscope slides (on a hot plate) to melting and alcohol-cleaned cover glasses pressed on the "melt" to form a capillary pathway from the edge of the cover glass to the cooled thermal protein preparations. As needed (three replications), a slide was placed on a sensortek TS-4 series thermal microscope stage set at 60° C. Two-three drops of water or 1% salt solution [NaCl; $FeCl_2$; $FeCl_3$; $MgCl_2$; $MnCl_2$; $CoCl_2$; and $Zn(C_2H_3O_2) \cdot 2H_2O$] were added to the margin of the cover glass while an edge of the thermal protein preparation was being viewed (oil-immersion light microscopy @ 1000x). Small (0.1-0.5 μm) protocells formed almost instantly (uncountable numbers visible within a few to 30 sec). This led us to believe that when this self-assembly reaction is carried out at 100° C, the results must be spontaneous. The more commonly studied protocells of about 3 μm diameter formed after ~3 min. in water and in 1.0% $MgCl_2$ and $MnCl_2$ solutions. These formed within 1-2 min. in 1.0% NaCl, $FeCl_2$, and $FeCl_3$ solutions. The formation of these in 1% $CoCl_2$ occurred after about 4-5 min. These times almost doubled when water or salt solutions were added to dialyzed products. Both sizes of protocells increased slightly with time (15-30 min. after the thermal protein products were moistened).

Protocells formed on the surfaces of thermal protein complex granules moistened by 50 μm droplets of distilled water (on alcohol-cleaned, 1 cm diameter microscope cover glasses incubated in a humid chamber 12 h at 23° C, and air dried at 23° C). Preparations were glued onto specimen stubs, coated with 50 nm palladium/gold alloy, and examined in a Hitachi S2460N scanning electron microscope (SEM) at 20 kV accelerating voltage. Protocells of various sizes were visible and the 3-μm sized protocells had formed linear and branched chains attached to the serface of the proteinoid with distal protocells usually being "buds"(Figures 2 and 3). The possibility that these were small protocells attached to the larger ones was set aside after comparing their sizes; "buds" being of intermediate diameters.

3. Results and Discussion.

We believe that the study of protocell formation and ultrastructure will help us to understand the environmental requirements and evolutionary sequences for the protocell to acquire modern mechanisms. In addition, it is thought that information from such studies may prove invaluable as juxtaposable data for expeditions to Europa and Mars. Understanding the effects of gravity on protocell formation is also important if we are to compare protocellular life found on Earth with that of other planets. This "thermal protein complex-sandwich" system to view the stages of formation at high magnification has the potential for extension. We have been able to dry the preparations and pry off intact cover glasses from the microscope slides. The next step is to view the surfaces of the cover glasses facing the thermal protein complex preparation in addition to the exposed protocells on the slide using SEM methods described above. This kind of high resolution study of protocells could be extended to the search for networks and matrices. If possible, these preparations also should be processed for transmission electron microscopy (TEM) matching structures visible with light and SEM with thin section data obtained using TEM. Analytical cytochemistry and cytofluorescence may be helpful in the study of events occurring during the origin of protocells (the transition from inanimate thermal protein complexes to biotic protocells) and their early evolution in the Domain Protolife (7-9).

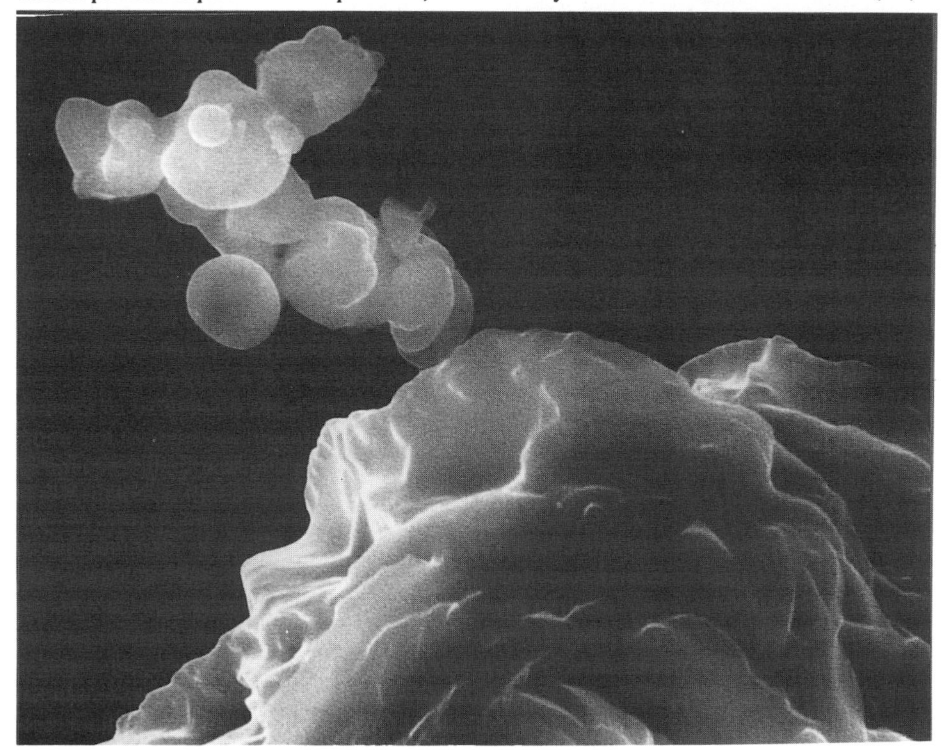

Figure 2. Scanning electron micrograph of protocells resulting from contact of aqueous 1.0% NaCl solution with a granule of TP9-Poly (asp, glu, trp; equimolar) incubated at 23 C.

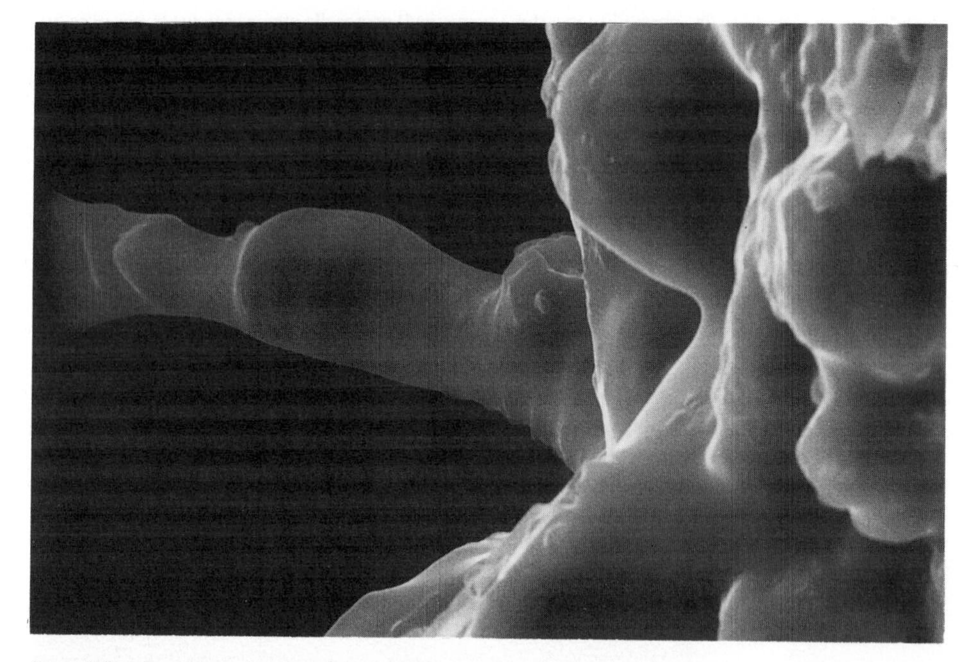

Figure 3. Scanning electron micrograph of protocells resulting from contact of aqueous 1.0% NaCl solution with a granule of TP9-Poly (asp, glu, trp; equimolar) incubated at 23 C. Note the attached base.

References

1. Fox, S. W. 1988. *The Emergence of Life; Darwinian Evolution from the Inside*. Basic Books, New York.
2. Fox, S. W. 1995. To cellular life and neurocellular assemblies. In *Evolutionary and Related Areas of Physicochemical Biology*. P. Poglasov, B. I. Kurganov, M. S. Kritsky, and K.L.Gladilin, Eds., Bach Institute of Biochemistry and ANKO, Moscow, pp. 161-175.
3. Fox, S. W. 1997. My scientific discussions of evolution for the Pope and his scientists. The Harbinger (World Wide Web Address-[http://entropy.me.usouthal.edu/harbinger/articles/rel_sci/fox.html]) 15:1, 6, 18, 19, 21.
4. Fox S.W. and Harada K. 1958. Thermal copolymerization of amino acids to a product resembling protein. *Science* 128 : 1214
5. Fox S.W. , Harada K., and Kendrick J. 1959. Production of spherules from synthetic proteinoids and hot water. *Science* 129; 1221-1223
6. Fox, S. W. and Middlebrook, M. (1954) Anhydrocopolymerization of amino acids under the influence of hypothetically primitive terrestrial conditions. *Federation Proc. 13,* 703.
7. Pappelis, A. and S. W. Fox. 1995. Domain Protolife: The Protocell Theory. In *Evolutionary Biochemistry and Related Areas of Physicochemical Biology*. B. F. Poglazov, B. I. Kurganov, M. S. Kritsky, and M. S. Gladilin, Eds., Bach Institute of Biochemistry and ANKO, Moscow, pp. 151-159.
8. Pappelis, A. and S. W. Fox. 1995. Domain Protolife: Protocells and metaprotocells within thermal protein matrices. In *Chemical Evolution: Structure and Model of the First Cell*. C. Ponnamperuma and J. Chela-Flores, Eds., Kluwer Academic Publishers, Dordrecht, pp. 129-132.
9. Pappelis, A. and S. W. Fox. 1996. Thermal peptides as the initial genetic system. In *Chemical Evolution: Physics of the Origin and Evolution of Life*. J. Chela-Flores and F. Raulin, Eds., Kluwer Academic Publishers, Dordrecht, pp. 157-165.

RADICALS, IONS AND MOLECULES IN CHEMICAL EVOLUTION

MOHINDRA S. CHADHA
C6 Beach House, Juhu,
Mumbai (Bombay) 400 049, India.
(e-mail: chadha@soochak.ncst.ernet.in*)*

1. Introduction

An area of great interest in chemical evolution research is the comparison of results of laboratory experiments with the more recent findings related to the chemistry of interstellar medium (ISM), comets, meteorites, planets and their satellites.
The occurrence and formation of hydrogen and nitrogen, methane and ammonia, hydrogen cyanide, carbon monoxide, formaldehyde, cyanoacetylene, aminonitriles etc and reactions of some of these molecules as such or in combination was presented at the 4th Trieste Conference on Chemical Evolution (Chadha, 1996).
An attempt is being made now to understand some of the mechanisms which may be operating in the formation of organic components in the ISM and relate these to the chemistry of comets, meteorites, other planets, satellites and the laboratory simulation experiments.

2. Nature of Interstellar Medium

The space between the stars in our Galaxy is almost entirely made up of hot (Temperature ~ 10^6K, and n ~ 10^4 nuclei/m^{-3}) and warm, somewhat denser gas (Temperature ~ 10^3–10^4K, n~10^6m^{-3}). The presence of highly ionized atomic oxygen (in hot gas) and H atoms (in warm gas) have been observed. Within this harsh environment, denser clouds of cooler gas are embedded, which contain almost the entire mass of the interstellar medium (Taylor and Williams, 1993).
Diffuse clouds: These are detectable in line of sight of bright stars, showing continuous absorption of star light from infrared to

J. Chela-Flores and F. Raulin (eds.),
Exobiology: Matter, Energy, and Information in the Origin and Evolution of Life in the Universe, 107–112.

ultraviolet wavelengths resulting from interstellar dust. These are cooler (Temperature ≅ 30-100°K) and have a higher number density ($n \cong 10^8$ hydrogen nuclei m^{-3}).
Dark clouds: These are opaque to visible light and have (Temperature ~ 10-30 K and densities of 10^9-$10^{10} m^{-3}$). These in association constitute giant molecular clouds with masses up to a million solar masses.

3. Chemical Evolution in the Interstellar Medium

A variety of molecular species have been identified by radioastronomers using IR and "radio" astronomical techniques which include observations in microwave, millimetre and sub-millimetre regions. The molecules detected range from simple diatomic species like H_2, CN, CO etc to polyatomic molecules like HCN, HCHO, NH_3, CH_3OH, CH_3CN, C_2H_5OH and even highly unsaturated species like CH_3C_4H, HC_9N, $HC_{11}N$ and CH_3C_5N.
The molecules which have been identified in outer space and have a direct relevance to Chemical Evolution and Origin of Life are shown in Fig. 1.

H_2	CO	HC≡C–CN
HCN	H_2O	H_2C=CH–CN
NH_3	CH_2O	H_2C=NH
CH_4	HCO_2H	H_3CCN, H_3CNH_2

Figure 1. Some interstellar molecules of direct relevance to Chemical Evolution.

There is evidence for the existence of charged species e.g. CH^+, HCO^+, HCS^+, $HOCO^+$, $HCNH^+$ etc. To date, over 100 molecular species (Irvine, 1995) have been identified, though the actual number is much larger if isotopic species are included. Taylor and Williams (1993) have summarised different aspects of interstellar chemistry and the importance of ion-molecule reactions has been highlighted.

4. Carbon Chemistry of Comets

Comets which consist of a nucleus, a coma and a tail demonstrate some interesting chemical composition. The comet nuclei contain simple and complex organic molecules and dust and rock embedded in a frozen water matrix. On heating, while approaching the Sun, the "parent" volatile compounds e.g. water, carbon dioxide, methane, acetylene, ammonia, hydrogen cyanide etc.) get ejected. In the resulting coma, on account of physico-chemical processes leading to partial or complete breakdown of the so-called parent molecules, daughter products are formed. The unchanged chemical entities are observed in the coma and the ionised (positive charged) species are observed in the tail.

Comets are believed to be made up from matter condensed and accreted in the outer regions of primitive solar nebula. There must be a relationship between the interstellar matter and the cometary components. There are similarities in the components identified in the interstellar medium and the coma and tail of comets.

5. Organic Matter in Meteorites

The organic matter in carbonaceous chondrites occurs in a variety of forms, which range from hydrocarbons (aliphatic and aromatic), carboxylic acids, amino acids, ketones and aldehydes to alcohols, amines, ureas, amidines and nitrogen containing heterocycles. One of the most thoroughly and extensively studied carbonaceous chondrites is the Murchison meteorite, (Kvenvolden *et al*, 1970). From these studies and those that followed, the nature of 2.0-2.5% total carbon in Murchison meteorite was obtained. The components identified were acids (monocarboxylic (C_2-C_8), dicarboxylic (C_2-C_9) and hydroxy (C_2-C_5); amino acids; alcohols (C_1-C_4); aldehydes (C_2-C_4), ketones (C_3-C_5), ureas, amines and heterocyclics.

The most exciting finding from these discoveries was the fact that the stereochemistry and the molecular structures of the amino acids and carboxylic acids has been unequivocally established to be of nonbiological and extra terrestrial origin.

6. Some Laboratory Studies and Possible Mechanisms

6.1 FORMATION OF HCN

Woeller and Ponnamperuma (1969) reported the formation of HCN in simulated Jovian atmosphere experiments and Gupta *et al* (1981) in the simulated Titan atmosphere.

Anders *et al* (1974) showed the formation of HCN and other nitriles during Fischer Tropsch reaction of CO, NH_3, H_2 at 500°C. Also its formation has been demonstrated by photolysis of a mixture of CH_4 and NH_3 by Ferris and Chen (1975) and photolysis of NH_3 in the presence of acetylene by Ferris and Ishikawa (1988).

They have proposed a free radial mechanisms for the formation of HCN.

6.2 PHOTOCHEMICAL PROCESSES IN INTERSTELLAR DUST FORMATION

Greenberg and Mendoza-Gomez (1993) have carried out pioneering work in providing insights into the nature of interstellar dust.
Briggs, Ferris, Greenberg and their collaborators (1992) have studied extensively, the products of photolysis of mixtures of CO, NH_3 and H_2O at 12°K. A large variety of products were identified by them which included C_2–C_3 hydroxy acids and hydroxy amides, glycerol, urea, glycine, hexamethylene tetramine, formamidine and ethanolamine in the product mixture. The middle molecular weight fraction was shown by Mendoza-Gomez and Greenberg (1993) to consist of highly unsaturated aromatic hydrocarbons as well as some oxygenated products. The volatile part of the photolytic products has been shown by Greenberg and Mendoza-Gomez (1993) to consist of CO_2, C_2H_4O, C_4H_6O, C_5H_8O, $C_6H_{10}O$, C_7H_5N, C_9H_8 etc.

6.3 FORMATION OF SIMPLE BIOMOLECULES IN INTERSTELLAR CLOUDS

Efforts are being made to study the mechanisms involved in the formation of a variety of compounds e.g., alcohols, aldehydes, ethers, ketones, acetylene derivatives, nitriles and amides. According to Steven Charnley of Ames, NASA (1995), such compounds can be formed by ion-molecule and neutral reactions in the cold, accretion of gas phase species on interstellar dust grains and ion-molecule and neutral-neutral reactions in evaporated molecules in hot cores.
A rich array of complex organic molecules exists in the hot cores. These are alcohols (methanol, ethanol), dimethyl ether, methyl formate, acetone, aldehydes (formaldehyde, acetaldehyde), formic acid, nitriles and amides.
The formation of the entire inventory of the observed organic molecules has been explained by invoking minimum number of reaction mechanisms on grain surfaces and the hot core gas.

The elaboration of the ketene ($H-\dot{C}=O$) to CH_3OH by H addition and higher alcohols by successive additions of C and H has been proposed.

6.4 PHOTOCHEMICAL MODELLING OF TITAN'S ATMOSPHERE

Raulin and collaborators (1994, 1995 and references cited therein) have carried out detailed and in depth studies on organic chemistry of Titan and comparison of components in Titan's atmosphere and those resulting from experimental and theoretical modelling.

7. Conclusions

An attempt has been made to highlight information available on the nature of organics in the ISM, in some comets, planetary satellites, meteorites (carbonaceous chondrites) and from chemical evolution experiments in the laboratory. The role of radicals and ions under different situations has been discussed. It may not be too presumptuous to conclude that a vast majority of gas phase simulation experiments carried out even at room temperature, leading to the formation of a variety of products (as such or on hydrolysis), could have gone through photolytic or radiation induced radicals or radical ions. Even in our early experiments (Chadha *et al* 1971a, b) when CH_4 and NH_3 were subjected to spark discharge some of these intermediates must have been involved in the formation of a large number of products.

REFERENCES

Anders, E., Hayatsu, R., and Studier, M. H. (1974) Interstellar Molecules: Origin by catalytic reactions on grain surfaces. Astrophysics. J. 192, L101-L105.

Briggs, R., Ertem, G., Ferris, J.P., Greenberg, J. M., McCain, P. J., Mendoza-Gomez, C. X., and Schutte, W. (1992) Comet Halley as an aggregate of interstellar dust and further evidence for the chemical formation of organics in the interstellar medium. Origins of Life and Evolution of the Biosphere 22, 287-307.

Chadha, M. S., Lawless, J. G., Flores, J. J., and Ponnamperuma, C. (1971a) Experiments in Jovian Atmosphere in Chemical Evolution and the Origin of Life. Eds., R. Buvet and C. Ponnamperuma. North Holland Pub. Co., Amsterdam, 143-151.

Chadha, M. S., Flores, J.J., Lawless, J. G., and Ponnamperuma, C. (1971b) Organic Synthesis in a Simulated Jovian Atmosphere - II Icarus 15, 37-44.

Chadha, M. S. (1996) Role of transient and stable molecules in Chemical Evolution: In Chemical Evolution: Physics of the Origin and Evolution of Life, Eds., J. Chake-Flores and F. Raulin, Kluwer Academic Publishers, Netherlands 107-122.

Charnley, S. B. (1995) Formation of simple biomolecules in interstellar clouds. Space Science Division Report, NASA Ames Research Centre, Calif. 30-33.

Ferris, J. P. and Chen, C. T. (1975) Photosynthesis of Organic Compounds in the Atmospheric of Jupiter, Nature (London) 268, 587-588.

Ferris, J. P. and Ishikawa, Y. (1988) The formation of HCN and Acetylene by photolysis of Ammonia in the presence of Acetylene: Application to the Atmosphere Chemistry of Jupiter, J. Am. Chem. Soc., 110, 4306-4312.

Greenberg, J. M. and Mendoza-Gomez, C. X. (1993) Interstellar Dust Reservoir: A reservoir of prebiotic molecules in the chemistry of life's origins. Eds. J. M. Greenberg *et al*, Kluwer Academic Publishers, Netherlands, 1-32.

Gupta, S., Ochia, E., and Ponnamperuma, C. (1981) Organic synthesis in the atmosphere of Titan, Nature 293, 725-727.

Irvine, W. M. (1995) Organic molecules in the gas phase of dense interstellar clouds. Adv. Space Res. 15, 35–43.

Kvenvolden, K. A., Lawless, J., Pering, K., Peterson, E., Flores, J., Ponnamperuma, C., Kaplan, I. R., and Moore, C. (1970) Evidence for Extraterrestrial Amino Acids and Hydrocarbons in the Murchison Meteorite, Nature, 288 923-926.

Mendoza-Gomez, C. X. and Greenberg, J. M. (1993) Laboratory simulation of organic grain mantles – Origins of Life and Evolution of the Biosphere 23, 23-28.

Raulin, F., Bruston, P., Coll, P., Cosica, D., Gazeau M-C., Guez, L., and de Vanssay, E. (1994) Exobiology of Titan, J. Biol. Phys. 20, 39-53.

Raulin, F., Aflalaye, A., Bruston, P., Coll, P., Gazeau, M-C, Sternberg, R., de Vanssay, E., Coscia, D., and Israel, G. (1996) In Chemical Evolution: Physics of the Origin and Evolution of Life, Eds., J. Chela-Flores and F. Raulin, Kluwer Academic Publishers, Netherlands, 357-371.

Taylor, S. and Williams, D. (1993) Star-studded chemistry, Chemistry in Britain 29(8), 680-683.

Woeller, F. and Ponnamperuma, C. (1969) Organic synthesis in a simulated Jovian atmosphere, Icarus 10, 386.

SOME ASPECTS OF THE IRRADIATION OF HCN AND RELATED COMPOUNDS AND THEIR ROLE IN PREBIOTIC CHEMISTRY.

A. Negrón-Mendoza and S. Ramos-Bernal
Instituto de Ciencias Nucleares, UNAM
México, D.F. México

1. Introduction.

Synthesis of organic compounds under possible Earth conditions is successful while the medium is reducing, but under oxidizing conditions, the products are negligible (Chang, 1971). There is no longer a consensus on the prevalence of reducing conditions in the early Earth atmosphere, and more evidences consider the atmosphere as a mixture of CO_2 CO, N_2 and H_2O. It is important to r-examine the prebiotic sources of organic compounds by pathways that fit geological/atmospheric requirements of a neutral atmosphere. These considerations reveled the need to evaluate penetrating sources of energy like ionizing radiation and sonochemistry.

Radiation-induced reactions have been proposed as mechanisms for synthetizing organic compounds under conditions that existed at the time life on Earth began. Simple precursors cause a large variety of compounds in radiation induced reactions.

CN-containing molecules are key compounds because of the high chemical reactivity of the CN group. The occurrence and formation of HCN came from several sources, terrestrials (like volcanic eruptions), and extraterrestrial (in comets and the interstellar space).

It is possible to synthetized cyanides eassily with simulated experiments (Chada et al., 1971, Toupace et al., 1978; Gupta et al., 1981 and Ehrenfreund et al., 1995). Implications for prebiotic chemistry are profound due to the fact that these compounds produce polymeric materials that upon hydrolysis release compounds of biological significance such amino acids, purines, carboxylic acids, etc.

Thermal polymers have been described very early in the literature, and their importance has been shown (Volker, 1960; Mattews and Moser, 1967; Ferris et al., 1981; Mathews et al., 1984). Much effort has been devoted to dilucidate the structure of HCN polymers. Yet, the thermal condensation reactions for HCN require very concentrate solutions, that seems unrealistic under simulating primitive Earth conditions.

J. Chela-Flores and F. Raulin (eds.),
Exobiology: Matter, Energy, and Information in the Origin and Evolution of Life in the Universe, 113–116.

In this paper we will present results from the formation of an oligomeric material starting from the gamma irradiation of very diluted solutions of HCN and mixtures of CN-containing compounds.

This research for the study of the oligomeric material formed by the action of ionizing radiation in aqueous solutions of HCN and ammonium cyanide, has followed two lines: a) To find the minimum concentration of HCN and doses of irradiation for the detection of oligomeric material. b) Study the effect of radiation dose in the formation of oligomers from HCN.

2. **Experimental Procedures**

2.1 PREPARATION OF CHEMICALS AND GLASSWARE.

All the chemicals were of the highest purity available. The water was triple distilled according to the standard techniques in Radiation Chemistry. Hydrogen cyanide and ammonium cyanide were generated in a special setup from KCN and sulfuric acid, and collected in a glass syringe in which it was then irradiated. The concentration used were from 0.001 to 0.2 moles dm^{-3} and the pH was 6-9. All the solutions were oxygen-free. Also mixtures of HCN (0.2 moles dm^{-3}), CH_3CN (0.04 moles dm^{-3}) and CH_3CH_2CN (0.02 moles dm^{-3}) were irradiated at doses up to 180 kGy.

The irradiation was carried out in a ^{60}Co source with 50 kCi nominal. The radiation doses were from 2.5 kGy to 3000 kGy.

The analysis was made according to the procedures followed by Draganic et al., 1985.

3. Results and Discussion.

3.1 ANALYSIS OF OLIGOMERIC MATERIAL

The irradiated samples presented visible changes of color, from colorless to a dark brown. Also some samples presented the presence of solid material. Important changes also occur with pH (from 6 to 9.6) due to the accumulation of ammonia and other basic products. The solution was lyophilized and the dry residue was analyzed looking for the formation of biologically important compounds.

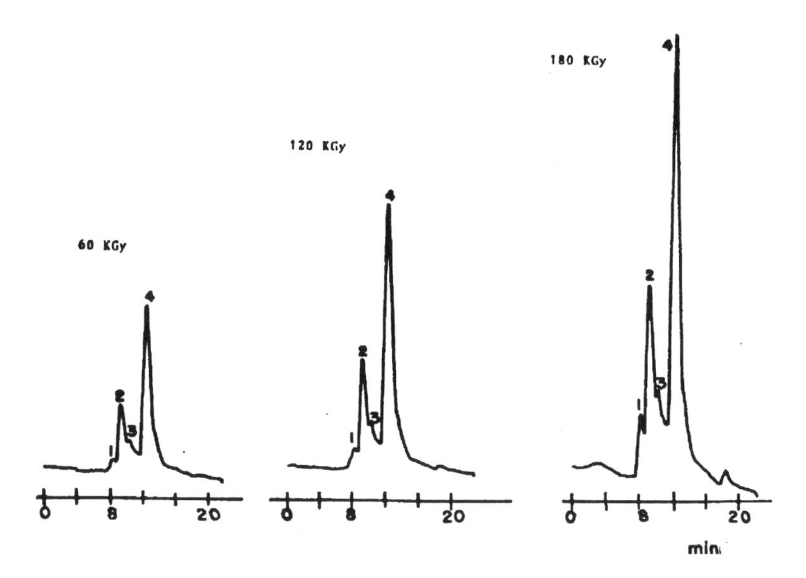

Figure 1. HPLC-chromatogram of CN-containing compounds exposed to different doses of gamma rays. : 1) oligomer with mw of 22000 daltons, 2) oligomer with mw of 16000 daltons, 3) oligomer with mw of 10000 daltons, 4) small molecules.

Figure 1 shows the HPLC chromatogram of mixture of HCN CH_3CN and CH_3CH_2CN; irradiated at different doses. The decomposition of HCN formed about 76% of an oligomeric material. This product constituted the principal constituent of the dry residue. The estimate of the molecular weight for such oligomers is up to 20000 daltons. The formation of the oligomers was observed in all the concentrations and doses studied. Oligomer concentration increased with the dose and the build up ends after the depletion of cyanide. Figure 2 shows the formation of the oligomer with molecular weight of 16,000 daltons, as function of the radiation dose and at three different initial concentrations of HCN. One interest of the present work was to evaluate the lower limits in the concentration of the target compound for the formation of oligomers. It was found that cyanide solutions of 10^{-3} moles dm^{-3} irradiated at 2.5 kGy are enough to produced these types of compounds.

Final Remarks

The formation of HCN oligomers was detected in very diluted solutions, and with very low radiation doses that probably were easily to accomplish in the early Earth. This results contrast with thermal polymerization of HCN in which high concentrations at basic pH are needed for the polymerization process. On the other hand, radiation-induced reactions are via free radicals, in which each CN molecule is added by steps.

Oligomers are produced with different structure than those formed in thermal polymerization. Characterization of the structure of these oligomers is in process. The data obtained up to now indicate that they are formed by polyamides and polyesters. These oligomers released compounds of biological significance like amino acids, purines and carboxylic acids when hydrolized.

The results found in this study remark the importance of radiation induced reactions as an energy source in terrestrial and extraterrestrial scenarios.

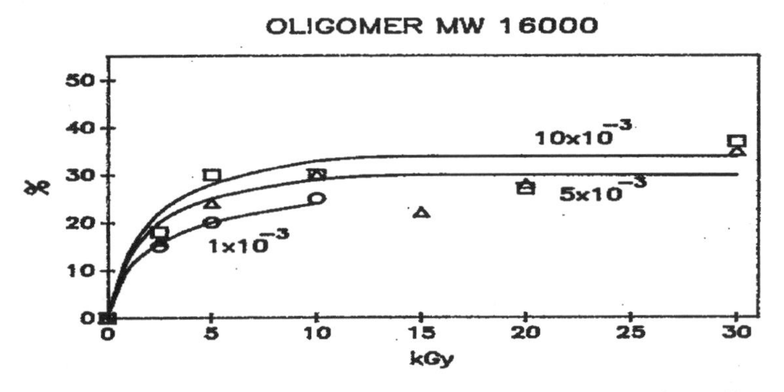

Figure 2. Effect of the irradiation dose in the formation of an olimomer at different initial concentrations HCN

References

Chada, M.S., Lawless, J.G. Thres, J.J., and Ponnamperuma), *Chemical evolution and the origin of life, R.* Buvet, C. Ponnamperuma eds. Amsterdam, North Holland, p.143-151, 1971.

Chang, S., DesMarais, D., Mack, R., Miller, S.L. and Strathearn, G.E. Prebiotic organic synthesis and the origin of life. In *Earth's earliest biosphere, its origin and evolution,* J.W. Schoppf, editor, Princeton University Press, New Jersey, 1983.

Draganic, Z., Draganic, I., Azamar, J.A. Vujosevic, S., Berber, M.D. and Negrón-Mendoza, A.., Radiation Chemistry of overirradiated aqueous solutions of hydrogen cyanide and ammonium cyanide,*J. Mol. Evol., 21,* (1985), 356-363.

Ehrenfreund, P., Boon, J., Commandeur, J., Sagan, C., Thomson, W.R., and Khare, B.), Analytical pyrolisis experiments of Titan aerosol analogues in preparation for cassini-Huygens mission, *Adv. Space Res.,*15 (1995), 335-342.

Ferris, J.P., Edelson, E.H., Auyeung, J.M. and Joshi, P.C., Structural Studies on HCN oligomers, *J. Mol. Evol.* 17 (1981), 69-77.

Gupta, S. Ochiai, E. And Ponnamperuma , C. Organic Synthesis in the atmosphere of Titan, *Nature,* 293 (1981), 725-727.

Matthews, C.N. and Moser, R.E., *Peptide synthesis from hydrogen cyanide and water, Nature*, 215 (1967), 1230-1234.

Mattews,C.N.,Ludicky, R.A., Schaefer, J., Stejskal, E.O. and McKay, R.A., Heteropolypeptides from hydrogen cyanide and water. Solid state ^{15}N NMR investigations. *Origin of life*, 14 (1984), 243-249.

Toupance, G., Bossard, A., Raulin, F., Far UV irradiation of model prebiotic atmospheres. In *Origins of life,* Noda, H., Ed., JSSP, Tokyo, p. 159-165. 1978.

Völker, T., Polymere Blausaure. *Angew Chem.* 72 (1960), 379-384.

THE ROLE PLAYED BY SOLID STATE SURFACES IN PREBIOTIC CHEMISTRY

S. Ramos-Bernal and A. Negron-Mendoza
Instituto de Ciencias Nucleares, UNAM
Mexico, D.F., 04510, Mexico

1. Introduction

A realistically simulated environment that could be the resemblance of the primitive Earth, would require the consideration of multiphase systems. Therefore, it is relevant to simulate the prebiotic Earth to consider the contribution of solid phases, since the enhance of chemical reactions by solid surfaces were by all means important. The solid state has also been considered as one of the most important prebiotic concentrators of organic molecules. This last property, in addition of its very well known reactivity that characterized the surfaces of some solids, made the solid state a potential catalyzer of very important reactions during the prebiotic Earth. In other words, after the solids adsorb organic molecules, they would then catalyze the synthesis of small molecules and oligomers into biomolecules. Also, to carry out these reactions an input of energy was necessary. One source that is proposed to be relevant in this scenery is ionizing radiation. Thus, to understand the relation of solids with organics in presence of energy, let us first see the interaction of radiation with the solid without any organic adsorbed, and then with an adsorbed molecule.

2. Interaction of Solids with Ionizing Energy

Research on the effects of ionizing radiation in solids has led not only to a greater understanding of solid state physics, but also for knowing better the radiation chemistry involved. In the solid state the ionized and excited species will be much more closely confined than in liquids. Thus, the distribution of this species in the solid state, will depend very much on the linear energy transfer (LET) . Furthermore, the atoms and molecules in the surroundings of the ionized and excited species are restricted in the solid state. The deposited energy is therefore confined to a very small region in the solid. With radiation of high LET, hot spots develop within very small volumes where intense excitations and ionizations are taking place, increasing about three times the magnitude of the temperature in this region for about one millionth of a second. Color centers will result in some solids where alternated ions carry opposite charges, which affects the absorption of light. Irradiation also causes electrons to be ejected from ions in the crystal and sometimes it happens that such electron becomes trapped in a negative ion vacancy.

J. Chela-Flores and F. Raulin (eds.),
Exobiology: Matter, Energy, and Information in the Origin and Evolution of Life in the Universe, 117–121.

3. Main Features about Radiation Catalytic Effect.

Since ionizing radiation is the principal source of energy that can penetrate solids, this is the type of energy considered here. The interaction of this type of energy with solids produces: excitation, degradation, as well as storage, and transfer of energy within the solid. Therefore, the role of solids as substrates for prebiological chemistry should be considered in addition with this high energy radiation. Moreover, chemical reaction occurring in the absence of radiation will be of no interest in this work.

Some general radiation induced effects on solids (without any organic adsorbed) are the following:

1) On irradiation, additional imperfections will appear on the surface of the solid and these new defects may be active in absorption and catalytic processes.

2) The changes induced in a solid by radiation can be distributed according to the duration of their relaxation time:

i) Effects that do not last after irradiation has stopped, like ionization by Compton effect and excitation of the crystal lattice.

ii) Effects that persist long after irradiation has stopped, and yet disappear at a considerable rate when the temperature is increased annealing these new defects. These apply to the formation of defects called color and V centers.

3) Defects caused by chemical conversions induced by radiation in the solid, such as redox reactions.

The crystal structure can be damaged only with particles of very high energy, it is when its energy is close to the displacement threshold of the atoms. Another types of damage are caused when interstitial atoms are formed in the closely packed row. These lengthens the chain of atoms over several interatomic distances causing clusters in which (n+1) atoms occupy n sites along the closely packed row. Such clusters exist together, and may move over many lattice sites so that surplus atoms with unsaturated valences interacting with the absorbed substance may appear to be present at the surface.

3.1 IRRADIATION OF SOLIDS WITH AN ADSORBED ORGANIC COMPOUND.

Organic chemical reactions within solids can be augmented via heterogeneous catalysis with the presence of ionizing radiation. The enhancement of these reactions can be accounted for: transfer processes, redox reactions, stabilization of intermediates, and finally by acidity behavior that can be brought from either Bronsted or Lewis type sites (Solomon D.H. 1968).

Since irradiation forms pairs of opposite charges, one of these charges must be trapped within the solid in order for one of the charge carriers to transfer its energy in some other zone of the solid. These carriers play a very important role within this radiation-induced catalysis.

A correlation between the changes of optical, and crystallographic properties of solids under irradiation, and the possible organic reactions inside solids may give some important clues in the synthesis of complex molecules during the prebiotic period of the Earth.

4. Charge Transfer

Irradiation on a solid creates pairs of opposite charges, that will desappear if recombination takes place. In order to avoid recombination, the non-equilibrium charge carries are trapped within the solid. This trapped charges play a very important role within the radiation induced catalysis. Organic compounds adsorbed on a surface of solids are subject to charge transfer adsorbed by the solid from radiation. This phenomenon is of particular interest because there are modes of energy transfer across interfaces that do not exist in homogeneous phases. In other words, solids can store energy from the gamma radiation and release it in various forms. A consequence of the energy transfer between the solid and the organic compound adsorbed on it, is that solid lattice should be able to serve as a moderator of energy.

It is known that energy transfer from solids to adsorbate is possible. Then it is likely that energy deposited in the bulk solid by penetrating energy sources, (such as ionizing radiation), should be well moderated to a minor energy form before transfer to a distant interfacially adsorbed reactant.

Energy deposited into the solid causes redistribution of energy among excited states, and changes in the energy-level structure due to material damage. The absorption of energy such as ionizing radiation is considerably less well understood than optical transitions, since the energy absorbed may be fractionated into: (a) heat; (b) luminescent decay; (c) chemical changes in the solid; (d) transference to an adsorbed reactant; or (e) energy stored as separated charge pairs.

The absorption and storage of electronic energy within a crystalline lattice substrate will have a significant impact on the chemistry of adsorbed organics. If the effect of bulk lattice excitation is to alter the surface electronic distribution, there must exist modes of energy transfer between the lattice substrate and the reacting medium. The role of electronic energy transfer in heterogeneous catalysis is not very well characterized. Experiments in energy transfer processes in solids have been done using doping with divalent ions (Ramos et al 1985), but further development in this topic is needed to understand its relevance in prebiotic chemistry.

The trapped separated charge pairs produced by high-energy radiation frequently have a long life, as evidenced by the capacity of many inorganic photoconductors for thermoluminiscence. Other manifestations of these metastable excited states are EPR signals and predicted optical absorption properties. The luminescence phenomena are indicative of energy storage and charge mobility in the solid and they are produced by solvatation/ desolvatation and mechanical stress. Since only very limited data are available about the wavelength distribution of the emitted light, it is not clear whether some or these luminescence phenomena derive from the same luminescence centers. The electronic energy level structure of the material will be described as characterized by optical absorption of the solid, or by EPR of doped solids before and after gamma irradiation (Ramos et al 1985)

5. Chemical Reactivity

The interaction between the organics and the solid surfaces can profoundly alter the electronic structure within adsorbed molecules. Thus, surfaces can serve as charge transducers from reactant to catalyst and viceversa. Molecules adsorbed on a surface are subject to energy as well as charge transfer to and from the solid. Besides ionizing radiation, triboelectronic energy can also produce chemical reactions in solids. Although triboelectronic processes are poorly understood, it is clear that surfaces serve as energy transducers. The energy deposited in the bulk solid should be well moderated to less high energy form before transferring to a distant interfacially adsorbed reactant.

Four aspects of the interactions of the radiation with solids with compounds adsorbed on it are important to chemical reactivity. These are the following:

5.1 ENERGY GAP

The interesting solids are those call insulators, which have a considerable gap between the valence and conduction bands. The so-call gap of energy is the difference in energy between these two bands. In these types of solids one of the principal defects produced by irradiation are the color centers, which are a product of the defects in the lattice and the mobility of the excited electrons (holes) that are transported to the trapping sites. Charge carriers will not be produced when transitions between localized energy levels (within the gap) occur.

5.2 TYPES OF DECAY

Any amount of energy larger than the lowest electronic levels of the solid will be dissipated as heat (vibrations of the crystal lattice). Since radiationless deactivation is fast in solids, the ionized effects of radiation will be moderated by the solid.

5.3.- ENERGY IN THE SOLID

The amount of energy stored is quantifiable in terms of:

i)-the band gap.

ii).-the energy required to reactivate the trapped charge to mobility.

iii)-the size of the quantum reaching the adsorbed reactant.

iv).-the number and degree of population of trapping centers in the solid.

The energy (apart from the radiationless decay), may be fractionated into: a.- heat b.- luminescence decay. c.-chemical changes d.-transference to the adsorbed reactant e.- energy stored in form of separated pairs. This energy stored is the responsible of thermoluminescence and others photophysical properties of solids. This stored energy deposited in trapped separated pairs is ready to be useable in a chemical reaction on the solid surfaces (Morrison 1980). If the solid is irradiated with light of a long-wavelength then this will reactivate trapped excitations, thus transporting bulk-trapped excitation to the surface to be usable in the chemical reactions.

5.4.TRANSFER

The role of electronic energy transfer in heterogeneous catalysis on solid surfaces is fundamental and makes necessary to include some spectroscopic properties of catalysts to see their reaction-promoting capabilities. In order for electronic excitation to produce a chemistry reaction effectively, is necessary an energy barrier between electron/hole pairs to prevent wasteful electron/hole recombination. The electron (hole) transfer in photoelectrochemical systems is the way of energy transfer. The latter is driven by a potential difference, between the electronic energy of the light-absorbing catalytic surface and the redox potential of the solution phase reactant. This barrier is provided by the space charge potential near the interface. Given these high excitation energies, impurity centers have a potentially larger impact on surface chemistry.

6. Final Remarks.

It was considered here the potential to intensify, via irradiation, the capabilities of solids to serve as substrates for prebiological chemistry. For supporting this broadly defined idea about the role of the interaction of high energy radiation with solids, our group has provided some results using clay minerals. (Negron-Mendoza and Ramos-Bernal 1992; Negron-Mendoza et al 1995). Ours results show, potential relationships between minerals, electronic excitation, and surface reactivity, as applied to chemical evolution. We have discovered and partially characterized thermoluminescent properties of these clays. They also show the energy storage and transfer processes in clays. And provide a basis for estimating the potential significance of solid surfaces role for catalyzing reactions relevant to prebiotic chemistry.

More specific experiments should still be made to test this hypothesis about the role played by solids and ionizing energy. Since this hypothesis is principally based on the coupled transport of electrical charge and electronic energy through the solid, which operates via production and mobility of electron/hole pairs, then the mobility of charge/electronic excitation between defect centers serves as the basis for a primordial inorganic electron transport chain.

References

Morrison, S.R. (1980)Electrochemistry of semiconductor and oxidated metal electrodes, Plenum, N.Y.

Negrón-Mendoza, A., Albarrán , G. and Ramos, S. (1993) Transformation of malonic acid adsorbed on a clay-mineral by gamma irradiation, *Radiat. Phys.Chem.*, **42**, 1003-1006.

Negrón-Mendoza, A. Ramos, S.. Albarrán, G. (1995) Enhance decarboxylation reaction of carboxylic acids in clay minerals.,*Radiat. Phys. Chem.* **46**, 565-568

Ramos S. , Hernandez J., Murrieta H., Rubio J. (1985) Model for F-center productionin alkali halides doped with divalent cation impurities that change their valence by irradiation. Phys Rev. B **31** 8164.

Ramos, S. and Negrón-Mendoza, A. (1992) Radiation Heterogeneous processes of ^{14}C-acetic acid adsorbed in Na-Montmorillonite. *J. Radianal. Nucl. Chem.* **160**, 487-492

Solomon, D.H. (1968) Clay minerals as electron acceptors and/or electron donors in organic reactions. *Clays Clay Miner.*, **16**, 31-39..

MUTUALLY CATALYTIC AMPHIPHILES: SIMULATED CHEMICAL EVOLUTION AND IMPLICATIONS TO EXOBIOLOGY

DANIEL SEGRÉ and DORON LANCET[1]
Department of Molecular Genetics, and [1]Head, Genome Center
The Weizmann Institute of Science, Rehovot 76100, Israel

Abstract. A description of the emergence of life should delineate a chemically rigorous gradual transition from random collections of simple organic molecules to spatially confined assemblies displaying rudimentary self-reproduction capacity. It has been suggested that large sets of mutually catalytic molecules, and not self-replicating information-carrying biopolymers, could have been the precursors of life. We present here a stochastic model in which the mutually catalytic molecules are spontaneously aggregating amphiphiles. When such amphiphiles exert on each other random catalytic effects, biased molecular compositions emerge, that are endowed with replication-like properties. This approach may have important consequences to the understanding of very early chemical evolution. It could also guide a search for extraterrestrial forms of very primitive life.

1. Introduction

Since the discovery of DNA and its important role in living systems, models for the origin of life have been mostly aimed at understanding properties such as evolution of self-replicating biopolymers [1, 2], mechanisms for coding [3, 4], and template-mediated oligomerization [5]. The successful results obtained in these fields still leave many unanswered questions about the very early steps of chemical evolution. It has been suggested [6] that, for understanding these stages of biogenesis, classical concepts, such as Oparin's coacervate droplets [7], should be reconsidered in the light of modern biological knowledge and with the aid of today's computer power. The mean-field model that Dyson himself suggested in 1982 [8], and Kauffman's results on sets of mutually catalytic polymers [9] are among the main theoretical models for analyzing prebiotic systems, in which the interactions among different kinds of molecules lead to the emergence of replication-like properties. One of the most important aspects of such models is that ensemble replication can happen even if none of its constituent molecules by themselves are endowed with the capacity to generate their own copies or to undergo autocatalysis. In a similar spirit, some authors [10, 11, 12, 13] have emphasized the potential importance of amphiphiles at the first stages of chemical evolution. It was proposed that spontaneous aggregation of such compounds to form micelles or vesicles in a prebiotic environment could have ensured high local concentrations, and the subsequent enhancement of catalyzed chemical reactions among the constituent or enclosed chemicals.
Our previously described Graded Autocatalysis Replication Domain (GARD) model [14, 15, 16, 17] which analyzes the explicit chemical kinetics of mutually catalytic sets, is extended here to exploring the emergence of self-reproducing molecular assemblies composed of amphiphiles. A further elaboration of the previous model is the use of a

J. Chela-Flores and F. Raulin (eds.),
Exobiology: Matter, Energy, and Information in the Origin and Evolution of Life in the Universe, 123–131.

stochastic simulation algorithm instead of solving sets of differential kinetic equations. The scenario we propose, based on former teachings [6, 7, 9, 10, 11], can be summarized in the following phases:
(i) Spontaneous formation of assemblies, by aggregation of molecular species (eg: amphiphiles) due to physicochemical interactions (eg: hydrophobic effects). The high local concentrations thus attained allow more effective catalysis.
(ii) Emergence of poorly organized metabolic processes, in the form of quantifiable mutually catalytic interactions among members of the assemblies. Assemblies, which may appear by chance, whose molecular compositions are more successful in catalytically recruiting additional monomeric species from the environment, have a selective advantage.
(iii) Growing and splitting of the assemblies initiate an evolution-like mechanism which leads to increased abundance of more efficient catalytic assemblies. The assemblies, when splitting, propagate their unique composition to daughter assemblies [10, 18], creating a primordial system for storing and transmitting information. A quantitative description of this scenario may help understand important aspects of very early self-organization, potentially responsible for the emergence of life on Earth and elsewhere in the universe.

2. Chemical dynamics of molecular assemblies: a Monte Carlo approach

We analyze possible dynamic patterns in what we would like to call the "primordial random chemistry era". For this, we developed a method for depicting the chemical kinetics of self-assembling molecular sets by realistic computer simulations. These are based on standard mass action chemical laws, and on stochastic dynamical rules [19]. This approach, that we named SMACC (Single Molecule Algorithms for Complex Chemistry) [16] is aimed at reproducing the global kinetics of chemical reactions, without loosing the capacity to resolve individual molecules. The computer simulations whose results are presented here, analyze the behavior of sets of amphiphiles subjected to spontaneous aggregation and to mutually catalytic interactions. Amphiphiles are assumed to include diverse chemical configurations sharing the general property of having a hydrophobic region ("tail") and a hydrophilic, charged or polar, region ("head"). As in other origin of life scenarios [20] it is assumed that primordial chemistry would provide an ample supply of such molecules [11, 21]. The crucial qualitative aspects of the model are: a) Amphiphilic monomers may form mixed assemblies (e.g. mixed micelles) in an energetically favorable reaction; b) The assemblies are dynamic [22], having "soft matter" properties, and may gain or lose individual monomers by insertion/removal reactions; c) The monomer addition/removal reactions may be catalyzed by configurations of neighboring monomers; d) When assemblies reach a critical size they undergo splitting to two smaller assemblies. While a variety of critical sizes and splitting pathways are conceivable, we consider here a simple case whereby all assemblies behave identically and both split progeny are identical in size. These constraints may be easily relieved in future analyses.
We consider the existence of a N_G global number of amphiphile types (A_i) whose concentrations in the surrounding aqueous medium are buffered by a constant external supply. For simplicity, all A_i concentrations are assumed to be equal.

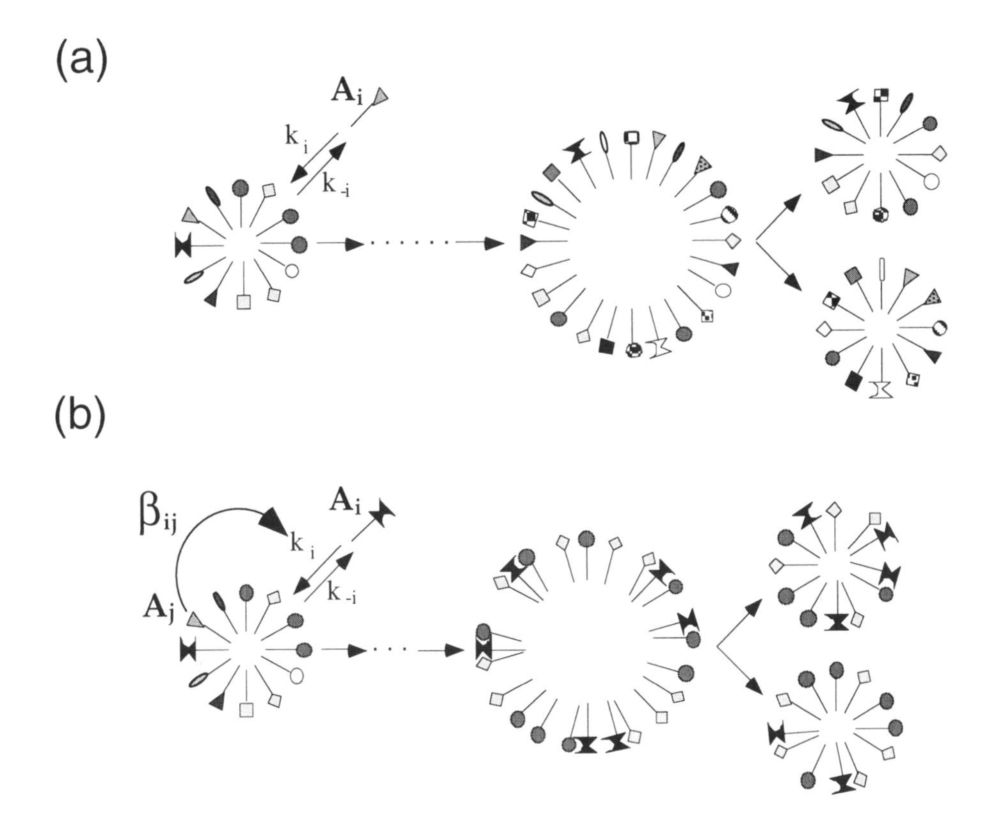

Figure 1. A schematic depiction of the reactions assumed to be responsible for growing and splitting of heterogeneous micelles. Amphiphiles are represented as having hydrophobic tails (sticks) and a variety of hydrophilic heads (geometrical shapes). In future simulations, tails could also be assumed to come in a large variety of types. Mutually catalytic interactions are symbolized by a geometrical complementarity between some neighboring species. Two scenarios are shown: **(a)** In absence of catalysis all the molecular species are equally likely to join/leave a micelle. Consequently, upon growth, if the number of possible species is much larger than the amount of molecules composing an assembly, no bias in the composition is expected to develop. When a micelle reaches a critical size (twice the initial one) its constituents molecules are randomly distributed into two new daughter assemblies. None of these new assemblies will probably resemble in the composition previously existed micelles. **(b)** When catalysis is exerted on joining/leaving of amphiphiles (arrow β_{ij}), species enhancing each other's reactions will tend to prevail after some time. When the composition becomes biased, and few species are present in large count, splitting generates assemblies whose composition resembles the one of the parent micelle.

The reaction by which a monomer A_i joins an assembly M_N composed of N monomers (n_1 of kind A_1, n_2 of kind A_2, etc.) can be represented formally as follows [23]:

$$M_N(n_1,n_2,\ldots,n_i,\ldots,n_{N_G}) + A_i \xrightarrow{k_i} M_{N+1}(n_1,n_2,\ldots,n_i+1,\ldots,n_{N_G}) \qquad (1)$$

Species A_i has a probability of insertion proportional to a basal forward rate k_i. Similarly, the reverse reactions are described by

$$M_{N+1}(n_1,n_2,...,n_i+1,...,n_{N_G}) \xrightarrow{k_{-i}} M_N(n_1,n_2,...,n_i,...,n_{N_G}) + A_i \qquad (2)$$

governed by a probability proportional to the backward basal rate k_{-i}.
It is postulated here as an additional simplification that, in absence of catalysis, the average time for a monomer to join the assembly does not depend on the specific monomer being inserted. In other words, that all k_i (and similarly all k_{-i}) are independent of i. In the presence of catalysis, it is assumed that the rates of the assembly growth/decline reactions are regulated by the kinetic terms $k_i(1+\langle n_j\beta_{ij}\rangle_j)$ and $k_{-i}(1+\langle n_j\beta_{ij}\rangle_j)$ for the forward and reverse reactions, respectively. The assembly average $\langle ...\rangle_j$ indicates the mean value of the catalysis exerted by the all molecules (A_j) present in the assembly at the time of insertion/deletion of the considered A_i. This averaging step is equivalent to an assumption of fast pre-equilibrium for the dynamic collisions between the incoming/outgoing A_i and all other members of the assembly. Similar mean-field assumptions have been used in other models [6].
The catalytic enhancement factors β_{ij} are the elements of an N_G X N_G matrix (the β matrix) with all the mutual catalysis values for the chemistry considered. The β matrix components are drawn from a probability distribution [15, 24, 25]. Based on these probabilities, a Monte Carlo simulation process determines the fates of the single molecules. The splitting event

$$M_{2N} \dashrightarrow M'_N + M''_N \qquad (3)$$

occurs whenever a micelle reaches twice the initial size.
Interesting effects can be seen only by simulating a large number of reactions. This makes it necessary to simplify the chemical rules in the described manner. More accurate mathematical models for the kinetics of amphiphiles assemblies [23] might be used in more advanced versions.

3. Random assemblies and the "vital bias".

The number of assemblies that can form by spontaneous aggregation of N molecules of N_G different kinds can provide a wide diversity of possible compositions. In particular, for $N_G >> N$, a vast majority of the species will be present in the assembly in a single copy, and most species will not be present in the assembly at all (figure 2A). Spontaneously formed assemblies will practically never have strong compositional biases, i.e. many copies of relatively few species types. In contrast, a crucial characteristics of present day life is that it contains a relatively small number of molecular types, as compared to the astronomical number of organic compounds that are generally conceivable. A common tendency in origins of life research is to try to explain how these exact compounds present in today's cellular life could have formed or could have been selected from the many possible substances. This process is often seen as a prerequisite needed for life to emerge. A different mechanism might be envisaged, in which the selection of the "right" building blocks coincides with the process by which life emerges.

The difference is subtle, but it has a very important implication: the latter view - unlike the former - just calls for a general decrease in the number of chemical species, without requiring that very specific compounds (e.g. nucleotides or amino acids) will be enriched. The involvement of non covalent assemblies in our model has significant consequences.While in the case of covalently linked biopolymers, changes in sequences seem to require laborious mechanisms involving sophisticated enzymatic catalysis, non covalent assemblies can easily change their composition by monomers joining/leaving. On the other hand it is often considered difficult to see how non-covalent assemblies may acquire the capacity of storing and propagating information, a property usually ascribed to biopolymers. However, as suggested by Morowitz [18], an assembly that grows and splits into two random halves, can propagate the specific composition to the daughter assemblies, provided that enough copies of each species are present before splitting. Precise relations between the size of the assembly and the number of species present in it can be deduced following this criterion [18]. In the results illustrated below, we show how random catalytic interactions within randomly seeded molecular assemblies could lead to the emergence of micelles capable of propagate their highly improbable biased compositions.

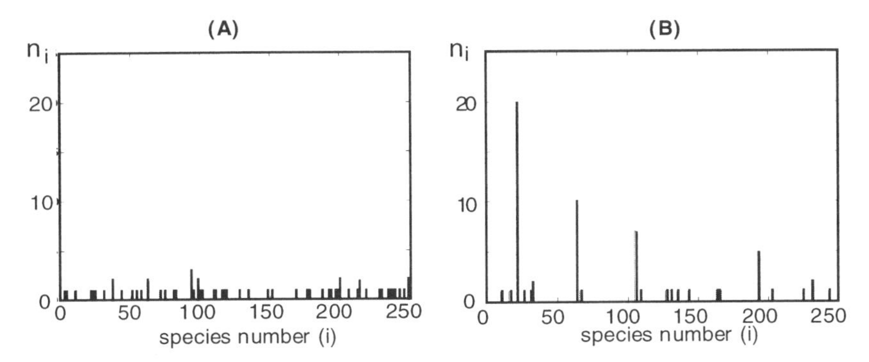

Figure 2. Histograms of the composition of two assemblies. For each species A_i, the number of individual molecules present in the assembly (n_i) is plotted. In general it is $\sum n_i = N$, with N between a minimal size (30) and the critical splitting size (60). The histograms are relative to micelles at their critical splitting size. **(A)** An assembly that underwent very little catalyzed joining/leaving reactions has no bias in its composition. This situation corresponds to an assembly of generation 1 (figure 3). **(B)** An assembly that underwent many steps of the catalyzed growing and splitting process (generation 15 in figure 3). The program for the Monte Carlo simulation of the reactions in an Amphiphilic GARD was implemented in Matlab on a DEC Alpha workstation.

4. Results of a computer experiment

Figure 2 shows some results from a computer experiment performed with an amphiphilic GARD as described above. After seeding randomly two micelles, we let them grow by mutually catalytic monomer joining reactions. The first that reaches the critical splitting size divides, and the two micelles born thereof start growing again in a similar way. The process is repeated cyclically, and biases in the compositions of the micelles are monitored. For quantifying the degree of bias we use a measure of compositional information I=-log(P) where P is derived from the multinomial probability distribution

$$P=(1/N_G)^N \cdot N!/(n_1!n_2!\dots n_{N_G}!). \quad (4)$$

The spontaneous increase of I (i.e. decreased probability of the assembly composition) as a function of generations is seen in figure 3. The maximal increase is roughly 20 standard deviations away from the mean of the distribution of randomly generated amphiphile assemblies (figure 3, inset).

These changes constitute a demonstration of the capacity of amphiphile assemblies to undergo an evolution-like change towards less probable structures. In the present model, progeny assemblies will practically never have the exact same composition of the original one.

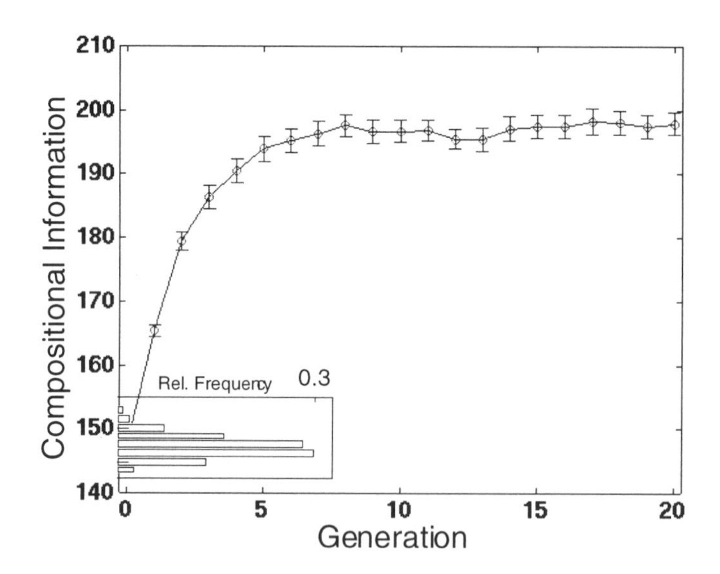

Figure 3. The curves shown are averages of 100 independent simulations, with bars indicating the standard deviation of the mean. Assemblies of size N=30 are randomly seeded at t=0 from an amphiphile chemistry of size N_G=250. In each generation two parallel assemblies undergo monomer addition/loss reactions governed by the kinetics described in Fig. 1. When one of the two assemblies reaches the critical splitting size N=60, its randomly sorted halves start the cycle again. This process is repeated for 20 generations. This amounts to following a trajectory on a tree by considering the faster-growing branch at each generation. For each generation, the value of the negative logarithm of the multinomial probability (I, see equation 4) is plotted. In the inset the distribution of I for spontaneously formed micelles is shown.

The fidelity of replication may actually be computed as a distance between "parent" and progeny using the metric

$$D=\sum |n_i'-n_i|/(n_i'+n_i), \quad (5)$$

where n_i', n_i are the counts for species A_i in the two generations considered. The similarity (defined as 1/D) becomes stronger for the higher generations.
The very high correlation coefficient (0.99) between the curves for the growth of I and of 1/D throughout the generations, indicates that when a more idiosyncratic composition appears via mutual catalysis, the assembly constitutes a more effective replicator. This may be intuitively understood by observing that as mutual catalysis results in an increased copy number of certain molecular species, the probability of equal copy number of such species in the two progeny assemblies increases. This kind of analysis seems therefore very appropriate for investigating in further detail the idea of compositional information [26].

5. Consequences of the amphiphile GARD for a Galactic Traveler in search for extraterrestrial life

One of the important future outcomes of origins of life studies should be the generation of guidelines for the search of life on other planets. Any experiment aimed at discovering signs of life would be biased by what we expect to find. This is particularly true in those cases where life has been "frozen" at very early stages of its emergence. The model described in the previous paragraphs allows one to delineate the following predictions on the nature of early life relics representing consecutive stages of prebiotic processes:
1) The occurrence of simple organic molecules of any kind, with no absolute requirement for specific molecular types, such as amino acids or nucleotides.
2) The occurrence of more complex organic molecules, up to several dozen carbon atoms, not necessarily oligomers of molecules known to constitute present day life.
3) The occurrence of amphiphilic molecules, i.e. those that have a polar part and a lipophilic part, without other constraints on chemical structure.
4) A high degree of diversity of the molecules described in (2) and (3): this is a necessary stage en route to the "narrowing" of molecular diversity that occurs later on.
5) The appearance of assemblies of molecules in the form of micelle-like structures.
6) Biased internal composition of such assemblies, whereby relatively few types are present. This criterion may apply to assemblies, but can also be discerned in disperse molecules if the assemblies did not survive.
7) A high degree of similarity in the internal composition of assemblies. If the assemblies did not survive, this may still manifest itself as macroscopic regions containing biased organic chemical composition.
8) Rudimentary mutually catalytic or photochemical properties of the chemicals present in assemblies or in spatial domains.
9) More advanced ultrastructure of assemblies, e.g. vesicles or higher structures.

Points 4 to 7 in this list have been explicitly modelled through the computer experiments illustrated in this work. Biased compositions were shown to be possible important hints for the presence of assemblies in which mutually catalytic interactions had become more and more efficient. Based on these results, one can speculate how an imaginary Galactic Traveler would plan a search for extraterrestrial "prebiotic fossils", i.e. relics of very early chemical evolution. For the Galactic Traveler, an unexpected chemical composition might sometimes justify a serious suspicion of past biogenic phenomena. Counter examples would be crystals or minerals, arrays of very few chemical species, which present an extremely biased chemical composition.

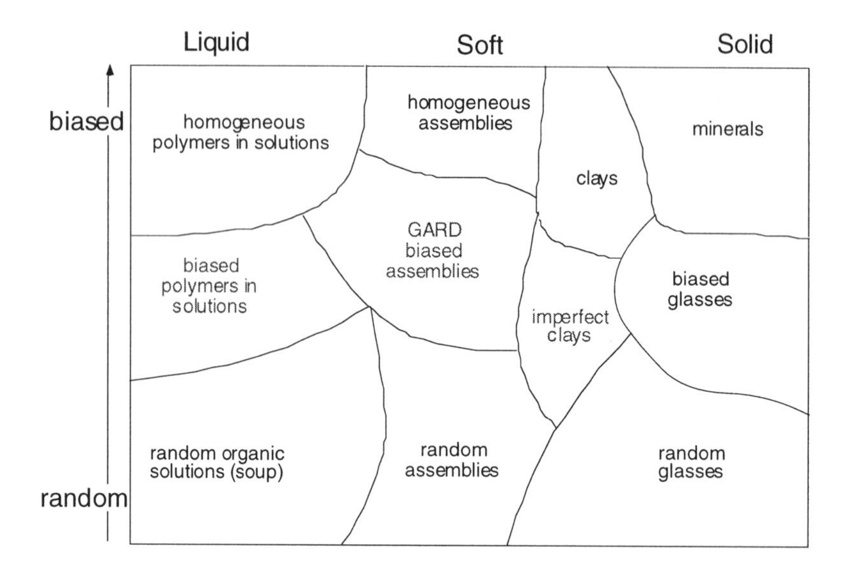

Figure 4. A diagram for a Galactic Traveler examining a sample of extraterrestrial material. The two attributes monitored here are the degree of compositional bias (vertical axis) and the compactness of the sample (horizontal axis). The idea that amphiphilic GARD-like assemblies could be good candidates for a primordial self-replicating system would suggest to survey with particular interest soft matter samples with some deviation from randomness in their composition. Such samples would be represented in the central region of the diagram.

These would usually not be considered as potential steps towards life. On the other hand minerals with imperfect crystal structures display a greater degree of chemical diversity and their life-like properties have been explored [27]. We suggest that an intermediate degree of chemical bias in soft matter (or what is left of it) would be indicative of very primitive prebiotic entities. Soft matter has the great advantage of affording the chemical dynamics needed for mutual catalysis and rudimentary replication, but at the same time allows effective chemical proximity which is not present in aqueous solutions. Thus the central region of the diagram (figure 4) is most likely to be interesting to a Galactic Traveler in search for the seeds of life.

6. Conclusions

A computer analysis of a model for the emergence of self-replicating units has been described, based on the idea that selective growth of micelle-like assemblies might result from random catalytic interactions among prebiotically formed amphiphiles of many different kinds. This idea can serve as a simple chemically realistic benchmark for performing *in silico* experiments that will help us understanding some aspects of early chemical evolution. As an alternative to scenarios focused mostly on present day life biopolymers, the model we suggest also indicates new possible viewpoints for exobiological explorations.

References

1. Eigen, M. and P. Schuster. "The Hypercycle." 1979 Springer Verlag. Berlin.
2. Orgel, L. E. "Molecular replication." Nature. **358**: 203-209, 1992.
3. Hartman, H. "Speculations on the origin of the genetic code." J. Mol. Evol. **40**: 541-544, 1995.
4. Nieselt-Struwe, K. and P. R. Wills. "The Emergence of Genetic Coding in Physical Systems." J. theor. Biol. **187**: 1-14, 1997.
5. Wright, M. C. and G. F. Joyce. "Continuous in Vitro Evolution of Catalytic Function." Science. **276**(25 april): 614-617, 1997.
6. Dyson, F. "Origins of Life." 1985 Cambridge University Press. Cambridge.
7. Oparin, A. I. "The origin of life." 1953 Dover Pub. N.Y.
8. Dyson, F. J. "A Model for the Origin of Life." Journal of Molecular Evolution. **18**: 344-350, 1982.
9. Kauffman, S. A. "The origins of order- Self-organization and selection in evolution." 1993 Oxford University Press. Oxford.
10. Morowitz, H. J., B. Heinz and D. W. Deamer. "The chemical logic of a minimum protocell." Origins of Life and Evolution of the Biosphere. **18**: 281-287, 1988.
11. Deamer, D. W. "Role of amphiphilic compounds in the evolution of membrane structure on the early earth." Origins of Life and Evolution of the Biosphere. **17**: 3-25, 1986.
12. Bachmann, P., P. Luisi and J. Lang. "Autocatalytic self-replicating micelles as models for prebiotic structures." Nature. **357**(7 May): 57-59, 1992.
13. Ourisson, G. and Y. Nakatani. "The terpenoid theory of the origin of cellular life: the evolution of terpenoids to choloesterol." Chemistry & Biology. **1**(September): 11-23, 1994.
14. Lancet, D., G. Glusman, D. Segré, O. Kedem and Y. Pilpel. "Self replication and chemical selection in primordial mutually catalytic sets." Origins of Life and Evolution of the Biosphere. **26**: 270-271, 1996.
15. Segré, D., Y. Pilpel, G. Glusman and D. Lancet. "Self-replication and Evolution in Primordial Mutually Catalytic Sets." Astronomical and Biochemical Origins and the Search for Life in the Universe, Proceedings of the 5th International Conference on Bioastronomy, IAU Colloquium N.161. Cosmovici, S.Bowyer and Werthimer ed. 1997 Editrice Compositori. Bologna.
16. Segré, D., Y. Pilpel and D. Lancet. "Mutual catalysis in sets of prebiotic organic molecules: evolution through computer simulated chemical kinetics." Physica A. **249**(1-4): 558-564, 1998.
17. Segré, D., D. Lancet, O. Kedem and Y. Pilpel. "Graded Autocatalysis Replication Domain (GARD): Kinetic Analysis of Self-Replication in Mutually Catalytic Sets." Origins of Life and Evolution of the Biosphere. **28** (in press): 1998.
18. Morowitz, H. J. "Biological Self-Replicating Systems." Progress in Theoretical Biology. Snell ed. 1967 Academic Press.
19. Bartholomay, A. F. "A stochastic approach to statistical kinetics with application to enzyme kinetics." Biochemistry. **1**(2): 223-230, 1962.
20. Miller, S. L. "A production of amino acids under possible primitive earth conditions." Science. **117**: 528-529, 1953.
21. Deamer, D. W. and G. R. Fleischaker. "Origins of life-The central concepts." 1994 Jones & Bartlett. Boston.
22. Lipowsky, R. "The conformation of membranes." **349**(7 February, 1991): 475-481, 1991.
23. Chizmadzhew, Y. A., M. Maestro and F. Mavelli. "A simplified kinetic model for an autopoietic synthesis of micelles." Chem. Phys. Lett. **226**: 56-62, 1994.
24. Lancet, D., O. Kedem and Y. Pilpel. "Emergence of order in small autocatalytic sets maintained far from equilibrium:application of receptor affinity distribution (RAD) model." Ber Bunsenges. Phys. Chem. **98**(9): 1166-1169, 1994.
25. Lancet, D., E. Sadovsky and E. Seidemann. "Probability model for molecular recognition in biological receptor repertoires: Significance to the olfactory system." Proc. Natl. Acad. USA. **90**: 3715-3719, 1993.
26. Morowitz, H. J. "Beginnings of Cellular Life." 1992 Yale University Press. London.
27. Cairns-Smith, A. G. "Genetic Takeover and the Mineral Origins of Life." 1982 The Cambridge University Press. Cambridge, U.K.

SALAM PHASE TRANSITION AND THE CAUSAL ORIGIN OF HOMOCHIRALITY

WENQING WANG, FANG YI
Department of Technical Physics , Peking University, Beijing 100871, P.R.China

Abstract. The origin of homochirality has been a basic and challenging problem of life science. Salam's phase transition hypothesis has been suggested. However, there are three problems implied in this putative phase transition of D- to L-amino acids. In this paper, we present our experiments involving the possible relevance of Salam's putative phase transition in the origin of homochirality.

Key words: Salam's hypothesis, phase transition, origin of homochirality,parity violation

1 Introduction

The first unifying principle in biochemistry is that the key biomolecules have the same handedness. The origin of optically active molecules in nature is a question that has intrigued scientists for well over a century. To account for the fact of homochirality, various proposals have been postulated. Ever since 1957, the search for the origin of homochirality from the view of the effects of the weak interactions has started.

2. Salam's phase transition

In 1991, A. Salam[1] proposed a new hypothesis: the subtle energy difference of chiral molecules induced by Z^0 interactions combined with Bose-Einstein condensation may cause biochirality among the 20 amino acids at critical temperature Tc, which is analogous to that of BCS superconductivity.Although Z effect is small beneath a low temperature, it may play an active role. Due to its effects, the hydrogens binding to α-carbon can give up their loose electrons, and act as metallic hydrogens. Based on the suggestion of Salam, the second order phase transition, due to parity violation of the third kind, causes a racemic mixture to become a chirality pure collection consisting of one enantiomer preferentially stabilized by the parity violation of weak neutral interactions.

When we attempt to lay the experimental bases of the origin of homochirality, some fundamental questions arise. Whether the putative second order phase transition of the type of Bose condensation may occur in amino acids? The range of the weak force is extremely short, how can we get some information about relatively long-range effects of weak force? In searching for effects of the weak force, it is advantageous to reduce the effect of the electromagnetic force, because that force can systematically mimic the parity violating effect of the weak force. A practical way to minimize the effect of the electromagnetic force is to work with a "forbidden" atomic transition, then the ratio of the strength of the weak force to that of the electromagnetic force is substantially increased. We found that under λ phase transition,the difference of D and L is amplified.

2.1 Specific heat measurement[2]

The temperature dependence of the specific heats for the D-, L-valine single crystals were shown in Ref.4. An obvious λ transition was observed at 270 ± 1K in the specific heat measurements of D- and L-valine by DSC. The phase transition was reversible.

J. Chela-Flores and F. Raulin (eds.),
Exobiology: Matter, Energy, and Information in the Origin and Evolution of Life in the Universe, 133–136.

It was shown in Ref.2. that the specific heat Cp value of D-valine was larger than that of L-valine and the same existed in the enantiomers of alanine, the peak showed 10^{-4} eV per molecule difference at the transition point. However, according to the quantum calculation by Mason and Tranter[3], the energy differences between D, L-amino acids are on the order of 3×10^{-19} eV per molecule, thus this result may lead us to such a conclusion that the ratio of the strength of the weak force to that of the electromagnetic force is substantially increased at the transition point.

2.2.2 DC magnetization measurements[4]

Magnetization of single crystals of D-, L-alanine and D-valine were measured as a function of temperature using the SQUID magnetometer in a field of 1kG. (Fig. 1)

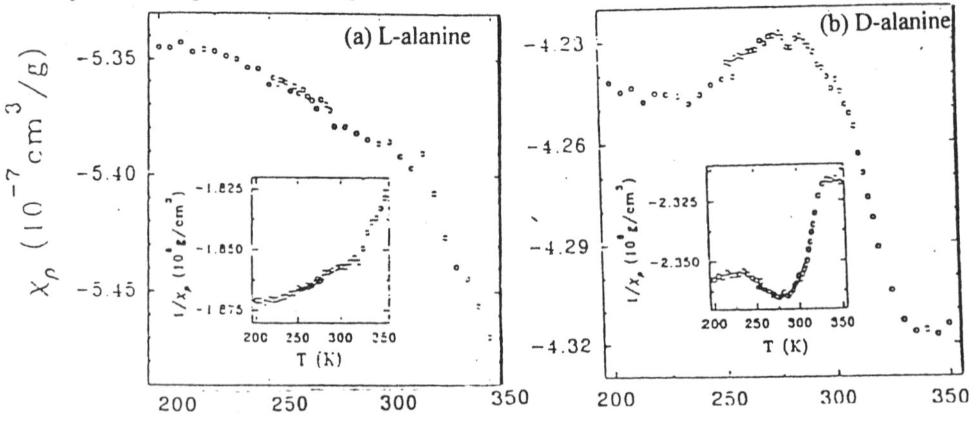

Fig.1 Mass susceptibility vs T of single crystals of D- and L-alanine

Their magnetic susceptibilities were negative ($\chi_\rho < 0$), the values were very low and absolute value about 10^{-7} cm^3/g.The magnetic transition temperature at which the discontinuity occured for D-alanine, valine was coincident with Tc of specific heat measurement. It was also shown that the χ_ρ vs T is different for D- and L-alanine.

Based on the above result, it might be proposed that the difference of the mass susceptibility χ_ρ ~ T curve between the D- and L-alanine is attributable to the variation of intramolecular geometry of chirality density, which is related to the parity violation energy shift of a chiral molecule and is a consequence of the short range of the weak interaction between the nuclei and electrons. The discontinuity in magnetic susceptibility might be induced by the occurrence of polarization phenomena. Quantum mechanical theory indicates that the magnetic susceptibility of a system with weakly interacting particles, for the case where the electron clouds of the particles are not spherically symmetrical, will be composed of two parts: (A) precession dismagnetism, (B) Van Vleck polarization paramagnetism. At elevated temperature, the role of polarization paramagnetism will be greater (in the D-case) , while in the L-case, the orientation paramagnetism which decreases with temperature predominantly affects the magnetism. Thus, different magnetic behaviour for D,L-amino acids has been observed in the experiments.

2.2.3 Laser Raman spectra measurements

Laser Raman spectra of single crystal of D-alanine at different low temperatures (100, 250, 260,

270,280,290K) are shown in Fig.2.

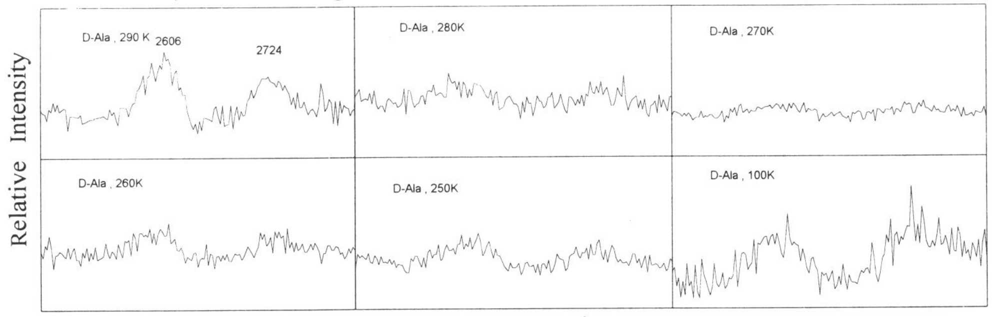

Fig.2 Raman spectrum of D- alanine single crystal at different temperatures

From Fig.2, among all the vibration modes, the bands at 2606,2724cm^{-1} should be assigned as second order C-H stretching frequencies[5]. The two peaks of D-alanine appeared to be most sensitive to temperature variation, they vanished at 270 K, but reappeared at 100K, with certain intensity enhancement when temperature decreased; while L-alanine had no such phenomenon.

The changes of integrated intensity at different temperatures were summarized in Fig.3.

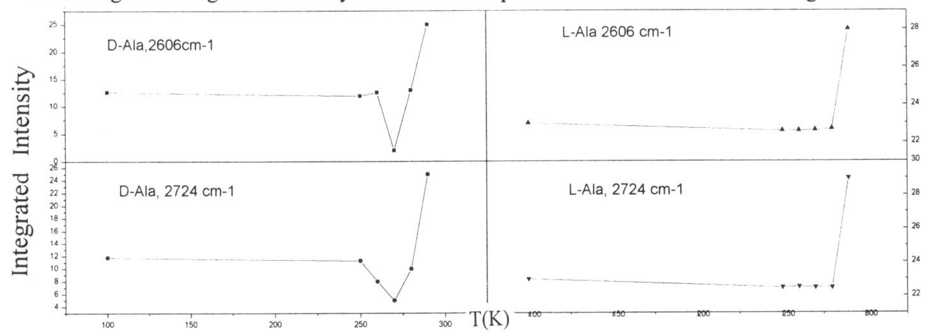

Fig.3 Integrated intensity of C-H stretching mode vs T

It was found that the integrated scattering intensity of L-alanine remained constant throughout most of the temperature region studied, while in the case of D-alanine, each curve underwent a sharp change, resulting in a minimum value at 270K, and gradually picked up intensity as temperature decreased.

Results from the mutual support of the above three different experiments lead us to such a possible conclusion: the anomalies in the curve for specific heat, magnetization measurements and Raman spectra might provide some insights about the phase transition underlying the origin of homochirality.

3. Problems of amplification mechanism

The suggestion of Salam that Bose-Einstein condensation may act as an amplification mechanism suffers from the difficulty of implying large activation energy barriers for the production of optically pure L-amino acids from D-amino acids, since the phase transition may introduce molecular changes that lead to the breaking of chemical bonds, so the possibility of alternative theory which is free from the above mentioned difficulties should be considered.

It is possible to account for this problem by assuming that the resonance hybrid of α- amino acids may be attributable to minimize this difficulty. Bonner and Lemmon[6] found that ionizing radiation led to racemization of optically active amino acids (both solid and dissolved in water).We will take alanine as an example for the interpretation. The unit cell of single crystal of alanine contains four zwitterions, the two ion groups of NH_3^+,COO^- are both attached to α-carbon. It's easy to produce α-radical through α-H abstraction by disturbance from surroundings (i.e. γ photon in crystalline amino acids, OH radical in liquid crystal state) We can draw three resonance hybrids of α-radical anion obtained from the zwitterion.

$CH_3-\dot{C}(NH_3^+)-C(=O)-O^- \rightleftharpoons CH_3-C(NH_3^+)=C(-\dot{O})-O^- \rightleftharpoons CH_3-\dot{C}(NH_3^+)-C(-O^-)=O$

(I) (II) (III)

Since I and III are nearly of the same stability, they contribute similarly to the stabilization of the structure, leading to enhanced stability of the resonance hybrid. This unhindered transition state undergoes the conformation change from the tetrahedral structure (sp^3 orbitals) to the planar trigonal intermediate (sp^2 orbitals).

$H^+ + NH_3^+-\dot{C}(CH_3)-C(=O)-O^- \longrightarrow$ (D-alanine) $+$ (L-alanine)

(D-alanine) (L-alanine)

Due to the weak-neutral current, which causes the difference in the susceptiveness of attack at different side, it may be possible to implement the transition from D- to L-amino acids which have lower ground state energies.

Based on these results, We conclude that at the point of λ phase transiton the difference of D- and L- is amplified. Working under λ phase transiton may minimize the effect of the electromagnetic force and the ratio of the strength of the weak force to that of the electromagnetic force is substantially increased.

References

1. Salam A., (1992), " Chirality, phase transition and their induction in amino acids " , *Phys. Lett.*,**B288**,153-160
2. Wenqing Wang, Xiangrong Sheng et al., (1994), " Low temperature specific heat anomaly of D-valine " , *J. Biol. Phys.*,**20**, 247-252
3. Mason S.F. and Tranter G.E., (1984), " The Parity-violation energy difference between enantiomeric molecules " , *Mol. Phys.* **53**, 1091-1111
4. Wenqing Wang ,Xiangrong Sheng et al., (1996), " Susceptibility behaviour and specific heat anomaly in single crystals of alanine and valine " , *J.Biol. Phys.*,**22**, 65-71
5. Wang C.H. and Storms R.D., (1971), " Temperature dependant Raman study and molecular motion in L-alanine single crystal " , *J.Chem. Phys.*, **55,** 3291-3298
6. Bonner W. A. and Richard M. Lemmon, (1978), " Radiolysis, racemization, and the origin of optical activity " , *Bioorganic Chem.*,**7**,175-187

PHYSICAL STUDIES OF ADSORBED BIO-ORGANIC SUBSTANCES IN Na-MONTMORILLONITE

A. Negrón-Mendoza[1], S. Ramos-Bernal[1], Dwight Acosta[2], and Isabel Gamboa de Buen [1]

[1] Instituto de Ciencias Nucleares, UNAM
México, D.F. 04510 México.
[2]Instituto de Física, UNAM,
México, D.F. 04510 México.

Abstract. Certain solids have been studied with respect to their ability to promote reactions in many organic systems. In particular, clay minerals have been implicated as a probable active surface for prebiological chemical reactions, occurring during the early evolution of the primitive Earth. It is frequently unknown, but very important to explain, how much a solid is involved in the chemical transformation that is taking place at its surface. To have some ideas of the former, some properties of the solid had been monitored and correlated with the adsorption of organic compounds.

In this work we present experimental results from ultraviolet spectroscopy, thermoluminescence, and electron microscopy obtained from samples with adenine and polyadenilic acid (Poly A) adsorbed in Na-montmorillonite. The results obtained indicate that both compounds are readily adsorbed into the clay. Adenine is adsorbed in the interlamellar space via ionic interchange and Poly A is binding mainly at the edges of the crystal. The thermoluminiscence signals allow us to support the proposal that defects in solids caused by natural radiation may serve as energy transfer for reactions in the clay.

2. Introduction

2.1. IMPORTANCE OF A SURFACE FOR PREBIOTIC SYNTHESIS

Chemical reaction may take place at hydrosphere/lithosphere, atmosphere/lithosphere interfaces, thus, a reliable simulation of relevant prebiotic environment would require the consideration of multiphase systems. In this sense, minerals are of particular interest for evolving molecules, because they can act as selective concentrating and internally binding agents. Without such concentration, most prebiotic scenarios come to halt, due to highly concentrated prebiotic environment may be implausible (Sillén, 1965) In particular, some significant fraction of monomers would have formed in mineral surface, or would have been adsorbed and concentrated and later condense in a complex molecule.

J. Chela-Flores and F. Raulin (eds.),
Exobiology: Matter, Energy, and Information in the Origin and Evolution of Life in the Universe, 137–142.

2.3. INFLUENCE OF MINERAL SURFACES ON THE CHEMICAL TRANSFORMATION OF ORGANIC COMPOUNDS.

Solid surfaces are important because the enhance chemical reactions by different mechanism. Clay minerals, carbonates, zeolites, silicates, sulfides, sulfates, and oxyhydroxydes are probably the most geologically relevant and abundant surfaces on the primitive Earth.

From the data published in the literature, the role of clays in chemical evolution as selective adsorbed, concentrator and catalyst for polymerization have been demons-treated. The adsorption of a large variety of biologically important compounds is well documented. For many compounds the adsorption is higher at an acid pH. Sometimes the binding of these types of compounds drops to almost null adsorption at pH 8. Since the pH of the primitive ocean is believed to be about 8, this may disqualify some results. Nevertheless, there are several natural conditions that attain higher acidities, and thus improve adsorption binding, supporting in this way the possible role of microenvironments in the primitive Earth.

In view of the perceived prebiotic importance of the nucleic acids, an extensive search has been made for the binding of nucleic acid bases and their derivatives in clay minerals. Some general observations are: The adsorption is pH dependent. The adsorption is more extensive in acid pH, and then it decreases steadily to a very low adsorption level at pH >6. Purines are adsorbed in larger extent than pyrimidines. Nucleotides and nucleosides are adsorbed in less extent than their corresponding bases (Perezgazga et al., 1998 and the references therein). There are two sites of binding, named the interlamellar channel and the edges of the clay. Bases are adsorbed mainly in the interlamellar channel. This conclusion was based in experiments blocking selectively the interlamellar channel with alkylamines or poisoning the edges with phosphates (Negrón-Mendoza and Ramos-Bernal, 1996; Perezgasga, et al., 1998).

When reactant species are adsorbed, in a mineral surface, their movement is restricted to two dimensions, rather than three dimensions of movement available in overlaying solution. Encounter frequencies between two diffusing species increase as the number of dimensions is decreased favoring reactions. (Hardt, 1979). Besides this enhancement, minerals on the early Earth were exposed to penetrating sources of energy such as ionizing radiation and ultrasound. The interaction of those natural energy sources, with mineral surfaces can influence the chemical transformation of organics; including such facets as excitation, degradation, storage and transfer. Energy transfer is of a particular interest because clays can store energy from the environment and release it in various forms. A consequence of the energy transfer between clays and organic compounds is that solids should be able to serve as moderators of energy. The interaction between the solid surface and the energy can be followed by changes in their microscopic structure and by the spectroscopic properties of the solid, such as its luminescence properties.

The questions posed in the present study related to the adsorption process of adenine and polyadenylic acid (Poly A); as model compounds for purines are: A) How the microscopic structure of the mineral is affected by the presence of an adsorbate? B) Can clays have thermoluminoscence properties? If so, can the accompany phenomena be used to explore the reactivity potential of the mineral? Is the luminescence signal of the non treated clay affected by the presence of an adsorbed molecule? To answer these questions we studied the adsorption of adenine and Poly A in Na-montmorillonite.

3. Experimental Procedures

The clays used in the present research were prepared from commercially available clays from Crook County, Wyoming, U.S.A. It was sterilized by autoclaving at 2 kgf/cm^2 for 20 minutes. The adsorption experiments were made with adenine (1 x 10^{-6} to 1 x10^{-3} mol dm^{-3} and Poly A 1 x 10^{-5} mol dm^{-3}), the preparation and analysis were made according to Perezgazga et al., 1998.

3.1 THERMOLUMINESCENCE MEASUREMENTS

The thermoluminescence (TL) consists in the light emission by a material, which has previously absorbed energy via ionizing radiation or UV light exposure when it is heated. When the material is naturally or artificially exposed to ionizing radiation, the electrons and holes produced during ionization are trapped at localized energy levels. As the temperature of the sample increases, the energy supplied was sufficient to enable the trapped electron to escape, and diffuse into the material until they recombine with a hole. The light emission is the result of the radiative recombination of electron-hole pairs. The graph of the TL intensity emitted as function of temperature is called the glow curve.

The natural TL response of the Na-montmorillonite without any treatment and with adenine and Poly A at different concentrations was studied. The natural TL is the TL intensity of a sample that had not been irradiated except those sources present in nature. Some samples with adenine were irradiated with ^{60}Co gamma-rays at a dose of 42 kGy.

The samples studied were made mixing the clay powder with tridestilled water to do a tablet, and then drying the sample at 40 °C. The reading of the TL signal was carried out using a Harshaw 2000 reader acoplated to a PC with a data acquisition system. The samples were heated from room temperature up to 450 °C, at a heating rate of 10 °C/s in a nitrogen atmosphere. The samples were read twice immediately after preparation, obtaining the TL signal and the residual signal (the sample holder and sample emission at high temperatures).

3.2. ELECTRON MICROSCOPY.

Transmission electron microscopy observations were carried out in JEOL 100CX and in JEOL 4000EX microscopes for conventional and high resolution modes respectively. Samples in each case were grinded in an agate mortar, dispersed in an ultrasonic bath and dropped on 200 mesh copper grids covered with a holey carbon film. Scanning electron microscopy observations were carried out in a JEOL 5200 microscope, the samples without grinding were covered with a sputtered metallic film before observations

4. Results and Discussion

4.1 ABSORPTION EXPERIMENTS

The results obtained of the adsorption of adenine and Poly A showed that these compounds are readily adsorbed in acid pH. The degree of binding decreases with an increased of pH. At pH 2 the samples are adsorbed 90-95%.

4.2 THERMOLUMINESCENCE RESULTS

TL response was defined as the difference between the TL signal and the residual signal normalized by the sample area. The intensity of the TL signal was dependent on the area of the sample and independent of the thickness to approximately 2 mm.

Figure 1 shows the natural TL glow curves obtained for the Na-montmorillonite without treatments and with adenine and Poly A adsorbed at different concentrations. The curves have not been normalized. These curves have a peak at 250 °C, and its intensity decreases when de pH of the adsorbed organic compound increases. Also, in Figure 1 is presented a glow curve of the clay with adenine exposed to gamma rays at 42 kGy, here the peak occurs at 170 °C.

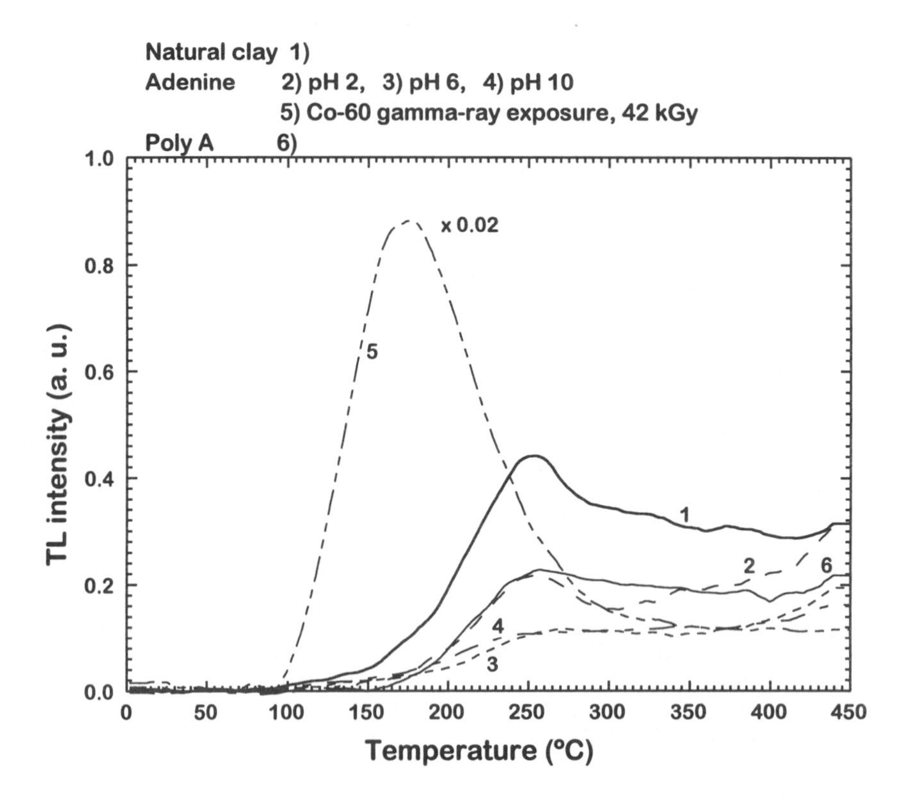

Figure 1. Glow curves for Na-montmorillonite, adenine + Na-montmorillonite, Poly A + Na-montmorillonite and Adenine +Na-montmorillonite irradiated at 42 kGy

The TL response of the samples showed, that there are significant differences in the response between the clay without, and with organic compounds, having a decrease in the response in the last case. Apparently there is no difference if adsorbate is adenine or Poly A. Also it was found that the response at pH 2 is larger than that at pH 6 and 10. For the irradiated sample, the response was approximately 180 times that obtained for the sample with adenine at the same pH.

Current research is going on to explain if with the TL measurements it is possible to find the concentration of adsorbed compounds. Because all clay samples presented thermoluminiscence signals imply that they have defects and traps associated with natural irradiation. This is very important due that it is kwon that heterogeneous catalysis can enhance some chemical reactions in compounds adsorbed in it. In nature, most of the energy produced by radioactive material is initially adsorbed in the solid phase. Consequently of this process, the solid phase may disipate the residual energy by different ways: heat, luminescence decay, chemical changes in the solid, transference to an adsorbed molecule or energy stored as separated charge pairs (holes or trapped electrons) namely non equilibrium charge carriers. An energized solid state, whether manifested by formation of chemically altered active sites or simply by separate trapped charges, might be expected to be more catalytic active that one not exposed to radiation field, because, most of the radioactive energy is initially adsorbed by the solid.

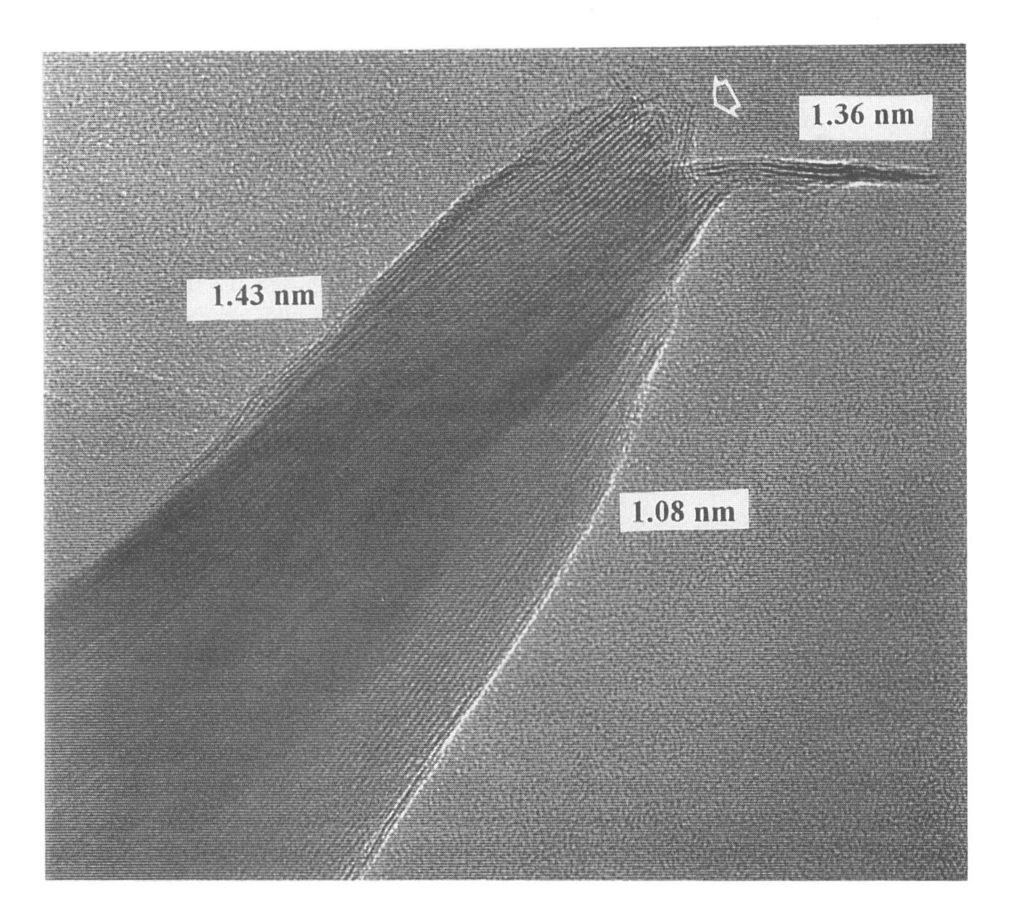

Figure 2. High resolution electron micrograph showing Na-montmorillonite lattice planes and what it might be an organic material (adenine) attached to the channel of the Na-montmorillonite material in the arrowed zone

The existence of stored energy, and the transduction process is manifested in a variety of physicochemical properties. These can be monitored by electron spin resonance, luminescence, and changes in the surface reactivity dependening on electronic energy input to the material.

4.3 ELECTRON MICROSCOPY

Conventional transmission and scanning electron microscopy did not reveal significative variations in morphology and structure in samples impregnated with organic molecules respect to the original ones. Yet, in several high resolution electron micrographs with lattice, and even atomic resolution, it is possible to detect material on and in Na-montmorillonite structure. In Figure 2, the Na-montmorillonite lattice distances runs from 1.08 to 1.43 nm. Also it is possible to observe in the arrowed zone, a crystalline material, showing atomic resolution, attached to the extreme of Na-montmorillonite planes. From this kind of micrograph it is concluded that foreign materials, organic molecules are attached or incorporated in the Na-montmorillonite structure, in particular, adenine is attached to the interlamellar channel.

5. Concluding Remarks

A reliable simulation of The primitive scenario on the Earth would require the consideration of multiphase systems. In this sense, minerals are of particular interest for evolving molecules. The capacity of these materials to carry out synthetic reactions can be directly related to charge and energy mobility associated with the defects of the crystal structure. These defects are formed by the interaction with penetrating energy sources. This interaction can be studied measuring some spectroscopic properties. The results obtained in this study with thermoluminescene an electron microscopy allow us to postulate that in nature, the energy release by radioactive material is initially adsorbed in the solid phase. Consequently of this process, there is an energy storage and transfer process in clays..

Acknowledgments

This work was partially supported by a CONACYT grant and DGAPA grant. We thank to Mr. S. Ham for doing the special glassware for these experiments.

References

Bernal, J.D., *The Physical Basis of Life*, Routledge and Kegan Paul, London, 1951.

Hardt, S.L. Rates of diffusion controlled reactions in one, two and three dimensions, *Biophysic Chem,* 10 (1979), 239-243.

Negrón-Mendoza, A., Ramos-Bernal, S. and Albarrán, G. Enhance decarboxylation reaction of carboxylic acids in clay minerals, *Radiat. Phys. Chem.* 46 (1995), 565-568.

Perezgaga, L. Negrón-Mendoza, A., Mosqueira, G. and De Pablo, L. Site of adsorption of purines, pyrimidines and its corresponding derivatives on sodium montmorillonite. To be published, 1998.

Sillén, L.G. Oxidation state of Earth's ocean and atmosphere I. A model calculation on earlier states. The mith of the "prebiotic soup" *Arkiv for Kemi* 24 (1965), 431-456.

On the Prevailing Direction of our Search into the Origin-of-Life -- "Second thoughts".

F.R. Eirich
Polytechnic University of New York
6 Metrotech Center
Brooklyn, NY 11201

1. Introduction

A polymer and colloid chemist, I became a participant in the Search for a plausible physicochemical mechanism that potentially led to the Origin of Life on our Planet, when Ahron Katchalsky drew me into his efforts to find a catalyst that would prevent the ready hydrolysis of amino acid phospho-adenylates and instead assist in the formation of polypeptides. Settling on Na-montmorillonite, he and his co-workers could achieve self-polycondensation of amino acids, to yield, for the first time, a spectrum (rather than a continuous distribution) of homo-polypeptides up to 80-mers[1]. After considerable work, Dr. Paecht and I were able to suggest a likely mechanism for these findings, including for the order in di- and tri-copolymers[2] though the corresponding polycondensation of nucleotides eluded us.

During the same period and since,a great deal of excellent theoretical and experimental work by a host of notable investigators[3] established pathways for the self-synthesis, under plausible prebiotic conditions, of the constituents (and their reactions) of putative primitive Life, but the self-assembly of presumed prebiotic entities to units which one could consider as"living", ie., the transition from in-animate to animate matter is not only still conjectural, but conceptually elusive[4]. Criticisms were also raised which claim principal difficulties in the way of the selfsynthesis of essential components under plausible circumstances[5].

2. What is Life, and Where and How Did It Start?

We are still haunted by Schroedinger's conjectural book. Considering his question, one is struck, by the absence of an agreement on the criteria for "living"[6], and secondly, by the uncertainty of the planetary condition in

J. Chela-Flores and F. Raulin (eds.),
Exobiology: Matter, Energy, and Information in the Origin and Evolution of Life in the Universe, 143–146.

general, and of the specific areas in particular, where the self-synthesis of "living" matter happened on our young planet.

The occurrence of self-organizing (irreversible) heat engines on a planet between its Sun and the heat sink of Space, thermodynamically favored, is exemplified eg., by the weather, ie, circulation within an irradiated atmosphere; another example are flames of combustion over outgassing chemical reactions. However, the coming of populations of Informed Heat Engine capable of evolving Life and Intelligence, surely needed the pre-existence of more subtle and intricate conditions. Exobiologically formed organic matter, whether from space or indigenous, must have evolved prebiotically to higher molecular weights with some enzymatic capability to break down energy-rich compounds. Thermodynamically favored, they evolved further to spawn Life in ways that are unknown and perhaps unknowable. For one conjecture, see Eirich (8).

We can only guess at the planetary locales that gave rise to molecules of a nature, and of sufficient molecular weight, to possess the catalytic ability and stability to produce, in their life span, the constituents of earliest life: volcanic vents on ocean floors, porous sediments, acid pools on cooling surface lava, drying water pools along ocean shores, freezing puddles above or below sea ice, subterranean mineral beds, et al, are the places thought of as potential spots for the synthesis of the required compounds. There is, further, our ignorance of the species and the number of molecules needed for the formation of "Life", whichever way defined, and of the way these components self-assembled.

3. Some Doubtful Assumptions

The only guidance concerning the constituents of early Life comes from the surmises of Biologists who contend that today's simplest organisms are by and large acceptable models for early Life. Their argument is based on the biological clock constituted by the protein-, or nucleotide-, family trees which indicate which organisms stayed less changed through the eons, so that they are good models for early living cells. But, even if one accepted this argument, in the face of the possibility that the biological clock might have ticked differently in earlier times along different branches of Life, there remains the likelihood that our ancestral line evolved for long times before it became the self-sustaining system that Biologists would call "alive". In other words, the unspoken assumption that we could learn of the Origin of Life, if only we could find plausible pathways for the components of Today's Life, appears now to me to have no logical basis; let alone the fact that, even if we had all components and, putting them together following the blueprint of today's cells, the assumption that they would become "alive" strikes me as optimistic.

These deliberations point to fundamental difficulties quite different from those claimed before (4, 5): ie., that, even should we succeed building, and recognizing, an entity in the line of descent of our progenitor, it would take billions of "evolving" generations and the same chain of "chances" that shaped our Life's evolutionary course, to arrive at equivalents of precambrian pro-or eu-caryoties. On the other hand, even if we tried and succeeded in a short-cut and produced precambrian-type cell-constituents "abiotically", we would still need the realization of the underlying assumption that putting the right components together in the right way, would, without priming, begin the reaction cycles whose sum we recognize as "living", and that Life from thereon would reproduce its parts "biologically" (by metabolism).

4. Looking for Alternatives

These difficulties do not militate against the good Science (on Chemical Evolution) that has developed while searching to understand the Origin of Life, but they constitute a reason for trying to build simpler models[9]. By building flying machines we came no closer to understanding flying Life, but we learned what flying entails. Equivalent benefits came from constructing speeding, diving, roboting and calculating machines,etc. Like Life, such machines are thermodynamically open systems, but have to be primed. Strangely, we have no machines which could be selfpriming, i.e, chemical ones (perhaps this is why Life uses a chemical machinery). The nearest chemical processes we have are emulsion polymerization which is self-priming, but also self-limiting, and self-igniting flames which can be self-propagating but are not self-encapsulating. Acting on the truism that "one can only understand what one can build", working with chemical models which exhibit features of "living" should place us in a better position to define "Life", and yield insights into the transition from non-living to living matter.

Since Life, whatever else it might be, is based on bounded irreversible processes, models which are micro flow-eactors for organic products are a logical choice. Our limited knowledge suggests that leaky heterotrophic liposomes (reagents diffusing in, wastes out), containing catalytically active building blocks of our biochemistry should be one practical starting point. Bio-engineers, cognizant of, but not necessarily influenced by their knowledge of working cells, should study the minimum requirements for the circular-periodic chemistry that enables their models to become minimal, self-maintaining and replicating systems. It would then make sense to study their formation under a variety of assumed prebiotic conditions.

5. Summary

It is arguable that much of today's research into the Origin Life amounts to an approach without a well defined target, since neither the conditions under which Life arose, nor the structure of the earliest "living" entities, nor the nature of the components which self-organized, can be known with certainty. It is a vague hope that, if a large number of the pieces known to be essential for Life after 4 billion years of prebiotic and biological evolution, were prepared, those essential for an energy dissapative activity will somehow fall into place. In addition to hopeful perseverance, notwithstanding its also formidable obstacles, one should, therefore, attempt to model systems that mimic functions of extant Life in order to become better acquainted with how selfmaintaining chemical systems will form spontaneously based on the interaction of plausible prebiotic building blocks.

6. References

1) Paecht-Horowitz, M., Berger, J.C.,and Katchalsky, A., Nature, 228, 639 (1970).
2) Paecht-Horowitz, M., and Firich, F.R., Origin of Life Evol. Biosphere, 18,359 (1988).
3) Deamer, D. W., and Fleischaker, G., "Origins of Life" (1994), Jones and Bartlett, Publ., Boston, MA.
4) Thaxton, C.B., Bradley, W.L., and Olsen, R.L. (1984),"The Mystery of Life's Origin", Philosoph. Library Inc., New York, NY.
5) Shapiro, R., "Origins" (1985), Summit Books, New York, NY.
6) Chyba, C.F., and McDonald, G.D., Ann. Rev. Earth Sci. 23,215 (1995).
7) Rothstein, J., in: "Prebiotic Selforganization of Matter" (1990), C. Ponnamperuma and F.R. Eirich, Eds., A.Deepak, Publ., Hampton,VA.
8) Eirich, F.R., in: "Chemical Evolution (1996), J.Chela-Flores and F. Raulin, Eds., Kluwer Academ. Publ., Dordrecht, Netherlands.
9 Eirich, F.R., in: "Chemical Evolution" (1995), Chadha, M., Negro-Mendoza, A., Oshima, T., and Chela-Flores, J., Eds., A. Deepak Publ., Hampton, VA.

Section 4.
Information

GENERIC PROPERTIES OF THE SEQUENCE-STRUCTURE RELATIONS OF BIOPOLYMERS

PETER F. STADLER
Institut für Theoretische Chemie und Strahlenchemie
Universität Wien, Währingerstr. 17, A-1090 Wien, Austria, &
Santa Fe Institute
1399 Hyde Park Rd., Santa Fe NM 87501, U.S.A.
Email: studla@tbi.univie.ac.at *or* stadler@santafe.edu

1. Modeling Molecular Evolution

Biological evolution is a highly sophisticated dynamical phenomenon, and its complexity is often confusing. For the purpose of analysis it may be partitioned into the four partial processes depicted in figure 1 [25]: population dynamics, (population) support dynamics, genotype-phenotype mapping, and fitness evaluation. These components are properly visualized as mappings between abstract metric spaces:

- the *concentration space* of biochemical reaction kinetics,
- the *sequence space* of polynucleotide sequences,
- the *shape space* of biopolymer structures, and
- the *parameter space* of kinetic rate constants.

Kinetics of chemical reactions or changes in populations are recorded in concentration space, the conventional space of chemical reaction kinetics. Variables count numbers of particles, molecules, virions, cells, or organisms, for example, and display their changes over time. Population dynamics is commonly described by differential equations, by difference equations in case of synchronized generations, or by a stochastic process in case of small particle numbers [12]. Concentration space deals only with the classes of genotypes actually present.

Sequence space (Q^n_α, d) is a metric point space containing all α^n possible genotypes of given chain length n, where α denotes the size of the polynucleotide or amino acid alphabet. Distances in sequence space are expressed by the minimum number of single nucleotide exchanges or point mutations

J. Chela-Flores and F. Raulin (eds.),
Exobiology: Matter, Energy, and Information in the Origin and Evolution of Life in the Universe, 149–156.

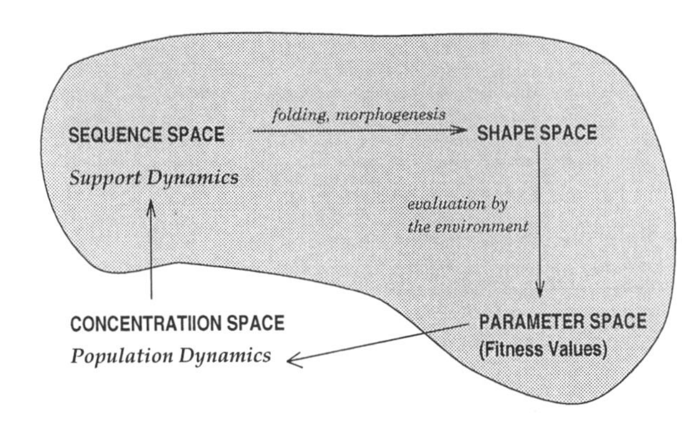

Figure 1. Evolutionary dynamics. *In vitro* evolution of biomolecules is partitioned into four components: (i) population dynamics, (ii) population support dynamics, (iii) genotype-phenotype mapping, and (iv) evaluation of the phenotype. Wright's notion of a *fitness landscape* is the composite of genotype-phenotype mapping and phenotype evaluation (shaded in gray). Population dynamics is tantamount to chemical reaction kinetics of replication, mutation and selection. Population support dynamics describes the migration of populations in the space of genotypes. Genotype-phenotype mapping unfolds biological information stored in polynucleotide sequences, which in turn is translated into kinetic constants for population dynamics.

converting one sequence into another: the Hamming distance d. The notion of a sequence space encapsulates a simple modeling of mutation frequencies because genotypes are grouped according to the minimum number of point mutation events that separate them.

The genotypes present in a population form a collection of points in sequence space. When a mutant is produced, an new point in sequence space is occupied and a new variable "appears" in concentration space; when a variant dies out, the corresponding point in sequence space becomes vacant and a variable "disappears" from the population dynamics picture. The cloud of occupied points in sequence space, the so-called *support* of a populations, therefore moves. Molecular phylogenies therefore can be viewed as reconstructions of the time evolution of the population support (or rather its center of mass, the consensus sequence) from the "snapshot" of the present-day sequences.

A complete model of evolutionary dynamics has to explain also the parameters (rate constants) entering the kinetic equations. These *fitness values* are the consequence of the species' *phenotypes*. The *shape space* $\mathcal{S}$ contains all phenotypes that are formed by processing genotypes in a given context. The genotype-phenotype mapping encapsulates all of morphogenesis in the case of higher organisms, in the context of molecular evolution, however, phenotypes are tantamount to the spatial structures of biopolymers. Sewall Wright [30] introduced the term *fitness landscape* for the combined mapping from genotypes to fitness values.

The dynamics of evolution is oftentimes simplified to adaptive walks on fitness landscapes [16], thereby disregarding concentrations. This is an acceptable shortcut as long as different phenotypes do not interact directly but only compete for common resources. Many simulation studies in this field use *ad hoc* substitutes for biological fitness landscapes such as spin glass Hamiltonians. It can be shown, however, that the properties of a fitness landscape are determined almost exclusively by its underlying sequence-structure map [28]. In the case of the most important biopolymers the landscapes deviate significantly from simple disordered systems.

2. RNA Secondary Structures

Meaningful notions of structure are of course context-depend: successful descriptions of active sites of enzymes (or ribozymes) or specific binding sites for regulatory elements on DNA, for example, require high precision at atomic resolution as provided by X-ray crystallography whereas studies on phylogenetic conservation of structures can be done much better on the coarse-grained level of protein *ribbon* diagrams or RNA secondary structures. Some levels of coarse-graining are not only of physical relevance but also suitable for mathematical modeling. RNA is a particularly fruitful system because the structure prediction problem can be solved efficiently at least at the approximate level of secondary structures.

Experimental energy parameters are available for the individual contributions as functions of type (stacked pair, interior loop, bulge, multi-stem loop), of size, of the type of the delimiting base pairs, and partly of the sequence of the unpaired subsequences, see e.g. [8]. Additivity of the energy contributions allows us to calculated minimum energy structure of an RNA sequence by means of dynamic programming [31]. An efficient implementation of this algorithm is part of the `Vienna RNA Package` [11] (available from `www.tbi.univie.ac.at`). The statistical properties of RNA secondary structures reported in this contribution depend very little on choices of algorithms and variations of the parameter set [29].

The number of acceptable secondary structures can be computed exactly. An asymptotic expansion shows that it grows like 1.8488^n, assuming that (i) the minimum stack length is two base pairs and (ii) the minimum size hairpin loop has three unpaired nucleotides. The number of different structures that are actually formed as minimum energy structure of some sequence is even smaller. Our best estimate is 1.65^n. In contrast, there are 4^n possible RNA sequences composed from the natural **AUGC** alphabet; thus many sequences must fold into the same structure. We call the set $S(\psi)$ of all sequences folding into the same structure ψ the *neutral set* of ψ.

The logic of RNA base pairing implies *a priori* restrictions on $S(\psi)$.

Only the combinations **GC**, **CG**, **AU**, **UA**, **GU**, and **UG** form stable base pairs. Therefore, if positions i and j are paired in a structure ψ, then only one of these six combinations of nucleotides may occur in the corresponding sequence positions. The sequences satisfying this condition form the set of *compatible sequences* $C(\psi)$. Clearly we have $S(\psi) \subset C(\psi)$. Proteins sequences apparently are not subject to a simple restriction of this form.

3. Protein Structures

Computational surveys of protein space similar to the explorations of the RNA world are precluded by the notorious complexity of the protein folding problem, and by the fact that there is no obvious biophysically meaningful and computationally simple coarse resolution of protein structures. For our purposes, however, a technique is sufficient that allows us to decide whether a sequence x folds into a particular structure ψ. This *inverse folding problem* is less demanding than predicting the unknown protein structure of a given amino acid sequence [3].

Our starting point is a potential function $W(x, \psi)$ evaluating the energy of a sequence x when folded into a structure ψ, the latter being defined defined by the spatial coordinates of its C^{α} and C^{β} atoms, respectively [1]. Recent studies using so-called knowledge based potentials, see e.g. [27], demonstrated that the energy of the native fold of a sequence x can be estimated from the distribution of the energy values of x in its conformation space. As a measure for the quality of fit of sequence x and structure ψ the *z-score* [4] $z(x, \psi) = [W(x, \psi) - \overline{W}(x)]/\sigma_W(x)$ is used. Here $\overline{W}(x)$ is the average energy of sequence x in all conformations in a database and $\sigma_W(x)$ is the standard deviation of the corresponding distribution. Empirically, native folds have z-scores in a narrow characteristic range. Hence one may assume that x is a member of the neutral set $S(\psi)$ if and only if the z-score of x in conformation ψ is in the native range. Of course, only native structures ψ that are already in a structure database can be explored by this method.

The number of possible protein sequences is enormous: for $n = 100$ residues there are 20^{100} sequences. On the other hand, the repertoire of stable native folds seems to be highly restricted or even vanishingly small [5]. For example, it has been frequently observed that seemingly unrelated sequences have essentially the same fold [13]. Again, there seem to be by far more sequences than structures.

4. The Topology of Neutral Sets

Data from both large samples of long RNA sequences [26] and from exhaustive folding of all short RNAs [9] suggest a distinction between *common* structures, which are formed by more than the average number of sequences, and *rare* structures. Common structures represent only a small fraction of all structures, but almost all RNA sequences fold into one of these few secondary structures. Lattice protein models exhibit the same behavior [18]. Only the common structures play a role in natural evolution and in evolutionary biotechnology [24].

For an understanding of molecular evolution we need to know how sequences folding into the same common structure are distributed in sequence space, because it is the topology of neutral sets that determines whether similar structures with distant sequences (a phenomenon commonly observed with both RNA and proteins) may have originated from a common ancestor, or whether they must be the result of convergent evolution. Three approaches have been applied so far to studying the topology of neutral sets: a mathematical model of genotype-phenotype mapping based on random graph theory [23], extensive sample statistics [26, 1] and exhaustive folding of all sequences with given chain length n [10]. It was noticed already in early work on RNA secondary structures [7] that a substantial fraction of point mutations are neutral, i.e., that many sequences differing only in a single position fold into the same secondary structure. The same seems to be true for proteins. On the other hand, sequences folding into the same structure are almost randomly distributed in sequence space (more precisely in the subspace of compatible sequences $C(\psi)$ in the case of RNA) [26]. The mathematical model therefore assumes that sequences forming the same structure are distributed randomly using the fraction λ of neutral neighbors as (the only) input parameter. If λ is large enough the model makes three rather surprising predictions [23, 21]:

• The connectivity of networks changes drastically when λ passes the threshold value $\lambda_{cr}(\alpha) = 1 - \sqrt[\alpha-1]{1/\alpha}$ where α is the size of the alphabet. Neutral sets consist of a single component that spans sequence space if $\lambda > \lambda_{cr}$. Below threshold, $\lambda < \lambda_{cr}$, the network is partitioned into a large number of components, in general, a giant component and many small ones. (For RNA it is necessary to split the random graph into two factors corresponding to unpaired bases and base pairs, respectively, and to use a different value of λ for each factor.) In the first case we refer to $S(\psi)$ as the *neutral network* of ψ. For both RNA and proteins λ seems to be well above the threshold. Extensive neutral networks in protein space discovered using knowledge-based potentials [1] meet a claim raised by Maynard Smith [20] for protein spaces that are suitable for efficient evolution. Lattice protein model also show neutral networks [19, 2]. For RNA secondary structures their existence is demonstrated in [10].

• There is *shape space covering* for $\lambda > \lambda_{cr}$, that is, in a moderate size ball centered at an arbitrary position in sequence space there is a sequence x

TABLE 1. Shape space covering radii and closest approach distances of neutral network for RNA secondary structures.

n	r^*	r_{cov}	$r_{cov} - r^*$	r_{ca}
30	3.1	5.2	2.0	
50	6.5	9.2	2.7	2.1
70	10.0	13.7	3.7	3.4
100	15.2	20.5	5.3	5.6

that folds into a an arbitrary prescribed secondary structure ψ. The radius $r_{cov} = \min\{h \mid B(h) \geq S_n\}$ of such a sphere, called the *covering radius*, can be estimated from simple probability arguments [24] in terms of the number of structures S_n and the number of sequences $B(h)$ contained in a ball of radius h. It is much smaller than the radius n of sequence space, see [10]. The average distance r^* of a random sequence to set of compatible sequences $C(\psi)$ of the target structure ψ is clearly a lower bound on the covering radius. Best upper bounds on r_{cov} listed in table 1. The covering sphere contains all common structures and therefore forms an evolutionarily representative part of shape space.

- The neutral networks of any two common structure ψ and ψ' "touch" each other if $\lambda > \lambda_{cr}$. The intersection theorem [22] guarantees that one can always find sequences that are compatible with both ψ and ψ' [22], i.e., $C(\psi) \cap C(\psi') \neq \emptyset$. Random graph theory then predicts that the distance between the neutral networks of two arbitrary common structures is $r_{ca} = 2$. This prediction has been tested computationally for RNA secondary structures [10]. One expects $r_{ca} \approx r_{cov} - r^*$ for long chains. Note that computational studies can only provide upper bounds for r_{cov} and r_{ca}. Preliminary results (Babajide, pers. comm.) suggest that there is shape space covering also for proteins.

5. Discussion

Algorithms for folding RNA sequences into secondary structures as well as inverse folding procedures for proteins based on knowledge-based potentials predict extended connected networks of sequences with identical structure and shape space covering. These observations have striking consequences for adaptation, based on a fairly realistic model of test-tube evolution [15]: (1) Acceptable starting points for evolution can be found everywhere in sequence space. (2) Finite populations diffuse along neutral networks, where their dynamics conforms the predictions of Kimura's neutral theory [17]. After a sufficiently long period of time all sequence information is lost, yet the phenotype is conserved. It is the maintenance of a phenotype, not of a genotype, which defines the mutation threshold beyond which adaptation breaks down. (3) On a single neutral network the population splits into well separated clusters. A population is not a single localized quasi-species [6] in sequence space, but rather a collection of different, independently diffusing quasi-species that share the same phenotype. (4) Neutral networks of different structures are interwoven. While drifting on a neutral network a population produces a fraction of mutants off the network and thereby explores new phenotypes [14]. A selection-induced transition between two structures, i.e., a phase of Darwinian selection, occurs in regions of sequence space where their networks come close to one another.

Neutral evolution, therefore, is not a dispensable addendum to evolutionary theory as it has often been suggested. On the contrary, neutral networks, arising as a consequence of the redundancy of sequence-structure (and possibly also structure-function) relationships of biopolymers, provide a powerful mechanism through which evolution becomes truely efficient.

Acknowledgments. The research on the sequence structure maps of biopolymers is an on-going joint effort with Peter Schuster, Walter Fontana, and Ivo Hofacker.

References

1. A. Babajide, I. L. Hofacker, M. J. Sippl, and P. F. Stadler. Neutral networks in protein space: A computational study based on knowledge-based potentials of mean force. *Folding & Design*, 2:261–269, 1997.
2. E. G. Bornberg-Bauer. How are model protein structures distributed in sequence space? *Biophys. J.*, 73, 1997. in press.
3. J. U. Bowie, R. Luthy, and D. Eisenberg. A method to identify protein sequences that fold into a known three-dimensional structure. *Science*, 253:164–170, 1991.
4. G. Casari and M. J. Sippl. Structure-derived hydrophobic potentials — hydrophobic potentials derived from X-ray structures of globular proteins is able to indentify native folds. *J. Mol. Biol.*, 224:725–732, 1992.
5. C. Chothia. Proteins. One thousand families for the molecular biologist. *Nature*, 357:543–544, 1992.
6. M. Eigen, J. McCaskill, and P. Schuster. The molecular Quasispecies. *Adv. Chem. Phys.*, 75:149–263, 1989.
7. W. Fontana and P. Schuster. A computer model of evolutionary optimization. *Biophys. Chem.*, 26:123–147, 1987.
8. S. M. Freier, R. Kierzek, J. A. Jaeger, N. Sugimoto, M. H. Caruthers, T. Neilson, and D. H. Turner. Improved free-energy parameters for predictions of RNA duplex stability. *Proc. Natl. Acad. Sci. (USA)*, 83:9373–9377, 1986.
9. W. Grüner, R. Giegerich, D. Strothmann, C. M. Reidys, J. Weber, I. L. Hofacker, P. F. Stadler, and P. Schuster. Analysis of RNA sequence structure maps by exhaustive enumeration. I. Neutral networks. *Monath. Chem.*, 127:355–374, 1996.
10. W. Grüner, R. Giegerich, D. Strothmann, C. M. Reidys, J. Weber, I. L. Hofacker, P. F. Stadler, and P. Schuster. Analysis of RNA sequence structure maps by ex-

haustive enumeration. II. Structures of neutral networks and shape space covering. *Monath. Chem.*, 127:375–389, 1996.

11. I. L. Hofacker, W. Fontana, P. F. Stadler, S. Bonhoeffer, M. Tacker, and P. Schuster. Fast folding and comparison of RNA secondary structures. *Monatsh. Chemie*, 125:167–188, 1994.
12. J. Hofbauer and K. Sigmund. *Dynamical Systems and the Theory of Evolution.* Cambridge University Press, Cambridge U.K., 1988.
13. L. Holm and C. Sander. Dali/FSSP classification of three-dimensional protein folds. *Nucl. Acids Res.*, 25:231–234, 1997.
14. M. A. Huynen. Exploring phenotype space through neutral evolution. *J. Mol. Evol.*, 43:165–169, 1996.
15. M. A. Huynen, P. F. Stadler, and W. Fontana. Smoothness within ruggedness: the role of neutrality in adaptation. *Proc. Natl. Acad. Sci. (USA)*, 93:397–401, 1996.
16. S. Kauffman. *The Origin of Order.* Oxford University Press, New York, Oxford, 1993.
17. M. Kimura. *The Neutral Theory of Molecular Evolution.* Cambridge University Press, Cambridge, UK, 1983.
18. H. Li, R. Helling, C. Tang, and N. Wingreen. Emergence of preferred structures in a simple model of protein folding. *Science*, 273:666–669, 1996.
19. D. J. Lipman and W. J. Wilbur. Modelling neutral and selective evolution of protein folding. *Proc. R. Soc. London B*, 245:7–11, 1991.
20. J. Maynard-Smith. Natural selection and the concept of a protein space. *Nature*, 225:563–564, 1970.
21. C. M. Reidys. Random induced subgraphs of generalized n-cubes. *Adv. Appl. Math.*, 1997. in press.
22. C. M. Reidys and P. F. Stadler. Bio-molecular shapes and algebraic structures. *Comp. & Chem.*, 20:85–94, 1996.
23. C. M. Reidys, P. F. Stadler, and P. Schuster. Generic properties of combinatory maps: Neural networks of RNA secondary structures. *Bull. Math. Biol.*, 59:339–397, 1997.
24. P. Schuster. How to search for RNA structures. Theoretical concepts in evolutionary biotechnology. *J. Biotechnology*, 41:239–257, 1995.
25. P. Schuster. Landscapes and molecular evolution. *Physica D*, 107:351–365, 1997.
26. P. Schuster, W. Fontana, P. F. Stadler, and I. L. Hofacker. From sequences to shapes and back: A case study in RNA secondary structures. *Proc. Roy. Soc. London B*, 255:279–284, 1994.
27. M. J. Sippl. Boltzmann's principle, knowledge-based mean fields and protein folding. an approach to the computational determination of protein structures. *J. Computer-Aided Molec. Design*, 7:473–501, 1993.
28. P. F. Stadler. Fitness landscapes arising from the sequence-structure maps of biopolymers. *J. Mol. Struct. (THEOCHEM)*, 1997. submitted.
29. M. Tacker, P. F. Stadler, E. G. Bornberg-Bauer, I. L. Hofacker, and P. Schuster. Algorithm independent properties of RNA secondary structure prediction. *Eur. Biophys. J.*, 25:115–130, 1996.
30. S. Wright. The roles of mutation, inbreeding, crossbreeeding and selection in evolution. In D. F. Jones, editor, *Int. Proceedings of the Sixth International Congress on Genetics*, volume 1, pages 356–366, 1932.
31. M. Zuker and D. Sankoff. RNA secondary structures and their prediction. *Bull. Math. Biol.*, 46:591–621, 1984.

GENETIC CODE: DINUCLEOTIDE TYPE, HYDROPATHIC AND AMINOACYL-tRNA SYNTHETASE CLASS ORGANIZATION

ROMEU CARDOSO GUIMARÃES
Departamento de Biologia Geral, Instituto de Ciências Biológicas,
Univ. Fed. Minas Gerais, 31270-901 Belo Horizonte MG Brazil
TelFax +55-31-274.4988, 499.2567; Email romeucg@icb.ufmg.br

Abstract. Correlations between hydropathies of anticodonic core dinucleotides (diN; 5' central→3' third base) and the correspondent amino acid (aa) proteic residues produced higher regularities than with hydropathies of aa molecules, indicating that protein phenotypes participated in shaping the genetic code. A double diagonal symmetry was recognized: assignments of RY, YR diN (of intermediate hydropathy, centered on the self-complementary ones) yielded a linear regression of the hydropathies, contained the punctuation signs (forming inverted couples), and corresponded to aminoacyl-tRNA synthetase (aaRS) class I; those of RR, YY diN (of extreme hydropathies, centered on homodinucleotides) corresponded to aaRS II. Dispersion of anticodonic type features along tRNA sequences, in concerted evolution with the correspondent aaRS binding sites, should account for this coherence. Deviations from the diagonal arrangement were 30% of the assignments, related to both anticodonic diN types and aaRS class subgroups.

1. Introduction

Hydropathic correlations in the genetic code established the hypothesis that assignments were based on physicochemical affinities between amino acids (aa) and nucleotides [see 5]. Correlations of hydropathies of anticodonic dinucleotide (diN; 5' central→3' third base) cores with those of the correspondent aa were limited due to utilization of properties of aa molecules [5], at a time when scales for properties of proteic aa residues were being developed. We explored scales of proteic aa residue hydropathies, discovering consistency also with aminoacyl-tRNA synthetase (aaRS) classes. Mutual coherence among properties of aa residues, tRNA structures at anticodonic or other sites, and aaRS class specificities indicated that these parameters converged in the aaRS microenvironments. Evolution of the genetic code was explained in part as derived from coherent properties of aa molecules and nucleotides, in an 'RNA world' [9], improved through hypercycling [11], where protein properties [see 2] fed back on the coding process. Parts of these data were presented in [3].

J. Chela-Flores and F. Raulin (eds.),
Exobiology: Matter, Energy, and Information in the Origin and Evolution of Life in the Universe, 157–160.

2. Materials and Methods

A diN hydropathy scale was averaged from [5] and correlated with 17 aa hydropathy scales. Nine of these were based mostly on properties of aa molecules: 7 taken from [5] and 2 from [8]. Eight scales were derived from the behavior of aa residues in proteins, normalized [10]. The hydrophilicity of mononucleotides [5; A<G<C<U] ordered the distribution of diN in the genetic code matrix. Terminator codons were read as UA and CA, and the secondary initiator codon (Val GUG) as AC. The leverage (Lh) introduced by an assignment into the linear regression obtained with the others was calculated [6; $P<0.05$ at Lh>1.96]. The significance value for Studentized residuals (St>1.33) from regressions with the 23 assignments were interpolated from plots of Lh and St.

3. Results

3.1. Diagonal symmetry of the hydropathic and dinucleotide type matrix. Correlations with all 23 assignments were not impressive (for the KD 'Kyte and Doolittle' scale [10]: diN=0.417-0.182aa, r=-0.744, msr=0.01) but an internal order was revealed by the St averages in the group of scales of hydropathies of aa residues (Fig. 1). Heterogeneity was higher with aa molecule scales (not shown). A 0.53 St divided the matrix into diagonal halves. The half of RY, YR diN (of intermediate hydropathies) contained 11 assignments, with low St and improved regression (for KD: diN=0.294-0.093aa, r=-0.867, msr=0.001); that of RR, YY diN (of extreme hydropathies) contained 12 assignments, with high St (Fig. 1). Internal consistency was 91%, with one exception in each half (Arg CG and Lys, the most basic and, in KD, the most hydrophilic aa). Eight assignments in the RR, YY diagonal (centered on the homodinucleotides) showed hydropathic coherence (Fig. 1) but the 3 small and hydroapathetic aa (Gly, Pro, Ser) were outliers from the regressions, corresponding to highly hydrophilic or hydrophobic diN. An inverted complementary symmetry was the only one consistently formed by this set of diN (GA/CU, GG/CC).

3.2. Physiological relationships.

3.2.1. Punctuation signs. These were included in the RY, YR diagonal, assigned to diN containing A and following consistent inverted symmetry (AU/UA, AC/CA). The former were in the extremes of the self-complementary diN axis.

3.2.2. Coherence of the diagonals with the aminoacyl-tRNA synthetase classes. The division of the genetic code into halves of aa by the aaRS classes [7] obeyed the hydropathic and diN type diagonals (Fig. 1). Coherence with aaRS I was 4/5 through the self-complementary diN axis and with aaRS II was 5/6 (overall 82% coherence) through the homodinucleotides. Coincidences fell to 70% through all boxes, with 3 deviations from each diagonal: Ala, Thr and His left the RY, YR diagonal to be charged by aaRS II; Leu, Glu and Arg CU, of the RR, YY diagonal, corresponded to aaRS I.

3.3. Evaluation of individual scales. The quality of scales was judged through the criteria of fitting to the symmetries and regularities described above. No 'best' scale could be chosen; those presenting the least deviations from the average of the 8 aa residue scales (Fig. 1) were, among the aa molecule scales, 'Grantham' and 'Woese et al.' [5], and KD, ROS 'Rose and Dworkin' and AMP 'average membrane preference of 7 scales', among the aa residue scales [10], seconded by 'Optimal Matching Hydrophobicities' and 'Membrane Propensity for Haemoproteins'.

AA Phe 0.69 II Leu 0.88 I	GA Ser 1.79 II	CA Cys 0.28 I Trp 0.25 I X	UA Tyr 0.38 I X
R	**R**	**Y**	**R**
AG Leu 0.62 I	GG Pro 1.79 II	CG Arg 1.08 I	UG His 0.20 II Gln 0.50 I
AC Val 0.03 I 'fMet'	GC Ala 0.24 II	CC Gly 2.12 II	UC Asp 0.55 II Glu 0.62 I
R	**Y**	**Y**	**Y**
AU Ile 0.06 I Met 0.06 I fMet	GU Thr 0.10 II	CU Ser 2.19 II Arg 0.79 I	UU Asn 0.89 II Lys 0.40 II

Fig. 1. Genetic dinucleotide and hydropathic anticode. Numbers are the average Studentized residuals of the 8 regressions utilizing amino acid residue scales; 'fMet', the wobbling initiator; aminoacyl-tRNA synthetase classes I, II. Highlighted are the **RY, YR** half (n=11; average 0.29), with values <0.5 (outlier Arg CG), and the **RR, YY** half (n=12), with higher values (average 1.11), 8 forming a 0.4-0.89 group and 4 (Gly, Pro, Ser) the significantly deviating assignments.

4. Discussion

4.1. Influence of protein phenotypes in shaping the genetic code. The improved regularities provided for by hydropathy scales of aa residues, as compared to scales of aa molecules, indicated that the hydrodynamic configuration of proteins participated in shaping the genetic code. Main refinements were the recognition of the hydroapatheticity of Pro and a correction on the value of Arg hydrophilicity. Our choices for ROS, AMP and KD coincided partially with indications of [10]. A consensus of the 8 aa residue scales utilized here will be presented elsewhere. The superiority of aa residues over aa molecules data indicated that properties of aa molecules were simpler organizing forces for the assignments, possibly in an 'RNA world' [9], later corrected and adjusted through the participation of protein phenotypes [systemic or 'informational' forces], in feeding back upon the coding processes through hypercyclic dynamics [11]. The consistency among aa residue hydropathy scales, which were developed through studies of other protein types, the genetic code organization and the aaRS class specificities for diN types integrated general protein folding properties with the aaRS multifaceted structures and functions.

4.2. Physiological correlations. The halves of the matrix described by the diN type symmetry were consistent internally and with the distribution of punctuation signs, and enabled the discovery that the aaRS class assignments followed the same rules. The distinct ways of approaching tRNA by the aaRS classes [7] provided a rationale for the physiological correlations: class I enzymes reach directly the identity elements of tRNA, via the minor groove, which is consistent with their association with diN of intermediate hydropathies and development of the fine tuning required for punctuation; class II enzymes reach tRNAs through the major groove, in consistency with their association with diN of extreme hydropathies. The diN type organization was discovered by examination of anticodonic assignments but an explanation for the coherences should account for tRNA features participating in aaRS binding being distributed in all arms, sometimes not including the anticodon loops. This can be achieved by the hypotheses of tRNA origins through replicative elongation of smaller precursors [1], which provide for dispersion of repeats of ancestral anticodonic type sequences along the molecule. Concerted evolution of tRNAs and aaRSs maintained the correspondent repeated binding properties and the coherence between aa and tRNA features. Data point to participation of both tRNA and aaRS in the departures from the strict obedience to the diagonal organization: the central anticodonic R is preferentially A for aaRS I and G for aaRS II; aaRS II subgroup of Gly, Pro and Ser was involved in both hydropathic non-coherence and in having pulled Ala, Thr and His out of the RY, YR diagonal; aaRS I subgroups of Val, Ile, Met and Cys, and of Arg CG and Gln, pulled Leu, and Arg CU and Glu, respectively, out of the RR, YY diagonal.

4.3. Early coding. It is difficult to conjecture on early tRNAs or mRNAs having homogeneous base composition or regularly alternating R and Y. There are indications that early nucleic acids were rich in A runs [4], which would favor starting with aaRS of coarser specificities, more similar to class II, in consistency with observations that this class corresponds to aa of lower sizes [7] and a larger variety of side chains.

Acknowledgments: CNPq, FAPEMIG, for financial support; R J Turner and J H Weiner, for offering the amino acid residue hydropathy scales; J E C Figueira, E A Reis, R M Assunção and A I C Persiano, for statistical assistance; E N Trifonov, D Moras, J Arnez and G Bernardi, for discussions and collaboration in progress; O V Davydov and O A Zhaxybayeva, for interest and discussions. This publication is dedicated to the memory of Prof. M Debrun, recently deceased.

References

1. Bloch, D. P., McArthur, B., Widdowson, R., Spector, D., Guimarães, R. C., Smith, J.: tRNA-rRNA sequence homologies: a model for the origin of a common ancestor molecule, and prospects for its reconstruction, *Orig. Life Evol. Biosph.* **14** (1984), 571-578; Nazarea, A. D., Bloch, D. P. and Semrau, A. C.: Detection of a fundamental modular format common to transfer and ribosomal RNAs: second order spectral analysis, *Proc. Nat. Acad. Sci. USA* **82** (1985), 5337-5341; Piccirilli, J. A., McConnel, T. S., Zaug, A. J., Noller, H. F. and Cech, T. R.: Aminoacyl esterase activity of the *Tetrahymena* ribozyme, *Science* **256** (1992), 1420-1424.
2. Di-Giulio, M.: The beta-sheets of proteins, the biosynthetic relationships between amino acids, and the origin of the genetic code, *Orig. Life Evol. Biosph.* **26** (1996), 589-609.
3. Guimarães, R. C.: Anticomplementary order in the genetic coding system, *Int. Conf. Orig. Life*, Int. Soc. Stud. Orig. Life, Orléans, 1996, pp. P100; The diagonal symmetry of the genetic code: hydropathy, dinucleotide type, and aminoacyl-tRNA synthetase classes. *Brazil. J. Genet.* **20** Supl. (1997), 339; Non-coding sequences original, genes derived, *Junk DNA: the role and the evolution of non-coding sequences*, Abstr. Int. Symp. Soc. Molec. Evol., Guanacaste (Costa Rica), 1997, pp. 6.
4. Guimarães, R. C.: Linguistics of biomolecules and the protein-first hypothesis for the origins of cells, in Ponnamperuma, C. and Chela-Flores, J. (eds.), *Chemical evolution: structure and model of the first cell*, Kluwer, Dordrecht, 1995, pp. 193-199; Guimarães, R. C. and Erdmann, V. A.: Evolution of adenine clustering in 5 S ribosomal RNA, *Endocyt. Cell Res.* **9** (1992), 13-45.
5. Lacey Jr., J. C. and Mullins Jr., D. W.: Experimental studies related to the origin of the genetic code and the process of protein synthesis: a review, *Orig. Life Evol. Biosph.* **13** (1983), 3-42.
6. Montgomery, D. C. and Peck, E. A.: *Introduction to linear regression analysis*, John Wiley and Sons, New York, 1992.
7. Moras, D.: Structural and functional considerations of the aminoacylation reaction, *Trends Biochem. Sci.* **22** (1997), 211-216.
8. Roseman, M. A.: Hydrophilicity of polar amino acid side chains is markedly reduced by flanking peptide bonds, *J. Mol. Biol.* **200** (1988), 513-522.
9. Saks, M. E. and Sampson, J. R.: Evolution of tRNA recognition systems and tRNA gene sequences. *J. Mol. Evol.* **40** (1995), 509-518.
10. Turner, R. J. and Weiner, J. H.: Evaluation of transmembrane helix prediction methods using the recently defined NMR structures of the coat proteins from bacteriophages M13 and Pf1, *Biochim. Biophys. Acta* **1202** (1993), 161-168; Personal communication, 1995.
11. Wills, P. R.: Self-organization of genetic coding, *J. Theor. Biol.* **162** (1993), 267-287.

COGNITIVE UNIVERSALS: ABSTRACT PSYCHOLOGY OF TERRESTRIAL AND EXTRATERRESTRIAL INTELLIGENCE

K. TAHIR SHAH
Dipartimento di Elettrotecnica Elettronica Informatica
Università di Trieste, Trieste, Italy

1. Introduction

Shall we be able to understand and communicate with aliens from an extraterrestrial Technologically Advanced Civilization (TAC)? Minsky [1984] argues that aliens may think like humans since all intelligent problem-solvers are subject to the same ultimate constraints - limitations on space, time, and materials. We propose that systematicity, meta-theoretical reasoning, theory of mind, language and descriptive granularity are pre-requisites to any kind of intelligence capable of developing a TAC. These are so called Cognitive Universals (CUs), a term used by Narens [1995] for capabilities such as comprehension of Arithmetic and Geometry. By definition, CUs are those cognitive capabilities which are common to all intelligent life forms, irrespective of their origin. CUs are necessary but not sufficient conditions for the development of a TAC. Other conditions and particular morphological characteristics such as appropriate body size, visual apparatus to scan the environment, an independent sensory apparatus (e.g., sound) to communicate with other beings and tactile sensory capability along with an appropriate size, form and number of "hands" capable of holding objects and modification also play a crucial role in tool making. Such morphological characteristics could possibly evolve anywhere not only to optimize the energy expense by the body, but also to optimize information processing resources by the brain.

We define a TAC as a civilization, which is capable of, at least, the following: (1) *Long-distance Interaction* (i.e., sending its missions anywhere within a reasonable traveling time, say, an intelligent being's life time). (2) *Hybrid biological-machine Intelligence*. (3) *Environmental modifications* anywhere within the traveling distance using hybrid life form (human)-android missions. The terrestrial civilization is on the threshold of being an advanced civilization in the above sense.

2. Abstract Psychology of Intelligence and Scientific Development

Any approach to CUs should be context-independent because extraterrestrial planetary environments are not necessarily like ours. That is, no matter what material is used by nature to evolve the brain under whatever external conditions, CUs remains invariant.

J. Chela-Flores and F. Raulin (eds.),
Exobiology: Matter, Energy, and Information in the Origin and Evolution of Life in the Universe, 161–164.

Different paradigms and physical processes provide underlying implementation support for the same set of CUs. What kind of information processing system, then, can be considered intelligent? What is the *essential and unchanging* nature of intelligence that can be used to characterize all its manifestations, terrestrial, extraterrestrial, and artificial? The answer, almost a universal consensus among AI researchers, is that virtually all aspects of intelligence are best modeled as information processing on representations. Moreover, intelligence is really not a single function implemented in a single architecture. Chandrasekaran [1990] defines intelligence as a "*coherent repertoire of generic information processing strategies, each of which solves a type of problem in a computationally efficient way, using knowledge of certain types, organized in a specific way, and using specific and locally appropriate control strategies.*" All kinds of intelligences are expected to face similar information processing problems, and the kinds of solutions that they are likely to adopt for these problem are to be essentially similar. This approach is referred to as *abstract psychology* because it does not discuss a particular intelligent agent or a class of agents (animals, humans, androids or aliens). We propose intelligence as a repertoire of higher level cognitive functions, namely CUs, but the level of abstraction used here is different than in other approaches.

However, we may come across the same kind of definitional-comparison problems in this context as when comparing our intelligence with terrestrial animals since each life form determines its version of reality, its *Umwelt*, depending upon its nervous system's architecture and limitations. A certain degree of communication and numerical competence such as relative numerousness or even proto-counting in some species of mammals [Davis & Pe'russe 1988] as well as planning and tactical deception capability in primates [Cheney & Seyfarth 1992] are not sufficient to develop even a primitive tool-making society. Moreover, animals do not possess an unbounded language generation capability or second-order intentionality (i.e., theory of mind).

Science is taken to originate from the following principles [Dilworth 1994]: (1) *The principle of the uniformity of nature* (e.g., the principle of space-time invariance). (2) *The principle of the perpetuity of substance*, which entails all conservation laws. (3) *The principle of causality*. Principles (1) and (3) entail: like causes have like effects. Our cognitive system and the brain evolved to deal with survival exigencies under the influence of these principles governing all kinds of phenomena. Our comprehension of the world is shaped by our cognitive capabilities. For example, if there was no serial, attention-based consciousness system, we would not have developed the notion of algorithm. Development of similar notions are also expected by any alien life form.

Simon's model [1992] of scientific discovery by any information processing system is founded on the physical symbol hypothesis which states that physical symbol systems, and only such systems, are capable of thinking. The possibility of machines making scientific discoveries confirms the teleological view that there are principles of rationality and a scientific method which is independent of social factors. On the other hand, the evolution of science is considered a social phenomena by Kuhn [1970], which is not predetermined by the nature of physical phenomena, but is the outcome of a process involving the goals and attitudes of the societies (e.g., priorities, political views, usefulness for military or other purposes and so on). In this paradigm, alien science can be similar to ours only if their social motives are similar to ours. However, evolutionary argument suggests that extraterrestrials are expected to have similar social attitudes as ours. This is because our complex social behavior is, nevertheless, a result of our struggle for individual and collective survival.

2.1 COGNITIVE UNIVERSALS

Adhering to the view that intelligence requires a symbolic-representational system, we propose the following CUs, all emanating from the principle of optimality.

1. *Systematicity*: Pylyshyn [1984] and Fodor & Pylyshyn [1988] argue that knowledge is systematic. It is a characteristic of thinking that if each concept in a set of potential constituent concepts is understood and relation structure can be instantiated by one assignment of the constituent concepts, then the thinker can also instantiate the relation structure with other permissible assignments of the concepts. Systematic reasoning with composable constituents requires symbols because as the diversity of knowledge a life form must represent and manipulate increases, it becomes increasingly difficult to find and to construct *de novo* specialized representations to provide appropriate encodings. Nature solved this optimization problem by evolving a system that, in addition to instantiation, enables more complex representations to be composed from simple ones. Therefore, information acquired in one task context has the potential to be made available in a different task context, which is exactly what is required for systematic reasoning with composable constituents. Systematicity is a key feature of rules of inference, judgments of perceptual similarity, and many other processes.

2. *Meta-Theoretical Reasoning*: This requires the ability to inspect and reason from higher levels (e.g., meta-level, meta-meta level) about objects, concepts etc., described at a lower level of abstraction. It is an essential part of our intelligence. Meta-theoretical reasoning entails many fundamental concepts such as algorithm, computation, semantics and truth, proof and so on. It is virtually impossible to develop mathematics or any scientific discipline without meta-theoretical reasoning capability. Examples of the meta-theoretical disciplines are: *philosophy of language* (reasoning and discussion about language), e*pistemology* (about the nature of knowledge), and *proof and validity theory* (meta-mathematics as reasoning and discussion about logical systems).

3. *Theory of Mind*: The development of a complex social behavior in a technological society requires the theory of mind (or the second-order intentionality) in order to have an effective communication among its members. It relates communication, cognition and social behavior. To be aware of other minds is a classical problem of the philosophy of mind. To say that mental states are *intentional* is a philosophical notion meaning that they are "about" something such as beliefs and desires (e.g., one believes that **p**, where **p** is a proposition). To have intentonality is referred to as the *first-order intentionality.* When an intentional being is aware of first-order intentionality of another being or its own, it is referred as to *second-order intentionality* or theory of mind. The accepted philosophical view is that human beings possess second-order intentionality. We tend to engage continuously about the mental states of others - a necessary process for effective interaction and communication with others.

4. Language: Ability to communicate thoughts requires generation and comprehension of an unlimited number of sentences in a language. The structure of terrestrial languages is very complex since it includes not only *syntax* (generation of unbounded number of legitimate expressions), *semantics* (meaning and truth) and pragmatics, but also such notions (introduced by Wittgenstein) as *private language*, distinction between the *actual and possible worlds* (e.g., counterfactuals) and *metaphors and analogies* as an integral part of thinking process. In the widely accepted view the basic structure of a language is *subject-predicate* correlated with *particular-universal*, the basic structure of reality. There are other aspects such as indirect speech acts, a kind of conversational implication that results from the performance of some other speech acts. Such speech acts save computational resources. The unbounded generative capacity exists in other systems too such as Numbers and distinct shapes and arrangements of geometrical objects. We believe that intelligent aliens need a language as complex as terrestrial languages. Anything less complex might be inadequate, too slow and computationally expensive to carry out communication of complex concepts necessary for the development of a TAC.

5. *Descriptive Granularity*: We define the ability to describe, extract and process information at multiple levels of abstraction as descriptive granularity. The ability to conceptualize external world at different granularities is a fundamental component of our intelligence. Any intelligent life form or machine must take into account granularity in its descriptive and reasoning processes. For instance, human perception is able to focus attention at various levels of detail and shift focus from one level to another. Focusing at a certain grain-size helps us ignore many confusing details and enables us to map the complexities of our *umwelt* into simpler model-of-the-world that are computationally tractable. In general, our knowledge consists of a global theory together with a large number of relatively simple, idealized grain-dependent, local theories, interrelated by articulation axioms [Hobbs 1985, McCalla et al., 1992].

4. References

Chandrasekaran, B., 1990: What kind of information processing is intelligence, in *The Foundations of Artificial Intelligence*, eds.: D. Partridge & Y. Wilkes, Cambridge University Press, pp. 14-46.

Cheney, D. L. & Seyfarth, R. M., 1992: Précis of how monkeys see the world, *Behav. & Brain Sci.*, **15**, 135

Davis, H. & Pe'russe, R., 1988: Numerical competence in animals, *Behav. & Brain Sci.*, **11**, 561-615.

Dilworth, C., 1994: Principles, Laws, Theories and the Metaphysics of Science, *Synthese*, **101**, 223-247.

Fodor, J.A. & Pylyshyn, Z.W., 1988: Connectionism and Cognitive Architecture, *Cognition,* **28**, 3-71.

Hobbs, J. R., Granularity, 1985: *Proc. of the IJCAI-85*, Los Angeles, CA. pp. 432-435.

Kuhn, Thomas S., 1970: *The Structure of Scientific Revolutions* (Second Edition), The University of Chicago Press, Chicago, USA.

McCalla, G., et al., 1992: Granularity Hierarchies, *Computers Math. Applic.*, **23**, 363-375.

Minsky, M., 1985: Why intelligent aliens will be intelligible in *Extraterrestrials, Science and Alien Intelligence*, ed.: E. Regis Jr., Cambridge University Press, Cambridge, UK, pp. 117-128

Narens, L., 1996: Surmising Cognitive Universals for Extraterrestrial Intelligences, in *Astrnomical and Biochemical Origins and the Search for Life in the Universe*, eds., C.B. Cosmovici, S. Bowyer and D. Werthimer, Editori Compositori, Bologna, 1997, pp. 561- 570.

Pylyshyn, Z. W., 1984: *Computation and Cognition: Towards a Foundation for Cognitive Science*, The MIT Press, Cambridge (Mass), USA.

Simon, H. A., 1992: Scientific discovery as problem solving, *Inter. Studies in the Phil. of Sci.*, **6**, 3-14.

Section 5.
Early evolution

EARLY METABOLIC EVOLUTION: INSIGHTS FROM COMPARATIVE CELLULAR GENOMICS

S. ISLAS, A. BECERRA, J. I. LEGUINA, and A. LAZCANO
Facultad de Ciencias-UNAM
Apdo. Postal 70-407, Cd. Universitaria
04510 México D. F., MÉXICO

1. Introduction

The use of small subunit rRNAs as molecular markers has led to universal phylogenies, in which all known organisms can be grouped in one of three major cell lineages, the eubacteria, the archaeabacteria, and the eukaryotic nucleocytoplasm, now referred to as the domains Bacteria, Archaea, and Eucarya, respectively (Woese *et al.*, 1990). A description of the last common ancestor (LCA, i.e., the cenancestor), of these three primary kingdoms may be inferred from the distribution of homologous characters among its descendants. In conjunction with the fragmentary information available from other organisms, the complete genome sequences now available in the public databases allow such characterizations, and in some cases can even provide insights into the nature of the cenancestor predecessors. Here we discuss the basic assumptions and strategies used in such approaches, and apply them to the understanding of the evolutionary assemblage of arginine biosynthesis. Additional aspects of the evolution of metabolic routes have been discussed in Peretó *et al.* (1997).

2. Some problems in comparative genomic analysis

The distribution of many biosynthetic enzymes found in all three primary lines of descent before complete genome sequences became available had already led to the idea that the cenancestor was comparable to modern prokaryotes in its biological complexity, ecological adaptability, and evolutionary potential (Lazcano, 1995). However, the differences in the metabolic repertoire and gene expression mechanisms among the three primary domains (cf. Olsen and Woese, 1997) demonstrate that the characterization of the LCA is an unfinished task, and that strong statements and broad generalizations should be avoided.

J. Chela-Flores and F. Raulin (eds.),
Exobiology: Matter, Energy, and Information in the Origin and Evolution of Life in the Universe, 167–174.

In principle, backtrack reconstructions of ancestral states can be achieved with a simple, straightforward methodology. Given the availability of complete genome sequences from the three primary domains, the cenancestor is defined by properties shared by all living organisms, minus those that are the outcome of convergent evolution and those acquired by horizontal transfer (Figure 1). However, cross-genomic analysis can be difficulted by unidentified proteins encoded by rapidly evolving sequences, as well as from the properties of a given genomic dataset. Inferences on the nature of the LCA can also be biased by the reduced DNA content of parasites and pathogens such as the mycoplasma, which have been selected as model organisms because of their small, compact genomes (Becerra *et al.*, 1997). Although the application of shotgun sequencing has led to an impressive growth of the databases in a very short time, larger volumes of complete genome sequences reflecting a broader cross-section of biological diversity are still required.

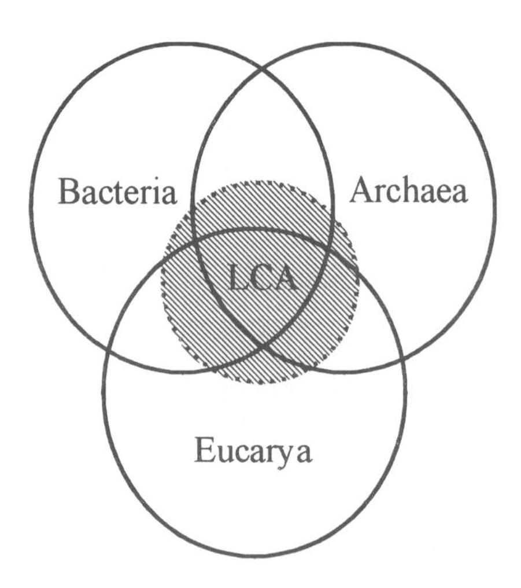

Figure 1. Intersection of the complete sequence spaces of the three domains defines the gene complement of the common ancestor (LCA). Identification of rapidly-evolving sequences would lead to a bigger set of ancestral genes (hatched areas).

The functions of many open reading frames (ORFs) derived from complete genome sequencing projects have been tentatively identified by computer searches based on structural similarities to known sequences in databases, but many more remain unidentified (30 to 50%, depending on the organism). Such databases are collections of the sequences that make up biological systems, but understanding how each component works is not enough for a proper description of how the entire system proceeds (Kanehisa, 1997). For instance, in the *Bacillus subtilis* tryptophan operon no sequence encodes the glutamine amido transferase required for anthranilate biosynthesis. This would pose a problem in comparative genomic-based metabolic reconstructions, had biochemical experimentation not demonstrated that in *B. subtilis* the required gene is shared with the folate biosynthetic route, in whose operon it is located (Crawford, 1989).

As summarized in Table I, understanding of the evolutionary development of metabolism can be obscured by a complex series of changes involving enzymatic additions, secondary losses, pathway replacements, and functional redundancies. Additional complications can result from (a) intraespecific enzyme substitutions involving paralogous proteins; (b) that possibility that extant enzymes may have participated in alternative routes which no longer exist or remain to be discovered (Zubay, 1993; Becerra and Lazcano, 1997); (c) homologous enzymes endowed with widely different catalytic properties (see below); and (d) intracellular horizontal transfer within nucleated cells (Embley *et al.*, 1997).

TABLE I. Some processes in metabolic evolution.

process	examples	reference
addition of enzymatic step(s)	oxygen-dependent cholesterol biosynthesis	Bloch (1994), Ourisson and Nakatani (1994)
	archaeal biosynthesis of 2,3-di-*O*-phytanil *sn*-glycerol	Stetter (1996)
loss of routes and enzymes	purine biosynthesis in parasites	Becerra *et al.* (1997)
pathway replacement	aerobic instead of anaerobic biosynthesis of monounsaturated fatty acids	Bloch (1994)
	fungal lysine biosynthesis	Vogel (1960)
functional redundacies	phosphatidylcholine biosynthesis	Bloch (1994)
	imidazole biosynthesis in purine and histidine biosyntheses	Peretó *et al.* (1997)

3. Did metabolism evolve backwards?

The first attempt to explain the emergence of metabolic pathways was developed by Horowitz (1945), who suggested that biosynthetic enzymes had been acquired via gene duplications that took place in reverse order as found in extant pathways. This idea, also known as the retrograde hypothesis, established an evolutionary connection between the primitive soup and the development of metabolic pathways, and is frequently invoked in descriptions of early biological evolution (cf. Peretó *et al.*, 1997). Prompted by the discovery of operons, Horowitz (1965) restated his model, arguing that it was supported not only by the overlap between the chemical structures of products and substrates of the enzymes catalyzing successive reactions, but also by the clustering of functionally related genes.

Although some operon-like gene clusters are found in both bacterial and archaeal genomes, whole genome comparisons between distant prokaryotes have shown that gene order can be easily eroded by extensive shuffling events (Mushegian and Koonin, 1996). This implies that the distribution in prokaryotic chromosomes of homologous genes encoding pathway enzymes cannot be used to (dis)prove the Horowitz hypothesis. However, if the enzymes catalyzing successive steps in a given metabolic pathway resulted from a series of gene duplication events (Horowitz, 1965), then they must share structural similarities. The known examples confirmed by sequence comparisons that satisfy this condition are limited to few pairs of enzymes and have been discussed elsewhere (cf. Peretó *et al.*, 1997).

4. The patchwork assemblage of biosynthetic routes

An alternative interpretation of role of gene duplication in the evolution of metabolism has been developed in the so-called patchwork hypothesis (cf. Jensen, 1976). According to this scheme, biosynthetic routes were assembled by primitive catalysts that could react with a wide range of chemically related substrates. The recruitment of enzymes from different metabolic pathways to serve novel catabolic routes under strong selective pressures is well document under laboratory conditions. Repeated occurrences of homologous enzymes in different pathways provide independent evidence of patchwork tinkering. Data derived from the ongoing genome projects has already demonstrated that a large portion of each organisms genes are related to each other as well as to genes in distantly related species. As discussed in the following section, the central role that gene duplication and recruitment have played in the assemblage of histidine anabolism (Alifano *et al.*, 1996) and purine nucleotide salvage pathways (Becerra and Lazcano, 1997) can also be extended to include arginine biosynthesis.

5. Gene duplication and arginine anabolism

The phylogenetic distribution of arginine biosynthetic genes suggest that this route was already present in the LCA. Hence, its absence in both *Helicobacter pylori* and the mycoplasma probably reflects polyphyletic secondary losses. Although the same chemical steps involved in arginine biosynthesis have been found in all organisms studied, two different strategies for the deacetylation of the intermediate *N*-acetylornithine have been described. In enterobacteria, the genus *Bacillus*, and the archaeon *Sulfolobus solfataricus* this reaction is catalyzed by *N*-acetylornithinase, the

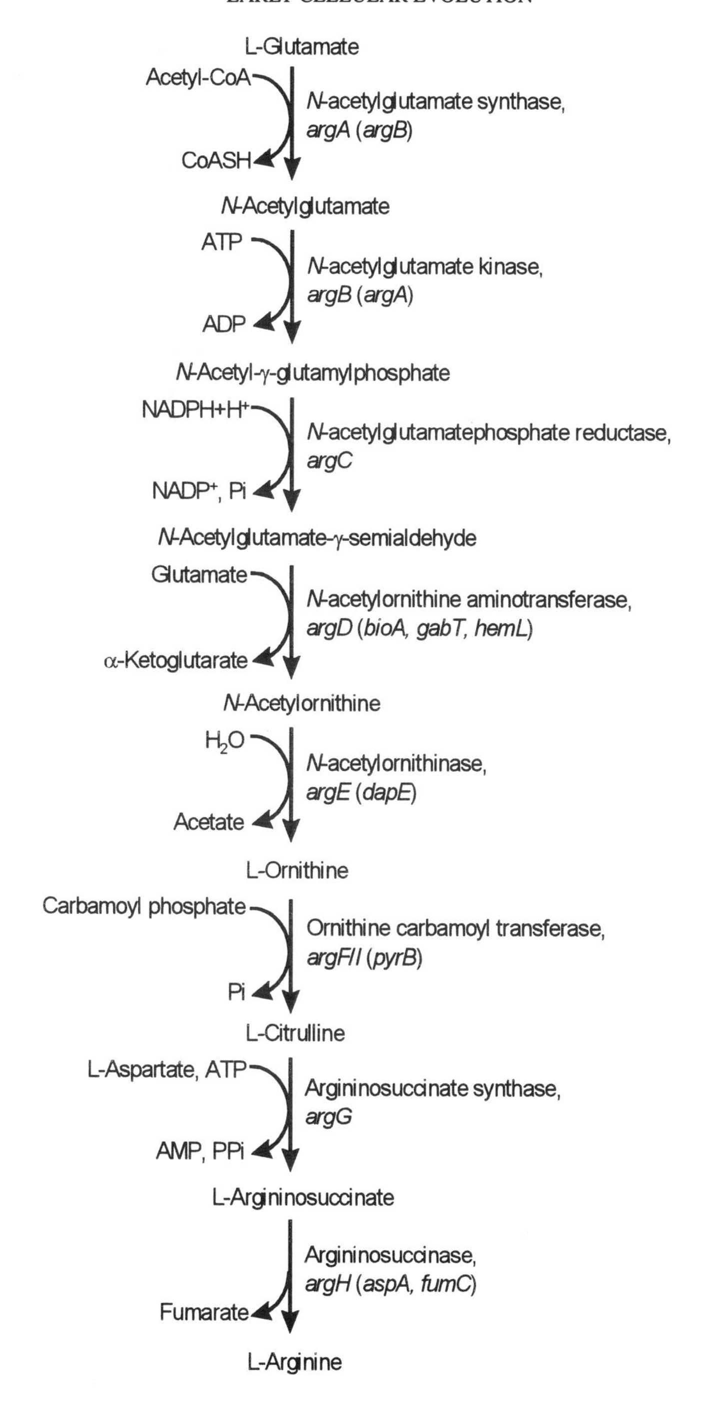

Figure 2. Arginine biosynthesis. The arginine biosynthetic genes paralogs are indicated within parenthesis.

gene product of *argE* (Figure 2), while in other prokaryotes and in fungi the acetyl group is removed by ornithine-glutamate acetyltransferase. There is no evidence of phylogenetic relationship between these two different enzymes. Another variation in this pathway occurs in the *E. coli* K12 strain, where two homologous genes (*argI, argF*) encode a family of four trimeric isoenzymes, that bind to L-ornithine and carbamoyl-phosphate to produce L-citrulline (Glansdorff, 1996).

Arginine biosynthesis consists of eight steps, five of which are mediated by enzymes that belong to different paralogous families (Figure 2). The list includes the pairs *argA/argB, argE/dapE,* and *argI/argF*, and the three- and four member families *argH/aspA/fumC* and *argD/bioA/gabT/hemL* (Riley and Labedan, 1996). Although the first two consecutive reactions in the pathway are catalyzed by the gene products of homologous sequences (*argA* and *argB*), we do not consider this as conclusive proof of the retrograde mechanism. Both reactions are chemically equivalent, and during the early evolution of this route they may have been catalyzed by an ancestral less-specific enzyme. Arginine biosynthesis thus provides additional evidence of the role of enzyme recruitment in metabolic evolution.

6. Homologous enzymes can have different catalytic properties

With the exception of proteins in which the evolutionary accretion of a functional motif or module has led new catalytic or binding properties, all enzymes encoded by paralogous genes can be expected to be endowed with comparable biochemical properties. However, reports on the existence of homologous enzymes that catalyze separate and mechanistically different reactions (Neidhart *et al.*, 1990) prompted us to search for additional examples in the available databases.

This analysis was performed using the database assembled by Riley and Labedan (1996), who compared the *E. coli* 1,862 protein sequences available as of April 1996 in the SwissProt databank. They concluded that 52.17% of all studied protein sequences had resulted from gene duplications, and classified them in paralogous families defined by sequence similarity. Their list includes 112 small families with only two sequences, 38 with three, 41 with three to seven, and 13 large families. As noted by Riley and Labedan, most of the members of paralogous families share comparable biochemical properties, with a scarce 1.23% of homologous protein pairs displaying what appear to be different functions.

We have repeated this analysis by looking exhaustively at all the characterized paralogous genes, and excluding from our sample 88 ORFs reported as hypothetical proteins. The resulting set was cross-checked with experimental data and the corresponding Enzyme Comission (EC) number. We have found a higher number of homologous genes with different EC numbers, which will be described elsewhere. An example is shown in Figure 3. It includes argininosuccinate lyase, which catalyzes the last step in arginine biosynthesis (Figure 2), and its homologs aspartate ammonia-lyase (that takes part in the synthesis and interconversion of aspartate and asparagine), and fumarate hydratase (which participates in the tricarboxylic acid cycle). As denoted by their corresponding EC number, these enzymes catalyze different reversible reactions (non-hydrolytic cleavage, (de)amination, and a hydration reaction, respectively). However, all three of them use fumarate as substrate, which suggest that the structural basis for their sequence similarity may be a large homologous binding site for this compound.

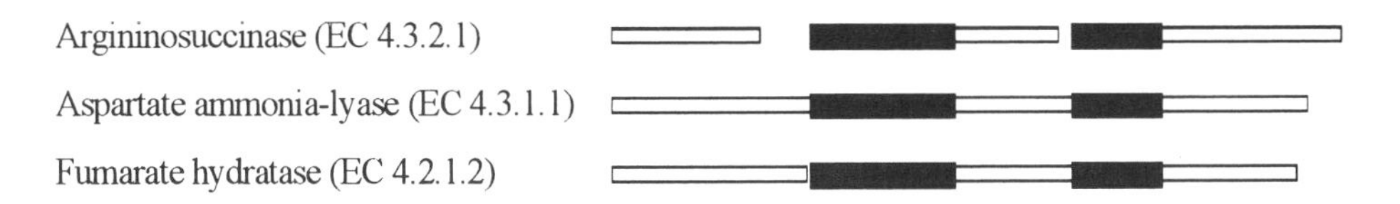

Figure 3. A three-member family of *E. coli* paralogous enzymes which different catalytic properties. The sequences were aligned using the Macaw program. The regions with statistically significant sequence similarity are shown in black.

7. Conclusions

The discovery that homologous enzymes that catalyze similar biochemical reactions are found in many different anabolic pathways supports the idea that enzyme recruitment took place at a massive scale during the early development of anabolic pathways. This conclusion is supported by analysis of the available genomic databases, which suggest that approximately 50% of cellular DNA is the outcome of paralogous duplications that may have preceded the divergence of the three primary domains. Such high levels of redundancy suggest that the wealth of phylogenetic information older than the cenancestor itself may be larger than realized, and that this information may provide fresh insights into a crucial but largely unexplored stage of early biological evolution.

Acknowledgements

The work of J. I. L. has been supported in part by the Consejo Superior de Investigaciones Cientificas (CSIC, Madrid, Spain). This paper was completed during a leave of absence of one of us (A. L.) as Visiting Professor at the Institut Pasteur (Paris), during which he enjoyed the hospitality of Professor Henri Buc and his associates at the Unité de Physicochimie des Macromolécules Biologiques. Support from the Manlio Cantarini Foundation (A. L.) is gratefully acknowledged. A.L. is an Affiliate of the NSCORT (NASA Specialized Center for Research and Training) in Exobiology at the University of California, San Diego.

References

Alifano, P., Fani, R., Lio, P., Lazcano, A., Bazzicalupo, M., Carlomagno, M. S., and Bruni, C. B. (1996) Histidine biosynthetic pathway and genes:structure, regulation, and evolution, *Microbiol. Rev.* **60**, 44-69.

Becerra, A. and Lazcano, A. (1997) The role of gene duplication in the evolution of purine nucleotide salvage pathways, *Origins Life Evol. Biosph.* (in press)

Becerra, A., Islas, S., Leguina, J. I., Silva, E., and Lazcano, A. (1997) Polyphyletic gene losses can bias backtrack characterizations of the cenancestor, *J. Mol. Evol.* **45**, 115 118.

Bloch, K. (1994) *Blondes in Venetian Paintings, the Nine Banded Armadillo, and other Essays in Biochemistry,* Yale University Press, New Haven.

Crawford, I. P. (1989) Evolution of a biosynthetic pathway: the tryptophan paradigm, *Annu Rev. Microbiol.* **43**, 567-600.

Embley, T. M., Horner, D. A., and Hirt, R. P. (1997) Anaerobic eukaryote evolution: hydrogenosomes as biochemically modified mitochondria? *TREE* **12**, 437-441.

Glansdorff, N. (1996) Biosynthesis of arginine and polyamines, in F.C. Neidhardt (ed.), *Escherichia coli and Salmonella typhimurium: Cellular and Molecular Biology*, AMS Press, Washington, D C, pp. 408-433.

Horowitz, N. H. (1945) On the evolution of biochemical synthesis, *Proc. Natl. Acad. Sci. USA* **31**, 153-157.

Horowitz, N. H. (1965) The evolution of biochemical synthesis -retrospect and prospect, in V. Bryson and H. J. Vogel (eds.), *Evolving Genes and Proteins*, Academic Press, New York, pp. 15-23.

Jensen, R. A. (1976) Enzyme recruitment in the evolution of new function, *Annu. Rev. Microbiol.* **30**, 409-425.

Kanehisa, M. (1997) A database for post-genome analysis, *Trends Genet.* **13**, 375-376.

Lazcano, A. (1995) Cellular evolution during the early Archean: what happened between the progenote and the cenancestor? *Microbiologia SEM* **11**, 185-198.

Mushegian, A. R. and Koonin, E. V. (1996) Gene order is not conserved in bacterial evolution, *TIGS* **12**, 289-290.

Neidhart, D. J., Kenyon, G. L., Gertl, J. A., and Petsko, G. A. (1990) Mandelate racemase and muconate lactonizing enzyme are mechanistically different and structurally homologous, *Nature* **347**, 692-694.

Olsen, G. J. and Woese, C. R. (1997) Archaeal genomics: an overview, *Cell* **89**, 991-994.

Ourisson, G. and Nakatani, Y. (1994) The terpenoid theory of the origin of cellular life: the evolution of terpenoids to cholesterol, *Chemistry & Biology* **1**, 11-23.

Peretó, J., Fani, R., Leguina, J. I., and Lazcano, A. (1997) Enzyme evolution and the development of metabolic pathways, in A. Cornish-Bowden (ed.), *Cell-Free Fermentation and the Growth of Biochemistry: Essays in Honour of Eduard Buchner*, Publicacions de la Universitat de Valencia, Valencia, Spain, (in press)

Riley, M. and Labedan, B. (1996) *Escherichia coli* gene products: physiological functions and common ancestries, in F.C. Neidhardt (ed.), *Escherichia coli and Salmonella typhimurium: Cellular and Molecular Biology*, AMS Press, Washington, D.C., pp. 2118-2202.

Stetter, K. O. (1996) Hyperthermophiles in the history of life, in G. R. Bock and J. A. Goode (eds.) *Evolution of Hydrothermal Ecosystems on Earth (and Mars?)*, Wiley, Chichester, pp. 1-10.

Vogel, H. J. (1960) Two modes of biosynthesis among lower fungi: evolutionary significance, *Biochem. Biophys. Acta* **41**, 172-173.

Woese, C. R., Kandler, O., and Wheelis, M. L. (1990) Towards a natural system of organisms: proposal for the domains Archaea, Bacteria, and Eucarya, *Proc. Natl. Acad. Sci. USA* **87**, 4576-4579.

Zubay, G. (1993) To what extent do biochemical pathways mimic prebiotic pathways?, *Chemtracts - Biochem. Mol. Biol.* **4**, 317-323.

IMPACTS OF DARWIN'S THEORY ON « ORIGINS OF LIFE » AND « EXTRATERRESTRIAL LIFE » DEBATES, and some wider topics

F. RAULIN-CERCEAU[1], M. C. MAUREL[2] and J. SCHNEIDER[3]

[1]*Grande Galerie de l'Evolution, Muséum national d'Histoire naturelle, 36 rue Geoffroy Saint Hilaire - 75005 - Paris - France- e-mail: raulin@mnhn.fr*

[2]*Institut Jacques Monod, Tour 43, 2 place Jussieu - 75251 Paris Cedex 05 - France- e-mail:Marie-Christine.Maurel@ijm.jussieu.fr*

[3]*Observatoire de Paris-Meudon - Place Jules Janssen, 92195 Meudon Cedex - France- e-mail: Jean.Schneider@obspm.fr*

1. Darwin and the Origin of Life :

Some sentences written by the British naturalist Darwin (1809-1882) in his publications or letters show his interest in this topic. Having expounded his evolutionary theory in the *Origin of Species*, Darwin concluded that « *There is grandeur in this view of life, with its several powers, having been originally breathed by Creator into a few forms or into one; and that, whilst this planet has gone cycling on according to the fixed law of gravity, from so simple a beginning endless forms most beautiful and most wonderful have been, and are being evolved* »(Darwin, 1859, Dick, 1996). A dozen years later, in a letter written in 1871 to the British botanist Joseph Dalton Hooker (1817-1911), he added a passing remark about a « warm little pond » in which life might have originated : « *It is often said that all the conditions for the first production of a living organism are now present, which could ever have been present. But if (and oh ! what a big if !) we could conceive in some warm little pond, with all sorts of ammonia and phosphoric salts, light, heat, electricity, etc., present, that a proteine compound was chemically formed ready undergo still more complex changes, at the present day such matter would be instantly devoured or absorbed, which would not have been the case before living creatures were formed* » (Shklovskii & Sagan, 1966).

Hence, the question of Origins (of life, of matter) appeared to Darwin as a very important problem to be solved, now that the question of the origin of species had become clearer. Darwin wrote in 1863, in a previous letter to Hooker : « *It will be some time before we see « slime, protoplasm, etc, generating a new animal. But I have long regretted that I truckled to public opinion, and used the Pentateuchal term of creation, by which I really meant « appeared » by some wholly unknown process. It is mere rubbish, thinking at present of the origin of life ; one might as well think of the origin of matter* » (Darwin, March 29, 1863). Obviously, at the time of Darwin, it was still rather early to describe as accurately as possible these steps of evolution, even if the first organic synthesis was already realized since the first half of the nineteenth century. According to E.N.Pavlovskii, Darwin did not touch on this aspect of evolution, but it was not his fault : the state of natural-historical science in his time did not allow him to formulate this question in such a way that it could be treated in real terms (Pavlovskii, 1959). Furthermore, Darwin was a naturalist and his work was based on observations of nature : his scientific methods were not of the same kind as those useful to understand the mechanisms of the Origin of Life.

J. Chela-Flores and F. Raulin (eds.),
Exobiology: Matter, Energy, and Information in the Origin and Evolution of Life in the Universe, 175–180.

At the end of the nineteenth century, many representatives of idealistic metaphysics in Russia argued that Darwinism suffered from the common ailment of « natural science materialism » : it asked questions that science was not equipped to answer (Vucinich, 1988). However, from Darwinian concepts, it was generally accepted that transformations from inorganic to organic matter, and then to proteins and primitive living organisms, were attributed to the properties of carbon compounds. This type of theory was based on the materialistic view of life and the unified view of nature, directly inspired by Darwin's work : there were no impassable barriers in nature (Kamminga, 1988). Every natural phenomenon was considered as part of a causal link : nature is one.

2. The idea of Evolution among the pioneers (especially Oparin) of the Origin of Life :

The Russian biochemist Oparin (1894-1980) suggested that the Origin of Life couldn't be solved in isolation from a study of the whole course of the development of matter which preceded this origin: life, not separated from the inorganic world by an impassable gulf, arose as a new quality during the process of development of that world (Oparin, 1959). The concept of a slow progressive development of matter was clear since his first publications on Origins of Life (i.e. since 1924, see Oparin, 1924). He stressed the point that an evolutionary approach to the study of this problem would be able, therefore, to open up a wider vista of possibilities for its solution. Moreover, according to Kamminga, he thought that the chemistry of organisms was not unique, organic compounds obeying the same physicochemical laws as inorganic substances. Oparin even noted a phenomenon reminiscent of a struggle for existence in the realm of crystals: one form of crystals is sometimes eliminated while the other one is « victorious at the first clash » (Kamminga, 1980).

Darwin is mentioned in the first Oparin's book in English *The Origin of Life* (1938) : *Ch.Darwin in one of his letters to Wallace criticized Bastian's experiments* (who considered it possible that living things originated from inorganic substances without intervention of any specific vital force), *and considered them altogether improbable. While he admits the correctness of the idea of archeobiosis, he pointed out that spontaneous generation had not been proved. According to Darwin, life must have originated sometime, somehow, but how this happened was still unknown* (Oparin, 1938). Oparin attended lectures privately given by K.A. Timiryazev, a Russian scientist who had gone to England to collect material on Darwin and Darwinism (Lazcano, 1992). But it seems that his first precise ideas related to chemical evolution are rather based in detail on the works of other authors. He cited Haeckel, Osborn, and particularly Omeljanski (for the possibility of a spontaneous generation of life on Earth at some remote epoch) and Vernadski (for the geochemical environment of the primitive Earth) (Oparin, 1938). It is quite sure that Oparin has been inspired by the general mechanism of Darwin's struggle for life and natural selection, in order to explain the last chemical steps leading to life. For instance, influenced by Darwin's natural selection, he couldn't imagine the emergence of a already fully autotrophic cell. He thought that, on evolutionary grounds, the first organisms must have been heterotrophic (Lazcano, 1992). Anyway, it seems to be an established fact that Oparin has been above all influenced by Engels's *Dialectics of Nature*, even indirectly, in his general view of evolution of matter. According to Ruse, dialectics leads to consider the formation of living bodies and the formulation of biological laws as proceeding in indissoluble unity: initial steps towards life were not only submitted to physical and chemical laws, but also to incipient biological laws including prebiological natural selection (Ruse, 1997).

3. Chemical Evolution :

The biochemist Benjamin Moore used the term « chemical evolution » for the first time, in a very explicit way, in his book *The Nature and Origin of Life* (1913) (Kamminga, 1988). Therefore, half a century after the birth of the theory of biological evolution, a chemical extension was proposed. But this raised a few questions : was this type of evolution similar to the one proposed by Darwin, *i.e.* based on natural selection ? Or is the problem of the origin of life, seen as a process of chemical evolution, quite distinct from that of its subsequent evolution ? According to Dobzhansky in 1972, the answer to this last question is yes : biological evolution has properties which evolutionary changes in inorganic nature do not have, such as adaptability of species, and life is an utterly improbable state of matter (Dobzhansky, 1972).

Nowadays, the question « does chemical evolution follow the Darwinian principles of evolution ? » is not completely solved, but some specialists of chemical evolution attempt to give an answer. Ancestors to terrestrial life, and related forms of life elsewhere in the universe, could be seen as part of the cosmic development. It is also possible that the physicochemical reactions, and similar accounts of steps from chemical to Darwinian evolution overlook some still unknown factors (e.g. the quantum theory provided for the full understanding of chemical reactions), but the principle of the continuous development from non-living to living matter can be considered well supported (Eirich, 1996). It is interesting to notice that, even today, in some scientific fields connected to Origins of Life but different from biology, Darwinian principles are quoted as general mechanisms underlying a sort of « natural selection ». For instance, in the field of cosmic chemistry, the survivance of molecules in some specific locations in space might be regarded as a consequence of the Darwinian principles. According to Tielens, any molecule that can be made in the laboratory likely has a « Darwinian » niche in space where conditions are just right for its survival (Tielens & Charnley, 1997).

4. Chemical Evolution versus Extraterrestrial Evolution, in Oparin's written works :

Oparin wrote that « *there is every reason now to see in the origin of life not a « happy accident » but a completely regular phenomenon, an inherent component of the total evolutionary development of our planet. The search for life beyond Earth is thus only a part of the more general question which confronts science, of the origin of life in the universe* » (Oparin, 1975). Oparin did not believe in Panspermia theory, but he was convinced that this did not mean that life existed only on the Earth : *We still have too little information to deny completely the possibility of existence of organisms on some other planets, whirling around stars similar to our Sun. But there can be no doubt that these worlds inhabited by living organisms are much farther removed from our solar system than are the nearest stars* (Oparin, 1938). He cited Jeans's work (1927) who suggested from calculations that the probability of any star becoming a Sun surrounded by planets was about one per hundred thousand times, and who thought that it would be extremely difficult to imagine life of a higher order capable of developing on celestial bodies radically different from our own planet warmed by the Sun (Oparin, 1938). It seems clear that extraterrestrial life was not Oparin's research program, since the contemporary science did not yet permit a definite answer to the question of life on other planets (Dick, 1996). But one of his very large public book intitled « Life in Universe » shows that this topic gave him something to think about (Oparin, 1958).

5. Example of Planet Mars :

The demonstration by Darwin in 1859 and 1871 that the living world, including man, is a product of evolutionary development placed the problem of extraterrestrial life in a new perspective (Dobzhansky, 1972).Darwin thought that if man had surpassed in mind the other denizens of Earth, this did not mean that he might not someway, or somewhere, be surpassed himself. The conclusion that had remained implicit in *The Origin of Species*, published in 1859, was made explicit by the time Darwin published *The Descent of Man* in 1871 (Sheehan, 1988). One of the Darwin's conclusion is that man is descended from some less complex organised form and bears in his bodily frame the indelible stamp of his more simple origin (Darwin, 1871).

Darwin made humankind's position in the biological scheme relative less than two decades before Giovanni Virginio Schiapparelli's discovery of the Martian « canals ». The term « canals » came from the famous « canali » (Schiapparelli, 1894) that Schiapparelli discovered in 1877, but « canali » in the Italian language is closer to « channels » than to « canals », term which corresponds to an artificial construction : this confusion is at the origin of the myth of an « intelligent » martian life during the end of the nineteenth century. The American astronomer Percival Lowell (1855-1916), himself a Social Darwinist, was responsible for the idea that the markings discovered by Schiapparelli were modern « canals » that had been dug by the Martians to transfer water from polar caps, but that idea must be regarded just as a suggestion, not as a theory (Lowell, 1895). « Canals » seemed to appear and disappear on the surface of Mars, and this was freely interpreted as evidence of a technological advanced civilization using irrigation for its agriculture (Dobzhansky, 1972).

It seems that in that late Victorian age, which had witnessed the great achievements of Darwin, Wallace, Huxley, and Spencer, the « *Philosophy of cosmos, with illustrations from celestial mechanics* » written by Lowell in 1891, could hardly have been formulated on other grounds than evolutionary principles (Sheehan, 1988). In this book, the question of life on Mars fitted for Lowell within a wider framework, in which life was seen as an « inevitable detail » in the scheme of « cosmic evolution ». Lowell tied his Martian proposals to a broad evolutionary scheme of planetary development, that is for Mars as well as for Earth, a slow death by desiccation : a planetwide desert was the inevitable result of the evolution of worlds (Sheehan, 1988). Lowell wrote « *each body, under the same laws, conditioned only by size and position, inevitably evolves upon itself organic forms* » (Lowell, 1935).

Still in the 1930s, we found some quotations mentioning evolved life on Mars. Oparin stressed in his book of 1938 that H.von Klüber pointed out that some investigators believed that there was a vegetation on the surface of Mars (von Klüber, 1931), this belief being based on a number of observations (Oparin, 1938).

6. Changes of perspective :

It is finally interesting for our purpose to point out that, while the main stream of biology led to an increase of a purely « materialistic » point of view, a quite opposite point of view has at the same time emerged from fundamental physics and semiotics. As soon as the early days of quantum theory, one of its founders, Niels Bohr, adressed the question of the universal validity of the method of analyzing a system in simple elements. This question takes its roots in quantum mechanics itself. To analyse a system, we must inescapably make use of an apparatus. But what about the analysis of the apparatus itself ?

To analyse its components, we must make use of a second apparatus, and so on. As Bohr puts it, «By the word experiment, we refer to a situation where we can tell others what we have done and what we have learned » (Bohr, 1949). This has led to the idea introduced by von Neumann of two different types of evolution of systems: -(1) the « Schroedinger » evolution of a system by itself (obeing the Schroedinger equation) and -(2) the measurement evolution (often called the « state vector (or wave packet) reduction »). The great epistemological revolution of quantum theory is that the evolution (2) is irreductible to evolution (1), as if an apparatus as such would not follow the Schroedinger equation. This has cast serious doubts on the pertinence of « objective properties » of systems. This notion of objectivity is one way of defining « materialism ». This was expressed by von Neumann: « Experience only makes statements of this type : an observer has made a certain (subjective) observation ; and never any like this : a physical quantity has a certain value » (von Neumann, 1955). This position was pushed even further by London and Bauer who introduced the idea that the « wave packet reduction » is not a physical phenomenon, but is only due to the intervention of the observer's consciousness (London and Bauer, 1938). This point of view had gained some credibility with the experimental proof (using the so called Bell inequalities) that the value of the polarization is not an objective property of photons, but is « produced » at the instant at which an observer makes the measurement (Aspect, 1976).

Since the 1980s, the « materialists » try to save the « realistic » or objectivistic position by using the notion of « decoherence ». One would pass continuously from the quantum microscopic world to the classical macroscopic world thanks to the destructive interferences of the many microscopic wave functions associated to the elementary constituents of large systems. But this hope has failed since it cannot explain, first what the notion of an « observable » means, and secondly why the final, probabilistic, outcome of a measurement is unique (Omnes, 1995). Thus, as was prophetised by Bohr since the 1920s, it is not possible to recover the classical level from the quantum microscopic level. In quantum theory, the classical, phenomenal, level has its own, proper *a priori*, rights.

This perspective has led very soon Bohr, after his father who was a biologist, to suggest that « life cannot be entirely analysed, since, as soon as you analyse too accurately a living system, you kill it ». Of course, today we have many kinds of « non destructive » measurements which limit the practical validity of the quoted statement, but in its principle, the statement remains valid in quantum theory: if you analyse too accurately a macroscopic measurement apparatus, you destroy its quality as an « apparatus ».

Parallel to the questions about the origins of life, there was, in fact since Plato (in *Cratylus*), a great debate about the origin of language. As far as language is the main if not the unique characteristic of humanity, this debate is of importance for the origin of the *Human* species. After long debates, there are today two irreconciliable positions:
- language is nothing else than a correspondance between a reality which it describes and the internal physiology of speech and audition, writing and lecture. - language is a series of symbols irreductible as well as their physical support. The first point of view fails to explain experiences such as the enonciation of the word « now ». From the point of view of the second proposition, the question of the origin of language is an ill-formulated question. Indeed, the modern developements of this point of view have shown that time does not precede language. It is on the contrary language which precedes time: time is so intimately intricated with language that, somehow, language creates time.

Thus language creates the prerequisit of its own origin, and there cannot be an objective origin. After Derrida, there is only a « supplement of origin » (Derrida, 1976), i.e. the origin can only be a construction, not a « reality ». In a subsequent work, we will demonstrate (Schneider, 1997) that this point of view is supported by several different researches (e.g. Cassirer, 1953).

As a conclusion, we want to stress the importance of the search for extraterrestrial « intelligences » for all of these questions, even those of epistemological interest. The discovery of such intelligences would provide new material to investigate these problems and perhaps to enforce one or another position. But this discovery is perhaps very far in time. In the meantime, it is worth pursueing the reflexion on the basis of human, and more generally biological, material.

References

Aspect, A. (1976) Proposed experiment to test the non separability of Quantum Mechanics, *Phys. Rev.* **D14**, 1944.

Bohr, N. (1949) Discussion with Einstein on Epistemological problems in Atomic Physics, in Schilpp (ed.), *Albert Einstein : Philosopher Scientist*, The library of Living Philosophers, Evanston.

Cassirer, E. (1953) *The Philosophy of symbolic forms*, Yale University Press, New Haven (1st ed., 1931)

Darwin, Ch. (1981) *On the Origin of Species, a facsimile to the first edition* (1859), Harvard University Press, Cambridge, p.490.

Darwin, Ch. (1863, March 29) C. Darwin to J.D. Hooker, in Francis Darwin (ed.), *The Life and Letters of Charles Darwin*, vol. III, John Murray, London, p. 18.

Darwin, Ch. (1871) *The Descent of Man*, Modern Library, New York, pp.919-920.

Derrida, J. (1976) *Of Grammatology*, John Hopkins University Press.

Dick, S.J. (1996) *The biological universe, the twentieth-century extraterrestrial life debate and the limits of science*, Cambridge University Press, Cambridge.

Dobzhansky, T. (1972) Darwinian Evolution and the Problem of Extraterrestrial Life, *Perspectives in Biology and Medecine* **15**, 2, 157-175.

Eirich, F.R. (1996) Prebiotic chemical evolution and Darwinian revolution, in J. Chela-Flores and F. Raulin (eds.), *Chemical Evolution: Physics of the Origin and Evolution of Life*, Kluwer Academic Publishers, p.303.

Kamminga, H. (1980) *Studies in the History of ideas on the Origin of Life from 1860*, PhD Thesis, Department of History and Philosophy of Science, Chelsea College, University of London.

Kamminga, H. (1988) Historical perspective: the problem of the origin of life in the context of developments in biology,*Origins of Life and Evol. Biosphere* **18**, 1-11.

von Klüber, H. (1931) *Das Vorkommen der chemischen Elemente im Kosmos*, Leipzig.

Lazcano, A. (1992) Origins of Life: the historical development of recent theories, in L.Margulis and L.Olendzenski (eds.), *Environmental Evolution*, The MIT Press, Cambridge (Ma), p.61.

London, F. and Bauer D. (1938) *La théorie de l'observation en Mécanique Quantique*, Hermann, translated in *The Quantum Theory of Measurement*, Wheeler and Zurek (eds.), Princeton University Press.

Lowell, P. (1895) *Mars*, Boston, p.153.

Lowell, A.L. (1935) *Biography of Percival Lowell*, New York, pp.68-69.

von Neumann, J. (1955) *Mathematical foundations of Quantum Mechanics*, Princeton University Press.

Omnès, R. (1995) *The quantum theory of measurement*, Princeton University Press, Princeton.

Oparin, A.I. (1924) *The Origin of Life*, translated by A.Synge in J.D.Bernal (ed.), *The Origin of Life*, London (1967), Appendix I, pp.197-234.

Oparin, A.I. (1938) *The Origin of Life*, The Macmillan Compagny, New York.

Oparin, A.I. and Fessenkov, V. (1958) *La vie dans l'univers*, Editions en langues étrangères, Moscou.

Oparin, A.I. (1959) in Oparin, Pasynskii, Braunshtein & Pavlovskaya (eds.), *The Origin of Life on the Earth*, I.U.B. Symposium Series, vol I, Pergamon Press, p. 2.

Oparin, A.I. (1975) in Calvin & Gazenko (eds), *Foundations of Space Biology and Medicine*, Washington D.C., vol. 1, pp. 321-367.

Pavlovskii, E.N. (1959) in Oparin, Pasynskii, Braunshtein & Pavlovskaya (eds.), *The Origin of Life on the Earth*, I.U.B. Symposium Series, vol.I, Pergamon Press, pp. 369-370.

Ruse, M. (1997) The Origin of Life: Philosophical Perspectives, *J. theor. Biol.* **187**, 473-482.

Schiapparelli, G. (1894) *The Planet Mars*, trans. W.H. Pickering, in Astronomy and Astro-Physics, **13**, pp. 635-640, pp. 714-723.

Schneider, J. (1997) *The Construction of Origins*, Contribution to the NASA International Origins Conference, Estes Park, May 19-23.

Sheehan, W. (1988) *Planets & Perception*, The University of Arizona Press, Tucson.

Shklovskii, I.S. and Sagan, C. (1966) *Intelligent Life in the Universe*, A Delta Book, New York, p.226.

Tielens, A.G.G.M. and Charnley, S.B. (1997) *Origins of Life and Evol. Biosphere*, **27**, 23-51.

Vucinich, A. (1988) *Darwin in Russian Thought*, University of California Press, Berkeley, p.258.

THE OLDEST FOSSIL MINERAL BACTERIA FROM THE EARLY ARCHEAN OF SOUTH AFRICA AND AUSTRALIA

FRANCES WESTALL

University of Bologna
DIPROVAL, via S. Giacomo 7, 40126 Bologna, Italy.
email: frances@geomin.unibo.it

Abstract

Finely biolaminated cherts from the 3.5-3.3 Ga Onverwacht Group of South Africa and a silicified stromatolite, ca. 3460 Ma in age, from the Warrawoona Group, NW Australia yield coccoid and bacillar structures, interpreted as possible and probable bacteria, preserved as fossils by mineral-replacement. The coccoid forms have an average diameter of 1 μm whereas the bacillar forms fall into two size classes: small, rod-shaped structures, 0.65-2 μm and some larger "sausage"-shaped rods, 2.0-3.8 μm in length. These bacteriomorphs are closely associated with a smooth/ropy to granular film coating, the bedding planes, interpreted as bacterial mat biofilm.

Introduction

Previous searches for bacterial fossils from the Early Archean of South Africa (the Swaziland Supergroup) and Australia (the Warrawoona Group) yielded filamentous structures of the order of tens of micrometers in length, interpreted as possible or probable filamentous bacteria and cyanobacteria (Schopf and Walter, 1983; Walsh and Lowe, 1985; Walsh, 1992; Schopf, 1993). Other, smaller coccoid structures of possible biogenic origin were described by Knoll and Barghoorn (1977). All these structures concerned carbonaceous, permineralised fossils, identified in thin section with a light microscope. Considering the small size of bacteria in general (most are between 0.3-2 μm in size, although the cyanophytes are generally considerably larger), scanning electron microscope (SEM) techniques offer more possibilities than the light microscope for studying the structure and mode of fossilisation of smaller fossil bacteria.

This paper describes a combined light microscope and scanning electron microscope study of fossil bacteria from the Early Archean of South Africa and Australia which, due to the better resolution attainable with the SEM, has substantially improved the documentation of and knowledge concerning Early Archean bacteria and their habitat.

Materials And Methods

A sequence of finely laminated, silicified volcaniclastic/evaporitic sediments from the Kromberg Formation (3416 ± 5 to 3334 ± 3 Ma) and the Hooggenoeg Formation (3472 ± 5 to 3345 ± 3 Ma), as well as a silicified stromatolite from the stromatolite type section from the Warrawoona Group at North Pole (3460 Ma), were selected for

J. Chela-Flores and F. Raulin (eds.),
Exobiology: Matter, Energy, and Information in the Origin and Evolution of Life in the Universe, 181–186.

thin section and SEM study. After optical microscope observation, the thin sections, together with chips displaying bedding plane surfaces, were gently etched in HF fumes to reveal the structures embedded in the quartz matrix. They were then coated with C and Au or Au/Pd for SEM observation with a Philips 515 (plus Edax EDX system) at the University of Bologna, and a Leo 440 (plus Kevex EDX system, and backscatter and cathodoluminescence detectors) at the University of Cape Town.

Results

The fine laminae of the cherts consist of dark brown, wavy, discontinuous wisps of the order of micrometers in thickness, creating packets of laminae tens of micrometers thick embedded in and separated by microcrystalline quartz (Fig. 1). Clusters of "dust" 50-100 μm in diameter are associated with the laminae. The clusters exhibit a radiating structure and comprise crystallites and spherules of the order of one to a few micrometers in size (Fig. 2). Many spherules appear to be joined together, or form straight or kinked chains, or occur in rectangular associations of four individuals. Both clusters and laminae are cross-cut by quartz veins, demonstrating that they are intrinsic to the rock and syngenetic with it.

The SEM observations show that the wispy laminae consist of extensive (up to 10,000 μm^2), often cracked, smooth or "ropy", to grainy sheets covering the surfaces of the bedding planes. The "ropy" sheets are formed of intertwined and bent fibers less than a micrometer in thickness. The sheets embed pseudomorph crystals of calcite, halite, gypsum and tourmaline, as well as rounded, elongated, "sausage"-shaped structures 2-3.8 μm in length (Fig. 5). The latter are curved, following the substrate, demonstrating that they originally consisted of soft, pliant material. Some of the "sausage"-shaped rods are attached to each other at their extremities. In contrast, all crystals exposed by HF etching display sharp-edged, euhedral characteristics and have not been rounded by the HF treatment. Pseudomorphs of aragonite also occur in the quartz.

EDX analysis of the 1 μm diameter spherules embedded in the quartz (Fig. 3) and pseudomorph aragonite shows that they contain trace amounts of Ca compared to the quartz matrix. They often occur together with small, 3-6 μm pseudomorphs of acicular tourmaline. Sectioned spherules are filled, not hollow structures. They exhibit organised association in small groups, pairs or straight to kinked linear associations. Linear associations may be coated by a common sheath exhibiting a wrinkled texture.

Short, rod-shaped structures, 0.65-1.9 μm in length and 0.32-1.1 μm in width,

---

Figure 1. Dark brown, wispy, wavy laminae in chert from the Hooggenoeg Formation, Onverwacht Group, South Africa. Scale 500 μm.
Figure 2. Detail of spherules in a cluster from the Kromberg Formation, Onverwacht Group. Note the joined (dividing?) spherules (arrows). Scale 10 μm.
Figure 3. Spherical bacteriomorphs (arrows) in the Kromberg Formation, revealed by HF etching. Note the paired and triplet associations linked by a common outer wall. Scale 1 μm.
Figure 4. Short rod-shaped bacteriomorphs looking like "rice grains" (arrows) and a larger, more oval bacteriomorph (center) from the Kromberg Formation. Scale 1 μm.
Figure 5. Large "sausage"-shaped bacteriomorphs (arrows) embedded in granular biofilm on a bedding surface of chert from the Hooggenoeg Formation. Note the curved attitudes of the rounded structures. Scale 1 μm.
Figure 6. Large, rod-shaped bacteriomorph in a round-ended hollow (large arrow) and a smaller rod-shaped bacteriomorph (small arrow) from the type stromatolite in the Warrawoona Group at North Pole, NW Australia. Scale 1 μm.

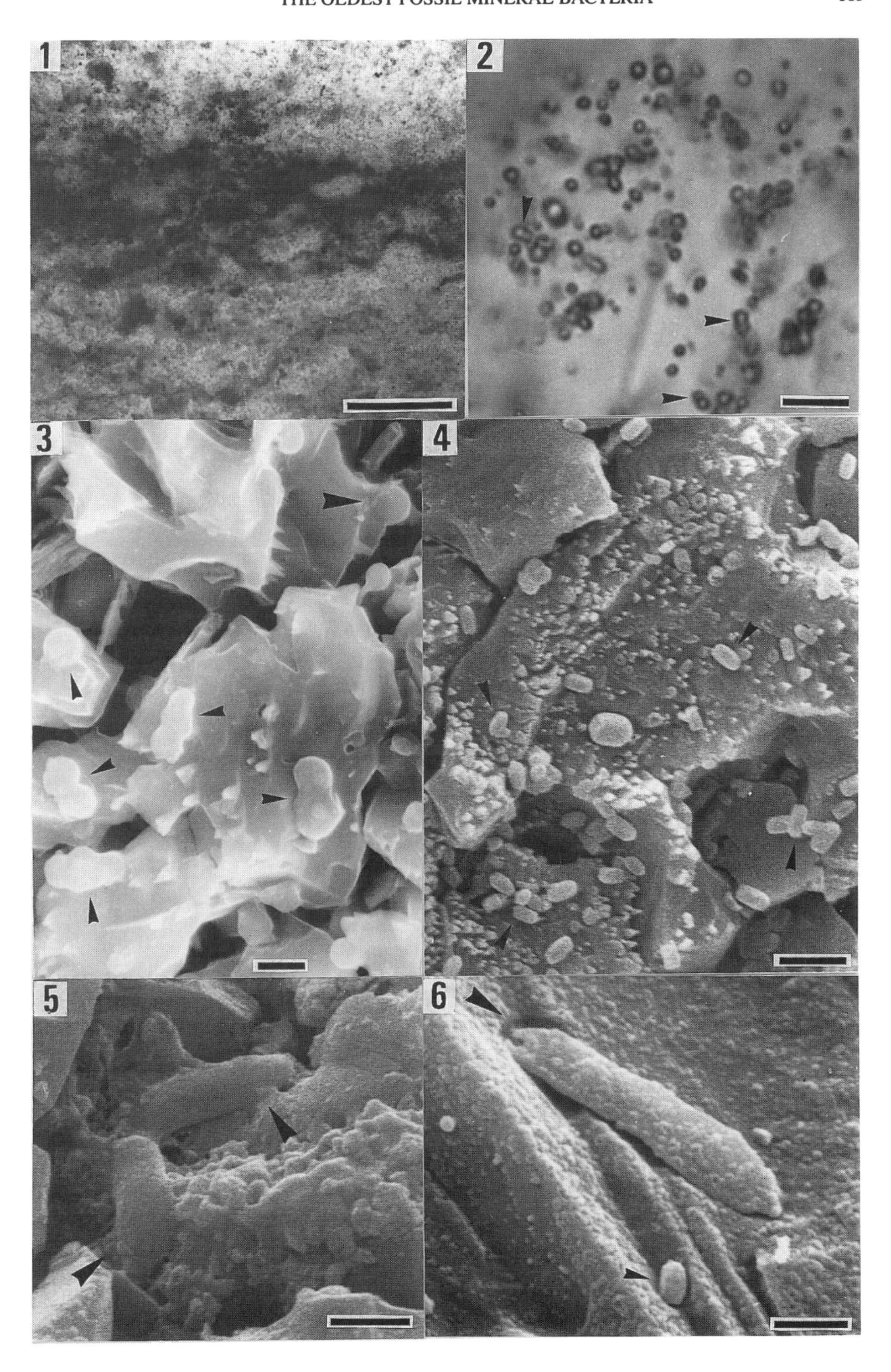
1
2
3
4
5
6

also occurring in clusters, were observed in chips of the cherts from the Onverwacht Group (Fig. 4). Some of the rods display a slight crystalline structure, whereas others are rounded in cross-section. The rods may be cemented together and some are joined at their extremities.

The sample from the Warrawoona stromatolite displays many similarities to those from the Onverwacht Group. The laminae consist of dark brown, elongated wisps and "dusty" clusters, sometimes with a radiating structure, are associated with them. Unlike the Onverwacht Group cherts, subrounded clasts of fine-grained sediments are trapped within the stromatolitic structure. Smooth biofilm can be seen embedded in the quartz matrix in the SEM images. Rare examples of filamentous structures and short rod-shaped structures occur in the silicified stromatolite (Fig. 6). The filamentous structures are 4 μm in length and 1 μm in width. Stereo photographs show that they are surrounded by a rounded hollow, indicating that the filament is a replacement cast of a previous, "sausage"-shaped structure. Occurring together with the filaments are short 0.6 μm long and 0.35 μm wide, rods very similar to those of the Onverwacht Group.

Interpretation

The laminations in the Onverwacht Group cherts are interpreted as bacteriological in origin on the basis of their morphological similarity with modern bacterial biofilm, and the fact that they occur on sediment bedding surfaces, as is common with modern bacterial biofilms in aqueous and littoral environments. The cracked texture of the biofilms on the Onverwacht bedding surfaces is reminiscent of subaerially exposed and dried modern intertidal biofilms. The structure of the wavy, wispy laminae of the Warrawoona stromatolite is very similar to that of the Onverwacht Group sediments, suggesting a similar interpretation. This is supported by SEM observation of biofilm embedded in the quartz.

Turning to the clusters of spherules and rod-shaped structures in the Onverwacht Group, there are strong similarities between the size, shape and organisation (in pairs, or linear associations) of the spherules and modern coccoid bacteria. The associations of one or more spherules joined to each other by an invagination or pinching of the outer common wall is very suggestive of bacterial cell division (Fig. 3). The outer walls of the spherules, with their nanometer scale roughness, is also similar to the outer walls of modern bacteria. The clusters are interpreted as colonies of bacteria. The larger, "sausage"-shaped rods (Fig. 5) are also very similar in size, shape and association (connection at their extremities) to modern, larger, rod-shaped bacillar bacteria, such as the cyanobacteria. Their curved attitudes and rounded cross sections contrast with the angularity of the pseudomorph calcite crystals also associated with the biofilm. Moreover, the associations of rods joined at their extremities is also suggestive of bacterial cell division.

The smaller rod-shaped structures (Fig. 4) are similar to modern heterotrophic bacteria in their size and shape, regularity of size and shape, organisation (joined at their extremities), and their occurrence in clusters. An argument against this interpretation is that some of the rods display crystallographic features or are cemented together. However, all the clearly identifiable crystals in the Onverwacht Group sediments display very distinct crystalline habits which cannot be confused with biogenic structures. Another explanation is that they may represent bacteria replaced by minerals, the growth of which has led to incipient crystallographic habits, as has previously been demonstrated (e.g. Vai and Ricci Lucchi, 1977; Castanier *et al.*, 1990).

At least some of the short, rod-shaped structures may, therefore, represent fossil bacteria.

Although few in number, the rod-shaped structures (Fig. 6) in the Warrawoona stromatolite are also interpreted as being of possible bacterial origin on the basis of their rounded morphology and size. The similarities between the clusters in the Warrawoona stromatolite and those consisting of probable and possible fossil bacteria from the Onverwacht Group lends further support to a biogenic interpretation for the Warrawoona stromatolite. This is the first time that microfossils have been described from the North Pole stromatolite locality.

The dried out bacterial mat biofilms and associated evaporitic minerals (including tourmaline), as well as the penecontemporaneous silicification by hydrothermal fluids, indicate periodically emergent shallow water sediments in an hydrothermally active environment. There is ample evidence for hydrothermal activity associated with the deposition of these volcaniclastic sediments interbedded with lavas (de Wit *et al.*, 1982; Byerly and Palmer, 1991; Westall *et al.*, submitted). The Ca traces in the microfossils from both the Onverwacht Group and the Warrawoona Group suggest that the bacteria were first replaced by calcite and subsequently silicified. These minerally-replaced fossils contrast with the previously described permineralised (silicified) carbonaceous remains (viz. Schopf and Walter, 1983; Walsh and Lowe, 1985; Walsh, 1992; Schopf, 1993): mineral replacement of bacteria has been demonstrated experimentally (Westall *et al.*, 1995) and is not uncommon in the rock record, especially in the Phanerozoic (e.g. Oehler, 1977; Wuttke, 1983; Chafetz and Folk, 1983; Knoll *et al.*, 1993; Liebig *et al.*, 1996).

The Overwacht Group and Warrawoona Group microfossils described here add valuable information concerning Early Archean fossil biota: (1) They are the first minerally-replaced bacteria to be described from the Early Archean; (2) They are of an order of magnitude smaller than all previously described microfossils from these formations (with the exception of the possibly biogenic spherules of Knoll and Barghoorn, 1977); (3) They are clearly associated with a shallow water, hydrothermal habitat. Considering the present discussion concerning hydrothermal vents as possible sites where life started (Russel and Hall, 1997), this latter observation is interesting.

Acknowledgements

Maarten de Wit, Jesse Dann, Dane Gerneke, Jan Högbom and Athos Fantazzini are gratefully acknowledged for their help and support.

References

Byerly, G.R. and Palmer, M.R. (1991) Tourmaline mineralisation in the Barberton greenstone belt, South Africa: early Archean metasomatism by evaporite-derived boron, *Contributions to Mineralogy and Petrololgy* **107**, 387-402.

Castanier, S., Maurin, A., and Perthuisot, J.-P. (1990) A trial to get dolomite in freshwater, *Geobios* **23**, 121-128.

Chafetz, H.S. and Folk, R.L. (1984) Travertines: depositional morphology and the bacterially constructed constituents, *J. Sedimentary Petrology* **54**, 289-316.

de Wit, M.J., Hart, R., Martin, A., and Abbott, P.(1982) Archean abiogenic and probable biogenic structures associated with mineralised hydrothermal vent systems and regional metasomatism with implications for greenstone belt studies, *Economic Geology* **77**, 1783-1802.

Knoll, A.H. and Barghoorn, E.S. (1977) Archean microfossils showing cell division from the Swaziland System of South Africa, *Science* **198**, 396-398.

Knoll, A.H., Fairchild, I.J. and Swett, K. (1993) Calcified microbes in Neoproterozoic carbonates: implications for our understanding of the Proterozoic/Cambrian transition, *Palaios* **8**, 512-525.

Liebig, K., Westall, F. and Schmitz, M (1995) A study of fossil microstructures from the Eocene Messel Formation using Transmision Electron Microscopy, *Neues Jahrbuch für Geologie und Paläontologie Monatsheft* **4**, 218-231.

Oehler, J.H. (1977) Microflora of the H.Y.C. Pyritic Shale Member of the Barney Creek Formation (McArthur Group), middle Proterozoic of northern Australia, *Alcheringia* **1**, 315-349.

Russell, M.J. and Hall, A.J. (1997) The emergence of life from iron monosulphide bubbles at a submarine hydrothermal redox and pH front, *J. Geological Society*, **154**, 377-403

Schopf, J.W. (1983) Microfossils of the Early Archean Apex Chert: new evidence of the antiquity of life, *Science* **260**, 640-646.

Schopf, J.W. and Walter, M.R. (1983) Archean Microfossils: new evidence of ancient microbes, in J.W. Schopf (ed.),*Earth's Earliest Biosphere*, Princeton University Press, New Jersey, pp. 214-239.

Vai, G. B. and Ricci Lucchi, F. (1977) Algal crusts, autochthonous and clastic gypsum in a cannibalistic evapotic basin: a case history from the Messinian of Northern Appenines, *Sedimentology*, **24**, 211-244.

Walsh, M.M. (1982) Microfossils and possible microfossils from the Early Archean Onverwacht Group, Barberton Mountain Land, South Africa, *Precambrian Research* **54**, 271-293.

Walsh, M.M. and Lowe, D.R. (1985) Filamentous microfossils from the 3,500-Myr-old Onverwacht Group, Barberton Mountain Land, South Africa, *Nature* **314**, 530-532.

Westall, F., Boni, L. and Guerzoni, M.E. (1995) The experimental silicification of microorganisms, *Palaeontology* **38**, 495-528.

Westall, F., de Wit, M.J., Dann, J., van der Gaast, S., de Ronde, C.E.J., and Gerneke, D. (submitted) Evidence linking Earth's oldest fossil bacteria to hydrothermal systems.

Wuttke, M. (1983) "Weichteilerhaltung" durch lithifizierte Mikroorganismen bei mittel-eozänen Vertebraten aus den Ölschiefern der "Grube Messel" bei Darmstadt, *Senckenbergiana lethaia* **65**, 509-527.

THE EVOLUTIONARY ASPECTS OF COENZYME PHOTOBIOCHEMISTRY

M.S. Kritsky[1], T.A. Lyudnikova[1], E.A. Mironov[2], and K.V. Neverov[1]
[1]A.N. Bach Institute of Biochemistry RAS, and [2]A.N. Nesmeyanov Institute of Organoelement Compounds RAS, Moscow, Russia
[1]E-mail <inbio@glas.apc.org>, [1]fax (7-095) 9542732

Photochemical processes driven by solar photons have strongly impacted prebiotic accumulation of organic molecules in outer space and on the Earth, and today photosynthesis is an ultimate source of organic matter for biosphere. Moreover, organisms are capable to perceive photons as informational signals governing their adaptation to the environment. The activity of this photobiological world is founded by photoreceptor molecules converting their excitation energy into the energy of chemical bonds or into an informational signal recognisable by cell metabolism regulators. These photoreceptors strongly vary by the apoprotein structure and by their chromophore molecules, which excitation by light starts the physicochemical events resulting in manifestation of physiological effect. All the variety of chromophores consists, in fact, of only three chemically distinct groups: tetrapyrrols (porphyrins including chlorophylls and the linear tetrapyrrols such as bilins), isoprenoids (carotenes and retinal), and heterocyclic compounds, the derivatives of isoalloxasine and pteridine. The few exceptions fall into very limited taxonomic niches.

Considering the diversity of structural and photochemical properties, the search of evolutionary roots of the processes driven by each chromophore group forms an individual problem. There is, probably, no much hope to reveal the abiotic ancestors of modern rhodopsin- or bacteriorhodopsin-mediated photoprocesses from the studies of free retinal in photochemical models. The excitation of its free molecules cannot be implemented into physiological processes observed in organisms, since the activity of biological retinal-bound photoreceptors involves conformational changes of its apoprotein followed in the case of bacteriorhodopsin by a H^+ translocation across the lipid membrane. The photoexcited free porphyrins can sensitise the electron transfer what was basically important for the development of photosynthetic mechanism. Nevertheless, the core of chlorophyll-mediated photosynthesis, its reaction centres (PRC) could not arise before apoproteins precisely arranging the chromophores to form these centres had appeared (for origin of PRC see [1]).

The isoalloxasine and pteridine heterocycles derivatives, flavins and pterins, are present in all cells. Many of them such as flavin mononucleo-

J. Chela-Flores and F. Raulin (eds.),
Exobiology: Matter, Energy, and Information in the Origin and Evolution of Life in the Universe, 187–190.

tide, flavin adenine dinucleotide (FAD), biopterin, molybdopterin and a family of folates are important coenzymes of 'dark' biochemistry. The results of chemical modelling suggest the abiotic origin of flavins and the pteridine derivatives [2]. This notion finds a support from the protein evolution studies, which have revealed the availability of flavins at the time of metabolism emergence about 4.0 Gy ago [3].

Pterins and flavins are chromophores of photoreceptors in organisms from archea and eubacteria up to higher eukaryotes. The biochemical systems under the control of these photoreceptors and the final physiological responses are widely varied, but the major role of these processes in organisms is to gain benefits from a changed illumination condition or to avoid stress hazards by adapting metabolism to light. The flavin and pterin photoreceptors studied today at molecular level are enzymes DNA-photolyase and eukaryotic nitrate reductase, and the photoreceptor pigment(s) cryptochrome(s). Besides FAD, 1,5-dihydroflavin adenine dinucleotide ($FADH_2$), N_5-substituted flavin, 8-hydroxy-5-deazaFAD (deazaFAD) and a pterin derivative, 5,10-methenyl-tetrahydrofolate (MTHF) functioning in these photoreceptors [4-6], there is a reason to suspect the participation of biopterin, folate and molybdopterin in biological sensing of light [7-9].

DNA-photolyase catalyses elimination from DNA of pyrimidine dimers formed after its irradiation by UV-photons (λ>300 nm). The enzyme widespread among pro- and eukaryotes binds these dimers and splits them in a photon-dependent reaction hence forming the original pyrimidine monomers. In all these organisms photoinduced excision of pyrimidine dimers proceeds by close molecular mechanisms that may point to evolutionary antiquity of the process. Photolyase contains two chromophore cofactors. One is always $FADH_2$ and the other can be either MTHF or deazaFAD. Dimer repair proceeds only after the light excites one of the chromophore groups. It is excited $FADH_2$ that directly participates in an intramolecular electron transfer leading to splitting of C-C bonds in dimer, and pterin and deazaflavin moieties play a role of light harvester that activates flavin by energy transfer [5].

Nitrate reductase, the first enzyme of the NO_3^- assimilation in microorganisms and plants, catalyses its reduction to NO_2^-. The electron transfer chain of eukaryotic nitrate reductase contains FAD and a pterin, molybdopterin, coordinating the Mo(V) atom of the active centre. The enzyme can exist in two interconvertable forms, one oxidised/active and the other reduced/inactive, and light plays a major role in their interconversion. Excitation of the FAD molecule in the inactive reduced enzyme causes reoxidation of the regulatory centres and restores its activity [6].

Cryptochrome(s) mediate(s) in plants and microorganisms adaptation of metabolism, behaviour and development caused by the UV-AB and a short-wave visible radiation [10]. Two molecular types of cryptochrome have been recently identified in plants. The first one is a family of proteins encoded by a fusion product of the DNA photolyase and tropomyosin genes. In modern cryptochrome both the photolyase and a tropomyosin parts have lost the sites responsible for these ancestral functions. The photolyase-like moiety of cryptochrome binds FAD and a pterin, possibly MTHF, and these chromophores have been suggested to act in electron phototransfer [4]. The other protein participating in the UV-AB and visible light reception is a protein kinase, which binds a FAD molecule and another still not identified chromophore [11].

Flavins and pterins absorb the UV-AB and blue light-photons, and the originated excited singlet and triplet molecules are active in the energy and electron transfer processes [10,12-14] (Fig.1). The tripet molecules efficiently generate $^1O_2$ with all chemical consequences of this photoprocess [13,15]. Under anoxygenic conditions and in presence of a suitable electron donor the strongly positive molecules of excited flavins and pterins convert to reduced forms (dihydroflavins and dihydro- and tetrahydropterins). The excited states of some reduced coenzymes such as $FADH_2$ and dihydropterins are also active in electron transfer. Flavins and pterins photocatalyse the electron transfer from a high potential donor to a substantially more negative acceptor with accumulation of free energy [10,14]. Among the donors are the compounds available from chemical evolution, for instance, ethylene and its derivatives such as ethylenediamine diacetate and ethylenediamine monoacetate [16]. In solution and in some enzymes, i.e. in nitrate reductase and in dihydrofolate reductase, the electron transfer is sensitised by triplet flavins and pterins [6,7,10,14]. In DNA photolyase the arrangement of chromophore close to its electron transfer partners permits their interaction with a singlet-excited $FADH_2$ [5].

$$Fl + h\nu \rightarrow {}^1Fl \rightarrow {}^3Fl \quad (1) \qquad Pt + h\nu \rightarrow {}^1Pt \rightarrow {}^3Pt \quad (1')$$

$${}^3Fl + O_2 \rightarrow Fl + {}^1O_2 \quad (2) \qquad {}^3Pt + O_2 \rightarrow Pt + {}^1O_2 \quad (2')$$

$${}^1PtH_4 + FlH_2 \rightarrow Pt\,H_4 + {}^1FlH_2 \quad (3)$$

$${}^1FlH_2 + Pyr<>Pyr \rightarrow FlH_2 + 2Pyr \quad (4)$$

$${}^3Fl + D \rightarrow FlH_2 + D_{ox} \quad (5) \qquad {}^3Pt + D \rightarrow PtH_2 + D_{ox} \quad (5')$$

$${}^3Fl + D \rightarrow FlH^{\cdot} + D^{\cdot}_{ox} \quad (6) \qquad {}^3Pt + D \rightarrow PtH^{\cdot} + D^{\cdot}_{ox} \quad (6')$$

$${}^3PtH_2 + D \rightarrow PtH_4 + D_{ox} \quad (7')$$

$${}^3Fl + D + A \rightarrow Fl + D_{ox} + A_{red} \quad (8) \qquad {}^3Pt + D + A \rightarrow Pt + D_{ox} + A_{red} \quad (8')$$

Figure 1. Photophysical and photochemical reactions of flavins (Fl) and pterins (Pt). The equations 1 (1') illustrate transition of molecules to the excited singlet and triplet states [10,12,13]. The 2 (2') is a generation of $^1O_2$ by the triplet coenzyme molecules [13,15]. Equations 3 and 4 show the energy transfer from a pterin to the flavin chromophore in DNA photolyase and the $FADH_2$-catalysed electron transfer inside the enzyme-substrate complex resulting in a splitting of pyrimidine dimer (Pyr<>Pyr) [5]. In 5 (5'), 6 (6') and 7 (7') excited flavins and pterins undergo 1e or 2e reduction and are converted into their reduced forms, dihydroflavins and dihydropterins (FlH_2, PtH_2) and tetrahydropterins (PtH_4) [10,13]. The 8 (8') reactions are the photocatalysed electron transfer from donor molecule (D) to acceptor (A) to form oxidised donor (D_{ox}) and the reduced acceptor (A_{red}) [6,10,14]. The donors with a E'_o values as high as +0.4 V (EDTA) and the acceptors with the E'_o as low as −0.4 V (methyl viologen) are active. Reactions 3,4,5,6' and 8 have been demonstrated in enzyme systems; and the 5 and 8 in solution, in artificial lipid membrane and in enzyme.

The above-listed evidences point to coenzymes as to probable primitive photoreceptors. Their abiotic availability suggests these molecules may have joined metabolism at the onset of evolution, and the photochemical properties are sufficient to sensitise the conservation of solar energy by prebiotic and primitive biological systems. The antiquity of flavin-binding proteins and of the enzymatic photorepair of DNA, as well as the spread of coenzyme-sensitised photobiological processes among a wide range of modern organisms also supports such a possibility.

Supported by grant (295-04-12092-a) of the Russian Foundation for Basic Research.

References

1. Blankenship, R.E.: Origin and early evolution of photosynthesis, *Photosynth. Res.*, **2** (1992), 91-111.
2. Heinz, B., Ried, W., and Dose, K.: Thermische Erzeugung von Pteridinen und Flavinen aus Aminosauregemischen, *Angew. Chem.*, **91** (1979), 510-511.
3. Rossman, M.G., Moras, D., and Olsen, K.W.: Chemical and biological evolution of a nucleotide-binding protein, *Nature*, **250** (1974), 194-199.
4. Ahmad, M., and Cashmore, A.R.: Seeing blue: the discovery of cryptochrome, *Plant. Mol. Biol.*, **30** (1996), 851-861.
5. Sancar, A.: Strucrure and function of DNA photolyase, *Biochemistry*, **33** (1994), 2-9.
6. De la Rosa, M.A., Roncel, M., and Navarro, J.A.: Flavin-mediated photoregulation of nitrate reductase. A key point of control in inorganic nitrogen photosynthetic metabolism, *Bioelectrochem. Bioenerg.*, **27** (1989), 355-364.
7. Ledbetter, J.W., Jr., Pflederer, W., and Freisheim, J.H.: Photosensitised rtduction of L-biopterin in the active ternary complex of dihydrofolate reductase, *Photochem. Photobiol.*, **42** (1995), 71-81.
8. Cremer-Bartels, G.H., Gerding, H., and Krause, K.: Increase of tetrahydropterins in cell-free retinal extracts in response to light exposure, in: J.E. Ayling *et al.* (eds.), *Chemistry and Biology of Pteridines and Folates*, Plenum Press, New York (1993), pp. 339-342.
9. Ninneman, H.: Participation of molybdenum cofactor of nitrate reductase from *Neurospora crassa* in light-promoted conidiation, *J. Plant. Physiol.*, **137** (1991), 677-682.
10. Presti, D.E.: The photobiology of carotenes and flavins, in D.E.Cosens and D.Vince-Price (eds.), *The Biology of Photoreception,* Soc. Experim. Biol. (1983), pp. 133-180.
11. Short, T.W., and Briggs, W.R.: The transduction of blue light signals in higher plants, *Ann. Rev. Plant Physiol.& Plant Mol. Biol.*, **45** (1994), 143-171.
12. Chahidi, C., Aubailly, M., Momzikoff, A., Bazin, M., and Santus, R.: Photophysical and photosensitizing properties of 2-amino-4 pteridinone: a natural product, *Photochem. Photobiol.,* **33** (1981), 641-649.
13. Neverov, K.V., Mironov, E.A., Lyudnikova, T.A., Krasnovsky, A.A., Jr., and Kritsky, M.S.: Phosphorescence analysis of the triplet state of pterins in connection with their photoreceptor function in biochemical systems, *Biochemistry (Moscow)* **61** (1996), 1149-1155.
14. Kritsky, M.S., Lyudnikova, T.A., Mironov, E.A., and Moskaleva, I.V.: The UV radiation-driven reduction of pterins in aqueous solution, *J. Photochem. Photobiol., B: Biol.*, **48** (1997), 43-48.
15. Krasnovsky, A.A., Jr.: Mechanism of formation and the role of singlet oxygen in photobiological processes, in: A.B. Rubin (ed.), *Molecular Mechanisms of Biological Effects of Optical Radiation*, Nauka, Moscow (1988), pp.23-41 (in Russian).
16. Nelson, K.E., and Miller, S.L.: The prebiotic synthesis of ethylenediamine monoacetic acid, the repeating unit of peptide nucleic acid, in: *ISSOL'96, 8th ISSOL Meeting, 11th Int. Conf. Origin of Life, Book of Program & Abstracts*, Orleans (1996), p.72.

ORIGIN OF THE PROTO–CELL MEMBRANE

– Great importance of phospholipid bilayer –

HAKOBU NAKAMURA
Biological Institute, Faculty of Science
Konan University, Kobe 658, Japan

The unit of all living systems is the cell. Cells are surrounded by cell (or plasma) membranes which are constructed from phospholipid bilayer (abbreviated to lipid bilayer hereafter) as skeleton.

Therefore, the living system is essentially a closed system which is border on the environment with a phospholipid bilayer, though much material and information is exchanged between the inside and outside of the system. This communication deals with the origin and evolution of the lipid bilayer as the skeleton of cell membrane. This is the central problem of the biological world on the earth, because all of the organisms are made of the cell (s).

1.The appearance of phospholipids on the earth

Lipids comprise a diverse class of chemical substances characterized by their solubility in non-polar organic solvents and low solubility in water. The solubility characteristics of lipids can be mainly attributed to the fact that they possess non-polar and low dielectric regions, namely fatty acids as long hydrocarbon chains and thus are extensively hydrophobic, although their carboxylic acid groups strongly polar because it can ionize by losing a proton in water. Consequently, fatty acids have both hydrophobic and hydrophilic characteristics.

On the other hand, phospholipids contain phosphoric acid which is esterified by the free -0H group of diglyceride as phosphatidic acid. Most biologically important phospholipids are derivatives of phosphatidic acid and contain other substances, such as choline, L-serine, or ethanolamine as nitrogen-containlng bases, and glycerol or inositol as non-nitrogen compounds, linked to one of the phosphate hydroxy groups. A small amount of phospholipid can be spread over the surface of water to form a monolayer of the molecules; in this thin film, the hydrophobic tail regions pack together very closely

J. Chela-Flores and F. Raulin (eds.),
Exobiology: Matter, Energy, and Information in the Origin and Evolution of Life in the Universe, 191–194.

facing the air and the hydrophilic head groups are in contact with the water. However, when the phospholipid-water system is agitated, the phospholipid molecules form a configuration known as minicelle, some of which construct a phospholipid sandwich, or phospholipid bilayer. The bilayer naturally closes to form a sphere, or liposome, in which a liquid can be contained. Studies have demonstrated that liposome is the skeleton of the cell membrane. For example, Table 1 shows that in various physical characteristics, artificial membrane quite resembles biomembrane.

Table 1. Comparison of acrificial membrane and biomembrane (nakamura, 1988)

character	artificial membrane (36°C)	biomembarane (25°C)
1. electron-microscopic figure	three layers	three layers
2. dipth (nm)	6.0-7.5	6.0-10.0
3. electric capacity (μ farad/cm^2)	0.4-1.0	0.5-1.3
4. electric resistance (ohm/cm^2)	10^5	10^5
5. surface tension (dyne/cm)	0.5-2.0	1.0
6. water permeability (μ m/sec)	92	0.4-400
7. glycerol permeability (μ m/sec x 10^2)	4.6	0.003-27

How and when did phospholipid molecules first appear in the primitive sea? The molecules canbe divided into several chemical constituents; fatty acids, glycerol, phosphoric acid, and nitrogen-containing bases or non-nitrogen groups. The origin of the phospholipid molecule is as derivative of chemical syntheses through chemical evolution.

Simulation experiments of chemical evolution have demonstrated the following:
(1). A variety of fatty acids as components of the phospholipid are synthesized from carbon monoxide and hydrogen by the catalytic influence of simple metals, as follows:

$$8CO + 17H_2 \xrightarrow{\text{Fe, Ni}} C_8H_{18} + 8H_2O$$

(2). Sugars such as triose (C_3) of glycerol, are spontaneously generated through polymerization of formaldehyde (HCHO); a major compound synthesized in spark discharge experiments to reduce gases containing CH_4, NH_3, H_2O, and H_2. Although most large sugars are unstable in aqueous solution, glycerol is quite stable and thus would increase in concentration in drying and freezing lakes, especially in the presence of clay.
(3). Phosphoric acid (H_3PO_4) is a hydration product of phosphorus pentoxide (P_4O_{10}).

Therefore, phosphatidic acid, as the basis of phospholipids, had to be generated during chemical evolution. And related compounds formed the liposomes of the proto-cell membrane, as shown in Fig-1.

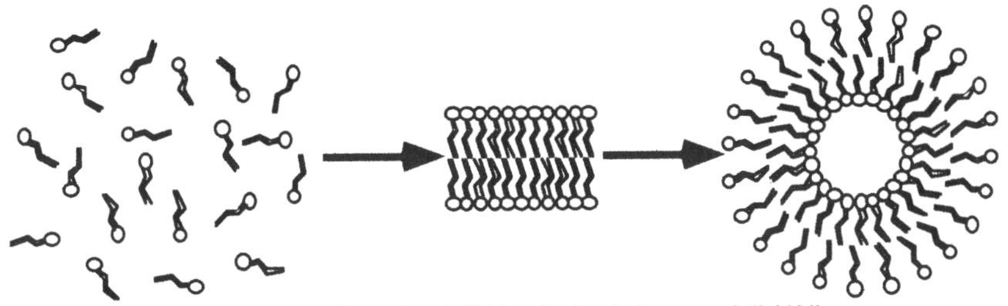

Fig. 1. Assemblages from phospholipid molecules to liposome vis lipid bilayer
These assemblages naturally occur in water, when agitated.

2. Formation of proto-cell

The proto-cell was the first living system that was generated through the chemical evolution. However, here, the detailed living mechanism of proto-cell is not discussed but its cell membrane.

(1). What is a biomembrane? All the membranes including the cell membrane, found in a single cell, contain a phospholipid bilayer as membrane skeleton. However, such biomenbranes function as more just liposomal membrane. Biomembrane contains a great variety of lipids and proteins, which differ greatly according to the type organelle, tissue, and species of the organism. Furthermore, there is differentiation even between the inner and outer layers of a lipid bilayer. Thus, the cell life as a whole can be considered the sum of its molecular functions among them.

(2). The lipid bilayer is impermeable to ions. The environment in which cell exists, for example sea water, contains various kinds of metalic and non-metalic ions, such as Na^+ and Cl^-, and the cell itself contains much protons (H^+) which actively move electron potential in the living state. However, the lipid bilayer possesses a remarkable characteristic, it is impermeable to ions. As a result, an ionic concentration difference, i. e., an ionic gradient, forms between the inside and out side of the lipid bilayer. The ionic gradient generates a spontaneously electro-chemical potential within the bilayer which is reffered to as membrane potential. The cell uses this potential as an energy source for various functions, including ATP synthesis.

Cellular energy is produced by the pumping of protons across mitochondrial cristae, which is driven by the respiratory chain, and /or across chloroplast thylakoids, which is driven by the photosystem. Moreover, excitation conduction in neuron fibers is energized by another type of ionic gradient of Na^+/K^+ and the highly electromotive forces produced by electric fishes such as *Narke japonica*, and *Tarpedo marmorata*, which reach the order of several hundred volts, are generated through an electric organ, the potential of which is strengthened by a number of ion-gradient membranes.

Proto-cells were acquired for energy to function from the electro-chemical potential of the cell membrane, in which the ionic gradient was spontaneously generated. Therefore, the lipid bilayer, as the plasma membrane, has functioned as an energy-generator from the very beginning.

3. Where are the membrane enzymes located in the lipid bilayer

Recent studies have established that metabolic carriers are alined according to the reaction sequence of the electron flow to synthesize ATP and NADPH, and the phospholipid bilayer of thylakoid membrane fixes these carriers in order. A similar phenomenon occurs in the membrane enzymes in the cell. The purpose of membrane differentiation is to copartmentalize metabolism. In fact, the membranous organelles have remarkably inproved the life efficiency of cells. For example, eukaryotic cells contain various intracellular membranes, each of which packages a specific metabolism: mitochondrion for respiratory metabolism (TCA cycle-respiratory chain), chloroplast for photosynthesis (photosystem- reduced pentose phosphate cycle), peroxisome (or microbody) for oxidative metabolism and so on. In summary, organelles serve to guarantee the (i) specificity and (ii) speed of a metabolism.

Bacteriorhodopsin, a typical transmembrane protein, crosses the lipid bilayer seven times, and serves as a light-activated proton pump in the cell membrane of the bacterium *Halobacterium halobium*. The cell membrane contains a specialized patch, the purple membrane. Each bacteriorhodopsin molecule contains a single light-harvesting group, a retinal, which gives the protein its purple color. When activated by a single proton of light, the excited retinal immediately changes its shape and causes a series of small conformational changes in the protein. The results in the transfer of one proton from the inside to the outside of the cell. In brightlight, a single bacteriorhodopsin molecule can pump several hundred protons per second, establishing a proton gradient across the cell membrane, which in turn drives the production of ATP by a second protein, ATP synthetase. This simple ATP-synthetic mechanism is important to the harvesting of energy from the environments. In higher cells, the proton pumping machinery is located in mitochondria or the chloroplasts.

References

Nakamura H. (1994) Origin of eukaryota from cyanobacterium:Membrane evolution theory. In Evolutionary Pathway and Enigmatic Algae (J. Seckbach), Kluver Academic Publishers, Dordorecht, pp. 3-18.

Nakamura, H. (1996) Origin and eevolution of the related organelles:mitochondrion, chloroplast, and peroxidase. Viva Origino 24, 137-158.

ANIMATE PROTOCELLS FROM INANIMATE THERMAL PROTEINS
Visualization of the Process

ARISTOTEL PAPPELIS[1], SIDNEY W. FOX[2], RANDALL GRUBBS[2], AND JOHN BOZZOLA[1]
1. Department of Plant Biology, MC 6509, Southern Illinois University at Carbondale, Carbondale, IL 62901 U.S.A. 2. Department of Marine Sciences, University of South Alabama, Mobile, AL 36688 U.S.A.

Abstract. Seven thermal protein complexes and the dialyzed products from one of these produced typical protocells (0.1-10 µm in diameter) on their surfaces when moistened with water. Protocells (about 0.5 µm in diameter) were visualized (800-1200x): almost instantly at 60 C; within a few min at 23 C; and, after about 15 min at 4 C. Protocells of about 3.0 µm diameter were observed associated with the thermal protein surfaces: within 30 sec to a few min at 60 C; after 4-10 min at 23 C; and, after 9 hr at 4 C. In all cases, the small protocells were free or in loose aggregates. The large protocells were often: free; linked in chains (filaments) or dendritic structures (5-15 protocells in the branched structures); or, more rarely, multi-linked protocellular clusters (15 or more protocells). This method of observing protocell formation provides opportunities for the study of factors involved in the self-assembly process and in protocell survival (e. g., effects of nutritional requirements for growth, differentiation, and reproduction).

1. Introduction

The protein-like products of thermal copolymerization of amino acids were found to form protocells when moistened (most commonly in water or aqueous 1.0% NaCl at 100 C). The protocells were seen as: individuals; linked protocells in linear or branched filaments; and/or, multi-linked protocells. These often were observed within networks formed by extraprotocellular thermal proteins. The presence of metal ions modified the morphologies of these networks. After the protocells were shown to exhibit the attributes of life, they became the answers to the origin-of-cellular-life questions posed by biologists (1-5); each step revealed by the processes of experimental studies. Although protocells were simple to produce, the moment(s) of self-assembly was never observed. The purpose of this study was to develop a method to visualize that process.

J. Chela-Flores and F. Raulin (eds.),
Exobiology: Matter, Energy, and Information in the Origin and Evolution of Life in the Universe, 195–198.

2. Materials and Methods

Seven thermal protein complexes were synthesized (R. Grubbs) by heating powdered, dry mixtures of amino acids at 180 C for 6 hours in Pyrex test tubes (1). They were: TP9=Poly (asp, glu, trp; equimolar); TP64=Poly (asp, glu, lys: equimolar); Poly (glu, gly: 1.47:1.50 by weight); Poly (asp, glu, pro: 1:1:1 by weight); leu-rich equimolar thermal protein; Poly 18 amino acids (2.0 g 18 amino acids:2.0 g leu by weight); Poly (glu, leu, try: 1.5:1.3:1.8 by weight); and, Poly (asp, glu, DL dopa; 1:1:1 by weight). Aliquots were crushed (powdered) for dialysis (Spectra/por tubing of 1,000 MW cutoff for 24 hours at 4 C with four changes of deionized water; dried at 37 C). Hydrolyzed samples and paper chromatography indicated recovery of the original amino acids.

Particles of thermal protein complexes or dialyzed products were heated on alcohol-cleaned glass microscope slides to melting. To these, alcohol-cleaned cover glasses were applied (with slight pressure) so that a gap formed sufficient for capillary movement of water to the cooled, exposed thermal protein surfaces while they were being viewed at 800-1200 magnifications (epifluorescence-phase contrast, video microscopy, etc.). Slides of each type of thermal protein preparation (from dry storage at 22 C) were individually chilled or preheated on a Sensortek TS-4 series thermal microscope stage. At 60 C, 1-3 drops of boiling water were introduced at the edges of the cover glass to make contact with the thermal protein products (simultanous viewing at 800-1200 magnifications). Similar trials were conducted at 4 and 23 C (1-3 drops of water at 4 or 23 C). At 4 C, prechilled preparations were viewed and returned to cold storage and restudied at hourly intevals (limited to 14 hr). Results were compared with protocells obtained from preparations (thermal protein complexes in water) made at 100 C.

Variously sized granules of thermal protein complexes were placed in 50 μm droplets of distilled water on alcohol-cleaned 1 cm-diameter cover glasses, incubated overnight at 23 C in a humid chamber, removed, and air-dried at 23 C. The cover glass preparations were glued onto specimen stubs, coated with 50 nm palladium/gold alloy, and examined in a Hitachi S2460N scanning electron microscope at 20 kV accelerating voltage.

3. Results and Discussion

When dry thermal protein complexes and their dialyzed products were examined microscopically, no protocells were observed. We infer that small and large protocells formed instantly when thermal protein complex materials come into contact with water at 100 C. When moistened at 4, 23, or 60 C, yellow-white "clouds" of the small, spherical protocells (~ 0.5 μm) formed on and along the surfaces of these solids: almost instantly at 60 C; within a few min at 23 C; and, within 30 min at 4 C. Larger protocells (~ 3.0 μm) emerged on or along these surfaces: within 30 sec to 3 min at 60 C; within 5 to 15 min at 23 C; and, after 9 hr of incubation at 4 C. Only one of the crude complexes (TP9) was studied at 4 C. Some dendritic forms appearing to be attached to the solid surfaces. Both the small and large protocells continued to increase in number and volume (diameter) with time. We infer that the 0.5 μm protocells enlarge (grow) to become the 3.0 μm diameter protocells. Multilinked protocells formed clusters of 15 or more protocells (each with at least one distinct collar). Some of the developing clusters began to be enclosed in a matrix of thermal protein products. The confirmation of protocell production was made using scanning electron microscopy. The dendritic forms had small protocells on their distal ends as though growth by budding was occurring; each bud enlarging and budding again. In the case of linear chains of protcells, the matrix gave them the appearance of cyanobacterial filaments with a sequence of protocells increasing in diameter from one end of the filament to the other. How the linkages formed remains to be determined.

We infer that the thermal proteins that self-assemble into protocells are at the same time structural elements of the wall-membrane boundary, multienzymic elements involved in protometabolism, "genetically" active (**Thermal Proteomes** = thermal peptide and thermal protein "genetic" informational units), growth- and differentiation-organizing centers, and transduction elements (6). Protocells with thermal proteomes would conduct nonribosomal syntheses of oligo/polypeptides and synthesize oligo/polynucleotides [coevolving oligo/polypeptides (**Proteomes**) and oligo/polynucleotides (**RNA-DNAomes**)(5)]. These three genetically active products would support further sequential evolution by increasing and decreasing each type of catalytically-genetically active product in an overlapping rise-and-fall manner, beginning with Proteomes dominating and ending with DNAomes dominating the others. The second to last common ancestor called the "progenote" (7) would be a metaprotocell heavily dependent on RNAomes. This would yield the "cenancestor" ("last common ancestor") (8). It would be the most complex metaprotocell, heavily dependent on DNAomes. The gap-bridging events enabling metaprotocellular life to emerge as the "minimum prokaryotic cell" having 256 DNA genes (9,10) remain unknown but are expected to involve coordinated gene expression. The rudimentary thermal protein wall-membranes of protocells would evolve by incorporating polypeptides, proteins, and lipids (lipids like those in archaeal and bacterial cells). "Membranes would arise through the gradual insertion of lipids between what were to become the transmembrane segments of integral membrane proteins (11)." Siefert et al. (12) reported data supporting this proposal: "--although the last common ancestor may have had a DNA genome,"--and--"was likely modern in its essence"--"it likely was preceded by progenotes with an RNA genome." --still in the process of evolving the relationship between RNA-DNAome genotypes and phenotypes, in a manner like that suggested earlier for RNA and DNA by Woese and G.E. Fox (7). The time to search for RNAomes and their regulators has arrived.

No prokaryotes or eukaryotes are known to contain genetic information to synthesize thermal proteins. Yet, the amino acid sequences found in thermal proteins are "programmed" into modern proteins (13, 14). Becerra et al. (8) stated that amino acid sequences in modern proteins may not allow us to peer beyond the time when proteins themselves came into being. However, Ivanov and Fortsch (13) and Ivanov (14) have shown us how to do just that. It is obvious that this "molecular memory" in modern proteins had to exist in peptide-protein coding sequences in Proteomes, RNA-DNAomes, and DNA genes. These modern amino acid and nucleotide sequences are "virtual" fossils (relics) of the first successful protocell. Thus, these remnant pathways (amino acid and nucleotide sequences) can be studied by backtrackers of DNA, RNA, and proteins interested in phylogeny and the origin of life but only into metaprotocells where these macromolecules emerged by means of the actions of Thermal Proteomes. Thus, the Thermal Protein Theory of the Origin and Evolution of Early Cellular Life unifies, with testable predictions, the origin of life theories and eliminates distracting speculations.

References

1. Fox, S. W. 1988. *The Emergence of Life: Darwinian Evolution from the Inside.* Basic Books, New York.
2. Fox, S. W., P. R. Bahn, A. Pappelis, and B. Yu. 1996. Experimental retracement of terrestrial origin of an excitable cell: Was it predictable? In *Chemical Evolution: Physics of the Origin and Evolution of Life,* J. Chela-Flores and F. Raulin, Eds., Kluwer Academic Publishers, Dordrecht, pp. 21-32.
3. Pappelis, A. and S. W. Fox. 1995. Domain Protolife: The Protocell Theory. In *Evolutionary Biochemistry and Related Areas of Physicochemical Biology,* B. F. Poglazov, B. I. Kurganov, M. S. Kritsky, and K. L. Gladilin, Eds., Bach Institute of Biochemistry and ANKO, Moscow, pp. 151-159.
4. Pappelis, A. and S. W. Fox. 1995. Domain Protolife: Protocells and metaprotocells within thermal protein matrices. In *Chemical Evolution: Structure and Model of the First Cell,* C. Ponnamperuma and J. Chela-Flores, Eds., Kluwer Academic Publishers, Dordrecht, pp. 129-132.
5. Pappelis, A. and S. W. Fox. 1996. Thermal peptides as the initial genetic system. In *Chemical Evolution: Physics of the Origin and Evolution of Life,* J. Chela-Flores and F. Raulin, Eds., Kluwer Academic Publishers, Dordrecht, pp. 157-165.
6. Kolesnikov, M. P. 1991. Proteinoid microspheres and the process of prebiological photophosphorylation. Origins of Life and Evolution of the Biosphere 21: 31-37.
7. Woese, C. R. and G. E. Fox. 1977. The concept of cellular evolution. J. Mol. Evol. 10: 1-6.
8. Becerra, A., S. Islas, J. I. Leguina, E. Silva, and A. Lazcano. 1997. Polyphyletic gene losses can bias backtracking characterizations of the cenancestor. J. Mol. Evol. 45: 115-117.
9. Mushegian, A. R. and E. V. Koonin. 1996. A minimal gene set for cellular life derived by comparison of complete bacterial genomes. Proc. Natl. Acad. Sci. USA 93: 10268-10273.
10. Mushegian, A. R. and E. V. Koonin. 1997. Response. J. Mol. Evol. 45: 117-118.
11. de Duve, C. 1991. *Blueprint for a Cell: The Nature and Origin of Life.* Burlington, NC, Neil Patterson Publishing.
12. Siefert, J. L., K. A. Martin, F. Abdi, W. R. Widger, and G. E. Fox. 1997. Conserved gene clusters in bacterial genomes provide further support for the primacy of RNA. J. Mol. Evol. 45: 467-472.
13. Ivanov, O. C. and B. Fortsch. 1986. Universal regularities in protein primary structure: preferences in bonding and periodicity. Origins of Life and Evolution of the Biosphere 17: 35-49.
14. Ivanov, O. C. 1993. Some proteins keep 'living fossil' pre-sequences. Origins of Life and Evolution of the Biosphere 23: 115-124.

DID THE FIRST CELL EMERGE FROM A MICROSPHERE ?

M. RIZZOTTI*, M. CRISMA†, F. DE LUCA*,
P. IOBSTRAIBIZER‡, P. MAZZEI*
*Department of Biology, University of Padova
†Biopolymer Research Center, CNR, Padova
‡Department of Mineralogy and Petrology, University of Padova

1. A simple standard procedure for producing proteinoid microspheres

There are essentially three hypotheses on the original organic aggregates which gave rise to the first cells: they are based either on the polynucleotides (or their precursors) endowed with enzymatic activity of the so-called RNA world, or on liposomes, or on the coacervates usually called *proteinoid microspheres* (Fox *et al.*, 1959; Fox and Dose, 1977; Honda *et al.*, 1993; Fox, 1995). In the last few years, the third hypothesis has been somewhat neglected, in spite of the great difficulties the other two are facing (Rizzotti, 1994, 1997).

To synthesize *proteinoids*, we start with a mixture of amino acids containing 0.01 moles aspartic acid, 0.01 moles glutamic acid, 0.005 moles alanine, and 0.005 moles leucine. The molar ratio of the components is more suitable to investigate their reactions than their weight ratio. The anhydrous mixture is mixed in a mortar, poured into a 20-ml test-tube, and put into an oven at a fixed temperature of 180°C for 6 hours (180°C is reached in approximately 45 min). When the test-tube is taken from the oven and cooled at room temperature, the material which has formed solidifies on the bottom and along the internal walls of the test-tube. Its colour varies from dark amber to brown, and its weight is reduced by about 15%.

The possible presence of peptide bonds in the proteinoids is checked in the infrared (IR) by means of the FT technique (Fourier transform; Griffiths and de Haseth, 1986) on a Perkin-Elmer FT-IR 1720× instrument, in which the recorded spectrum is the average of 25 scans. The powder of proteinoids (or the initial mixture of amino acids) is pressed into pellets with KBr (0.5 mg powder + 60-70 parts of KBr). The resulting spectrum is compatible with a limited presence of amide (including peptide) bonds, as the weak band at 1680 cm^{-1} demonstrates (fig. 1). Moreover, a remarkable amount of imide groups appears (band at 1715 cm^{-1}), together with other groups difficult to identify (Bellarmy, 1958). In any case, the amide groups form quickly, as they are already present after 1.5 hours of heating (fig. 1), when the dicarboxylic acids begin to melt, while during the following hours they decrease and imides increase (Fox, 1965).

J. Chela-Flores and F. Raulin (eds.),
Exobiology: Matter, Energy, and Information in the Origin and Evolution of Life in the Universe, 199–202.

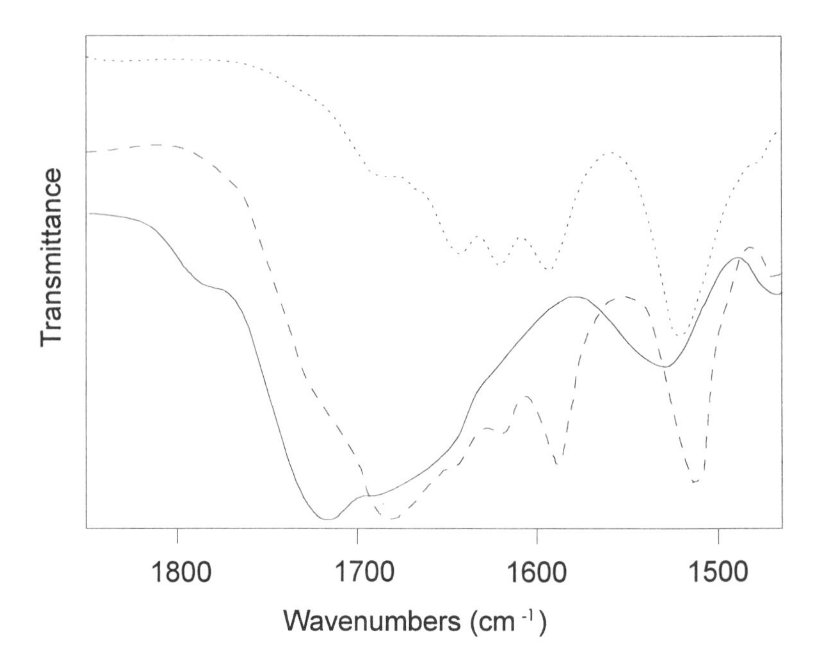

Fig. 1. Infrared spectra of the standard amino acid mixture at heating times of 0 (dotted line), 1.5 (dashed line) and 6.0 (solid line) hours. The band at 1680 cm^{-1} indicates amide groups, and that at 1715 cm^{-1} indicates imide groups.

To form the *microspheres*, the inside of the test-tube is scratched with a spatula and 0.2 g of the resulting powder are poured into a test-tube containing 4 ml of distilled water. The contents are mixed with a magnetic stirrer for 15 min and the suspension decanted for 30 min; the supernatant is apparently limpid.

If the presence of peptide bonds is checked in the suspension by means of some typical protein reactions (Gornall, Bradford, Amidoblack), contrasting results are obtained (Rizzotti and Mazzei, in press).

If a drop of the suspension is put on a microscope slide and observed at 125 magnifications, from tens to thousands of microspheres per field are noted. Their diameter is a few µm. No difference is found if heating is carried out in teflon test-tubes. Very few microspheres are visible before 1.5 h heating. Before decantation, larger carbon particles with vitreous aspect are visible and large spheroid objects are often noted along the boundary of the drop. If the suspension is boiled for times ranging between 1 min and 1 hour before decantation, a greater amount of material is brought into suspension. High speed centrifugation with an Eppendorf microfuge for 1 min is enough to sediment most of the microspheres (and all the larger suspended objects).

After the drop dries, the microspheres are no longer visible (even at 1000 magnifications). They become visible again if water or acetic acid are put on the slide: if acetic acid is used, the microspheres become permanently visible even after drying, although their aspect changes slightly. A similar but minor effect is also shown by ethanol. Instead, formalin, after drying, causes a sort of coagulation of the microspheres into larger roundish structures, with the aspect of filaments or clusters.

Under the polarizing microscope with crossed nicols, the preparation is always extinct, as expected from isotropic substances.

2. Variations to the standard procedure

If the amino acid mixture is heated in vials sealed under air or N_2, the resulting material is a viscous liquid, but it forms normal microspheres. If the vials are sealed under CO_2, they must be opened carefully because of the release of gas. The microspheres are more clearly visible than usual, less aggregated, and mixed with parallel fibers which are also visible with crossed nicols. Moreover, the IR spectra show a stronger amide signal and the Bradford reaction a much higher absorbance.

The most common ions in sea-water can affect the aggregation and stability of standard microspheres. In 1% NaCl, they show their usual aspect, but when the microscope preparation is dried, they align along straight segments resulting from the crystallization of the salt, and remain visible even after drying. In 1% $MgSO_4$ (but not in $CaCl_2$ 1%) they tend to associate among themselves, and in the dried preparation they collect in clusters.

The starting mixture can be changed in order to evaluate the effect of of hydroxy acids. When leucine is substituted with valine, and hydroxy acids (lactic, malic and thioglycolic) are added, microspheres are obtained which are apparently more stable, although not to drying. The same thing happens when leucine and alanine are substituted with glycine. The presence of lysine in place of leucine and alanine does not allow the formation of microspheres (Fox and Dose, 1977).

If the above three hydroxy acids are heated (in the absence of amino acids), the IR spectra of the resulting pale yellowish cream mainly show free groups and no (or very few) polyesters. Nevertheless, some polymerization does occur because normal microspheres form in 1% NaCl and larger ones in 1% $MgSO_4$ (whereas they do not form in water or 1% $CaCl_2$). Heating in teflon test-tubes does not affect the result.

3. Conclusions

Amide bonds form rapidly, but subsequently they decrease in favour of imide bonds, which mainly increase due to the condensation of aspartic acid (Fox, 1965). The formation of the microspheres seems to be correlated, to some extent, to the imide groups. As a consequence, the possible presence of peptide groups seems not to be so important in the microspheres. The presence of hydroxy acids, as in the mixtures found in meteorites and in the most popular organic synthesis simulation experiments (Miller, 1987), does not affect the formation of microspheres.

The scenario simulated in our experiments is that of the interface between dry rocky crust and atmosphere, with temperatures well above the freezing point of water, in the presence of amino acids and other organic molecules and in contact with aqueous solutions (rain, bodies of fresh water, seas). These environmental situations are not required to be constant, only transitory. The composition of the atmosphere and transitory hydrosphere do not seem to be critical at all, whereas the other current models of precellular aggregates require far more critical conditions.

From a neutral and rational standpoint, it is difficult to avoid the impression that, as soon as the above scenario appears on any planetary rocky surface, microspheres diffuse extensively. Thanks to their huge numbers and slight variability, microspheres may plausibly compete in several kinds of environment, undergo further chemical evolution owing to their selective chemical properties (Fox, 1965), and eventually switch to biological evolution.

Acknowledgements

The authors are particularly indebted to S. Fox for suggestions and discussions. They are also grateful for comments to L. J. Allamandola, V. Basiuk, M. Bressan, M. Greenberg, K. Harada, A. Pappelis.

References

Bellarmy L. J. (1958) *The infrared spectra of complex molecules*, Methuen, London.

Fox S. W. (1965) A theory of macromolecular and cellular origins. *Nature* **205**, 328-340.

Fox S. W. (1995) Thermal synthesis of amino acids and the origin of life. *Geochim. Cosmochim. Acta* **59**, 1213-1214.

Fox S. W., Dose K. (1977) *Molecular evolution and the origin of life*, Freeman, San Francisco.

Fox S. W., Harada K., Kendrick J. (1959) Production of spherules from synthetic proteinoid and hot water. *Science* **129**, 1221-1223.

Griffiths P. R., de Haseth J. A. (1986) *Fourier transform infrared spectrometry*, Wiley, New York.

Honda H., Maezawa M., Imai E., Matsuno K. (1993) Catalytic accretion of thermal heterocomplex molecules from amino acids in aqueous milieu. *Orig. Life* **23**, 177-183.

Miller S. L. (1987) Which organic compounds could have occurred on the prebiotic Earth? *Cold Spring Harbor Symp. Quant. Biol.* **LII**, 17-27.

Rizzotti M. (1994) Lipid vesicles: are they plausible primordial aggregates? *J. Biol. Physics* **20**, 155-162.

Rizzotti M. (1997) Did a precellular RNA world ever exist? in Cosmovici C. B., Bowyer S., Werthimer D. (eds.) *Astronomical and biochemical origins*, Compositori, Bologna, pp. 477-481.

Rizzotti M. , Mazzei P. (in press) Aggregati precellulari: variazioni sperimentali sulle microsfere. *Annales Biotheoretici* **2**.

Section 6.
Exobiology: General Perspectives

EXOBIOLOGY IN EARTH ORBIT

G. HORNECK

Deutsches Zentrum für Luft- und Raumfahrt, Institut für Luft- und Raumfahrtmedizin, Abteilung Strahlenbiologie
D-51170 Köln, Germany

Abstract

The primary goal of exobiological research is to reach a better understanding of the processes leading to the origin, evolution and distribution of life on Earth or elsewhere in the universe. In this endeavor, scientists from a wide variety of disciplines are involved, such as astronomy, planetary research, organic chemistry, paleontology and the various sub disciplines of biology including microbial ecology and molecular biology. Space technology plays an important part by offering the opportunity for exploring our solar system, for collecting extraterrestrial samples, and for utilizing the peculiar environment of space as a tool. Questions that have been tackled by exobiological experimentation in Earth orbit include the relevance of extraterrestrial organic molecules to the emergence of life on Earth or an any other planet, the role of solar UV radiation in evolutionary processes related to life, and the chances and limits for life to be transported from one body of our solar system to another or beyond. Studies on the responses of microorganisms to the environment of space have been carried out by use of balloons, rockets and spacecrafts. They have demonstrated the enormous lethal potential of extraterrestrial solar UV radiation. However, if shielded against the influx of solar UV, bacterial spores cope with the harsh environment of space for more than 5 years without any loss in viability. Future research e.g. on the International Space Station by use of the SEBA (Space Exposure Biological Assembly) facility will provide further clues to these important exobiological questions.

1. Introduction

Space technology offers the opportunities for *in situ* investigations in the environment of space. Utilizing space or special components of this environment as a tool, specific questions of exobiology can be tackled, such as of the role of stellar UV radiation in the evolution of potential precursors of life, of the role of UV radiation climate in prebiotic and biological evolution on Earth or any other celestial body, on the likelihood of interplanetary transfer and the limiting factors for Panspermia. The results of previous

J. Chela-Flores and F. Raulin (eds.),
Exobiology: Matter, Energy, and Information in the Origin and Evolution of Life in the Universe, 205–212.

exobiological experiments in Earth orbit and perspectives for future research will be discussed in this paper.

2. The Environment in Earth Orbit

The environment in Earth orbit is characterized by a high vacuum, an intense radiation climate of galactic and solar origin and extreme temperatures (Table 1). In free interplanetary space, pressures down to 10^{-14} Pa prevail. Within the vicinity of a body, the pressure may significantly increase due to outgassing. In a low Earth orbit, pressure reaches 10^{-4} to 10^{-6} Pa. The major constituents of this environment are molecular oxygen and nitrogen as well as highly reactive oxygen and nitrogen atoms. In the vicinity of a spacecraft, the pressure further increases, depending on the degree of outgassing. In the Shuttle cargo bay, a pressure of 3 x 10^{-5} Pa was measured, after bay doors were open for more than 100 h.

The radiation climate of our solar system is governed by components of galactic and solar origin. The galactic cosmic radiation, entering our solar system, is composed of protons (85 %), electrons, α-particles (14 %) and heavy ions (1 %) of charge Z>2, the so-called HZE particles. The solar particle radiation, emitted in solar wind and

TABLE. 1. Environment parameters in Earth orbit utilized in exobiological experiments

Space parameter	SL1[1)] D2	LDEF[2)]	EURECA[3)]	BIOPAN[4)]
Space vacuum				
Pressure (Pa)	≈10^{-4}	≈10^{-6}	≈10^{-5}	≈10^{-6}
Residual gas (cm^{-3})				
H	10^5	10^5	10^5	10^5
He	10^6	10^5	10^5	10^5
N	10^6	10^4	10^4	10^4
O	10^9	10^5	10^5	10^5
H_2O, organics	+	–	–	–
H_2O, N_2O, NO	+	–	+	–
Solar UV radiation				
Fluence (Jm^{-2})	≤10^3	≈10^9	≤3x10^8	≈10^7
Spectral range (nm)	>110	>50	>110	>170
	>170	>170	>170	
	>290		>280	
	>300		>295	
	212		220	
	223		230	
	230		260	
	260		290	
	290			
Cosmic radiation				
Dose (mGy)	0.7	4800	200-400	8200-7.5
Temperature (K)	243-290	264-302	295-318	253-288
Exposure time (d)	10	2107	336	15

1) Spacelab missions 1 and D2
2) Long Duration Exposure Facility
3) European Retrievable Carrier with the Exobiology Radiation Assembly ERA
4) BIOPAN on FOTON

during solar flares, is composed of 90-95 % protons, 5-10 % α-particles and a relatively small number of heavier ions. In the vicinity of the Earth, in the radiation belts, protons and electrons are trapped by the geomagnetic field.

The spectrum of solar electromagnetic radiation spans over several orders of magnitude, from short wavelength X-rays to radio frequencies. At the distance of the Earth (1 AU), solar irradiance amounts to 1360 W m^{-2}, the solar constant. Of this radiation, 45 % is attributed to the infrared fraction, 48 % to the visible fraction and only 7 % to the ultraviolet range. The extraterrestrial solar spectral UV irradiance has been measured during several space missions, such as Spacelab 1 and EURECA.

The temperature of a body in space, which is determined by the absorption and emission of energy, depends on its position towards the sun, and also on its surface, size and mass. In Earth orbit, the energy source are the solar radiation (1360 W m^{-2}), the Earth albedo (480 W m^{-2}) and terrestrial radiation (230 W m^{-2}). In Earth orbit, the temperature of a body can reach extreme values.

Exobiological experiments in Earth orbit have been performed in the last decade, utilizing the cargo bay of the Space Shuttle (Spacelab 1, D-2) or free flying satellites (LDEF, EURECA and FOTON). The parameters of space utilized during these missions are listed in Table 1 (also summarized in Horneck and Brack 1992).

3. Results of Exobiological Experiments in Earth Orbit

3.1. EXOBIOLOGY EXPERIMENTS SIMULATING THE CONDITIONS OF THE INTERSTELLAR MEDIUM

According to Greenberg's "cyclic evolutionary model for interstellar dust", interstellar grains consist of a silicate core and a mantle of relatively non-volatile higher molecular weight organic compounds (Greenberg 1983). The steps of their formation proceed in the following manner: (i) silicate particles of a mean radius of 0.05μm blown from cool stars into space form the seeds on which atoms and molecules of the interstellar gas condense; (ii) ultraviolet light from distant stars photolyzes the condensed mixture thereby creating new molecules and radicals; (iii) grain explosions replenish molecules in the gas phase. In this way, each grain accumulates a complex organic mantle that diffuses and/or regenerates during the passage of the grain through different regions of the interstellar medium. Each complete cycle takes about 10^8 years and the mean life time for an interstellar grain is about 5 x 10^9 years before it is consumed by star formation (Greenberg 1995).

During the EURECA mission (31 July 1992 - 1 July 1993) in the Exobiology Radiation Assembly ERA (Innocenti and Mesland 1995) such complex organic residues, produced in the laboratory, were exposed to the full spectrum of solar UV and vacuum-UV light in order to simulate the photolytic processes thought to occur during chemical evolution of interstellar grains (Greenberg 1995). With respect to the UV-radiation the exposure of this organic residue to solar UV for about 6 months provided a total fluence of 3.1 x 10^{22} hν cm^{-2}, which is equivalent to the photon dose received in about 10^7 years in the diffuse cloud medium of interstellar space. Hence the solar UV irradiated samples of the EURECA mission were expected to resemble the interstellar organic mantels which have undergone at least one complete evolutionary cycle. It is interesting to note, that the solar UV irradiated samples changed in color from yellow to brown. Their morphological structure appeared to be unchanged. The color change indicates an increase in the carbonization and associated greater visual absorption which results from photodissociation and depletion of oxygen, nitrogen and hydrogen from the samples (Greenberg 1995). Infrared spectra showed, that the 3.4 μm feature of the solar irradiated laboratory organic residues is very close to that of diffuse cloud interstellar dust, much closed than that of a wide variety of other suggested sources of organics. This precise match in the 3.4 μm infrared features provides experimental support to Greenberg's "cyclic evolutionary model for interstellar dust " as a most likely way of origin of the organic grain mantles on interstellar grains (Greenberg 1995).

3.2. EXOBIOLOGY EXPERIMENTS ON INTERPLANETARY TRANSFER OF LIFE

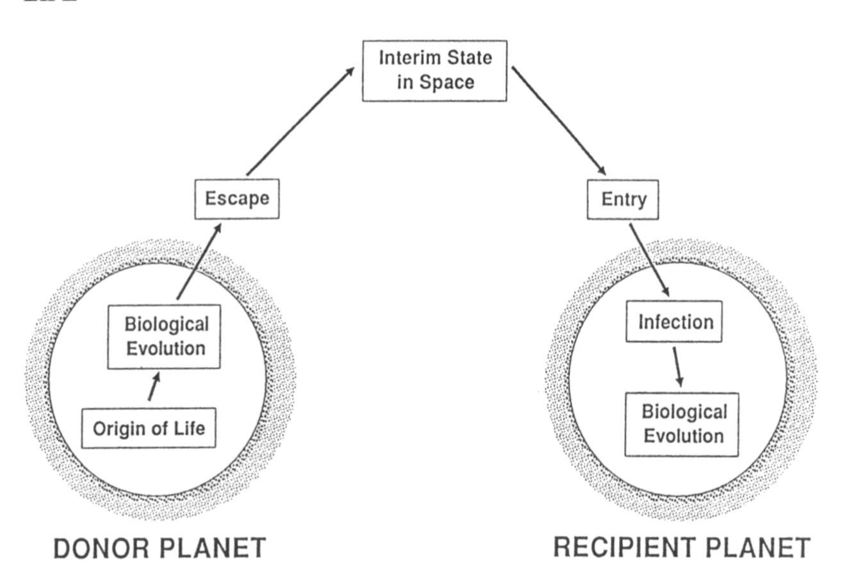

Figure. 1. Hypothetical scenario of an interplanetary transfer of life.

Although it will be difficult to prove that life could be transported through our solar system, estimates of the chances for the different steps of the process to occur can be obtained from measurements, calculations, and experiments. The fundamental phases and requirements of Panspermia are: (i) escape process, i.e. removal to space of biological material which has survived being lifted from the surface to high altitudes; (ii) interim state in space, i.e. survival of the biological material over timescales comparable with the interplanetary or interstellar passage; (iii) entry process, i.e. non-destructive deposition of the biological material on another planet (Figure 1). Recent discoveries, such as the detection of meteorites of Martian origin, the high survival of bacterial spores in space over extended periods of time, the high UV-resistance of micro-organisms at the low temperatures of deep space, the biochemical evidence of a common ancestor for all life forms on Earth, and the likelihood of artificial or directed transport by probes sent to other planets have given new support to the idea of Panspermia (summarized in Horneck and Brack 1992).

In order to study the survival of resistant microbial forms in the upper atmosphere or in free space, microbial samples were exposed *in-situ* by use of balloons, rockets or spacecrafts and their responses were investigated after recovery (reviewed in Horneck and Brack 1992, Horneck 1995). For this purpose, several facilities were developed, such as the exposure device on Gemini, MEED on Apollo, ES029 on SL1, ERA on EURECA, UV-RAD on Spacelab D2 and BIOPAN on FOTON. These investigations were supported by studies in the laboratory in which certain parameters of space (high and ultrahigh vacuum, extreme temperature, UV-radiation of different wavelengths, ionizing radiation) were simulated. The microbial responses (physiological, genetic and biochemical changes) to selected factors applied separately or in combination were determined.

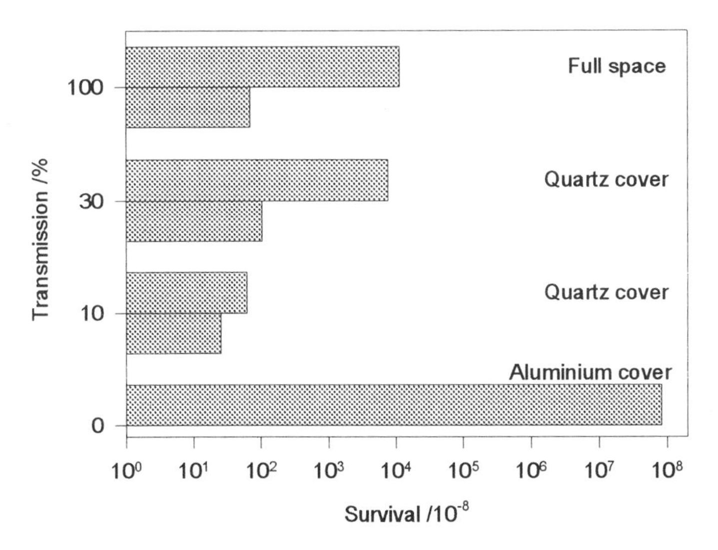

Figure. 2. Survival of spores of *Bacillus subtilis* after exposure to the space environment for nearly 6 years during the LDEF mission.

Early experiments in space have demonstrated that certain microorganisms survive short periods of exposure in space - several hours or days -, provided they are shielded against the intense solar ultraviolet (UV) radiation. To date, the maximum exposure time of microorganisms to the space environment has been the 2107 day duration of the LDEF mission (Horneck et al. 1994). Up to 70 % of *Bacillus subtilis* spores survive extended periods of time exposed to the vacuum of space, even if exposed as a monolayer (Figure 2) (Horneck 1993, Horneck et al. 1994). Nevertheless, even in the survivors, the genetic material is affected, indicated by an increased mutation rate, delayed germination, cross linking of DNA and protein, and the requirement of cellular repair processes to restore viability after vacuum exposure, respectively.

Solar UV-radiation has been found to be the most deleterious factor of space as tested with dried preparations of viruses, bacterial or fungal spores (Horneck 1993, Horneck et al. 1994). The incidence of the full spectrum of solar UV-light (> 170 nm) kills 90% of *B. subtilis* spores within seconds. To reach the same effect on Earth, the exposure to sunlight takes approximately 1000 times longer (Figure 3) (Horneck and Brack 1992). This difference is attributed to the ozone layer that protects the biosphere from the most harmful fraction of solar UV-radiation (< 295 nm). Action spectra of the solar photons (160 nm < λ < 320 nm) for killing bacteriophage T1 or *B. subtilis* spores (Horneck 1993, Horneck et al. 1995) correlate closely with the absorption spectrum of DNA. This indicates that the DNA is the critical target for lethality. If spores of *B. subtilis* are simultaneously exposed to solar UV-radiation and space vacuum they respond with an increased sensitivity, to a broad spectrum of solar UV-light (> 170 nm) as well as to selected wavelengths.

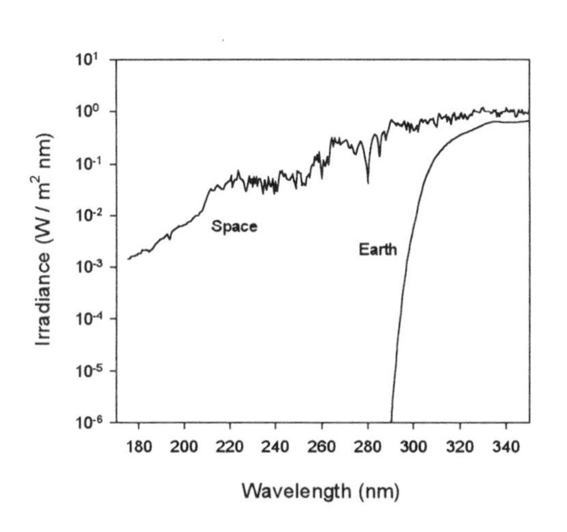

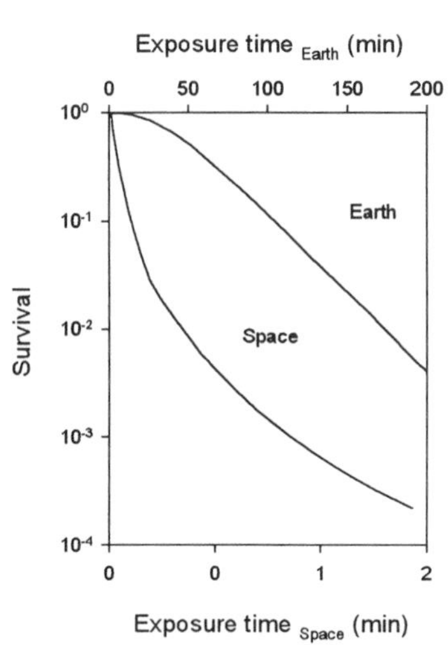

Figure. 3. Survival of spores of *Bacillus subtilis* after exposure to extraterrestrial solar UV radiation (>170 nm) (Space) or to global UV radiation (>290 nm) at the surface of the Earth (Earth).

During the major part of a hypothetical journey through space, if shielded from solar thermal radiation, micro-organisms will be confronted with the 4 K cold emptiness of space. Under these very cold conditions, thermodynamic and chemical reactions are nearly frozen. The photobiological response to solar UV-radiation may be completely different from room temperature conditions. Only the latter response (at room temperature) so far has been tested in space. Laboratory experiments under simulated interstellar medium conditions point to a remarkably less damaging effect of UV-radiation at these low temperatures. Treating *B. subtilis* spores with 3 simulated parameters of space simultaneously, i.e. UV-radiation (> 110 nm), vacuum, and low temperature (10 K), results in an unexpectedly high survival, even at very high UV fluences. At this low temperature range, the inactivation cross sections obtained are up to 2-3 orders of magnitude lower than at room temperature (Weber and Greenberg

1985). The temperature profile of bacterial spore UV-sensitivity shows a maximum at 190 K. From these data it has been estimated that, in the most general environment in space, spores may survive for hundreds of years (Weber and Greenberg 1985).

Among the ionizing components of radiation in space, the heavy primaries, the so-called HZE particles are the most effective species. To understand the ways by which particles of cosmic radiation interact with biological systems, methods have been developed to localize precisely the trajectory of an HZE particle relative to the biological object and to correlate the physical data of the particle relative to the observed biological effects along its path. By use of visual track detectors that were sandwiched between layers of biological objects in a resting state - this concept is known as the "Biostack" (Bücker and Horneck 1975) - in a variety of test systems including viruses, bacterial spores, plant seeds or shrimp cysts, injuries such as somatic mutations in plant seeds, development disturbances and malformations in insect and salt shrimp embryos, or inactivation in bacterial spores were traced back to the traversal of a single HZE particle (reviewed in Horneck 1992). Such HZE particles of cosmic radiation are conjectured to set the ultimate limit on the survival of spores in space since they penetrate even high shielding. The maximum time of a spore to escape a hit by a HZE particle (e.g. iron of LET >100 keV/μm), has been estimated to amount to 10^5-10^6 years.

To travel from one planet of our solar system to another, e.g. from Mars to Earth by random motion, a mean time of several hundred thousands to millions of years has been estimated for boulder-sized rocks, but only 2 months for microscopic particles. Weber and Greenberg (1985) have drawn the following picture for an interstellar transport: The clouds of gaseous and particulate matter between the stars move in a random fashion at speeds of about 10 km/s. If a bacterial spore is captured in such a cloud, it will be swept along with the gas. Given the distance between neighboring stars of 0.1-1 pc, this corresponds to a passage time of 10^5-10^6 years. Thus, for interstellar infection, micro-organisms would have to survive at least for 10^5-10^6 years in such a cloud.

4. Future Opportunities for Exobiology Experiments in Earth Orbit

For future research in the field of exobiology in Earth orbit, the International Space Station will provide a Space Exposure Biology Assembly (SEBA) which will be equipped with a sun pointing exposure tray for studies in the field of photoprocessing and photobiology. In addition, SEBA may allow the accommodation of a device for non-destructive dust collection and for atomic oxygen processing. Herewith, SEBA will provide the opportunities for further studying the relevance of extraterrestrial organic molecules to the emergence of life on Earth or any other planet, the role of solar UV radiation in evolutionary processes, related to life, the chances and limits for life to be transported from one body of our solar system to another or beyond, and of exobiological phenomena in aerosols and clouds.

5. References

1. Bücker, H., Horneck, G. (1975) Studies on the effects of cosmic HZE-particles on different biological systems in the Biostack I and II flown on board of Apollo 16 and 17, In: O.F. Nygaard, H.J. Adler, W.K. Sinclair (eds.), Radiation Research, Academic Press, New York pp. 1138-1151
2. Greenberg, J. M. (1983) The largest molecules in space: interstellar dust. In: C. Ponnamperuma (ed.) Cosmochemistry and the Origin of Life, D. Reidel, Dordrecht, pp. 71-112
3. Greenberg, J. M. (1995) Approaching the interstellar grain organic refractory component, The Astrophys. J., 455, L177
4. Horneck, G. (1992) Radiobiological experiments in space: a review. Nucl. Tracks Radiat. Meas., 20, 185-205
5. Horneck, G. (1993) Responses of *Bacillus subtilis* spores to space environment: results from experiments in space, Origins of Life, 23, 37-52
6. Horneck, G. (1995) Exobiology, the study of the origin, evolution and distribution of life within the context of cosmic evolution: a review, Planet. Space Sci., 43, 189-217
7. Horneck, G., Brack, A. (1992) Study of the origin, evolution and distribution of life with emphasis on exobiology experiments in Earth orbit. In: S.L.. Bonting (ed.), Advances in Space Biology and Medicine 2, JAI Press, Greenwich, CT, pp. 229-262
8. Horneck, G., Bücker, H., Reitz, G. (1994) Long-term survival of bacterial spores in space, Adv. Space Res., 14, (10)41-(10)45
9. Horneck, G., Eschweiler, U., Reitz, G., Wehner, J., Willimek, R., Strauch, K. (1995) Biological responses to space: results of the experiment „Exobiological Unit“ of ERA on EURECA I, Adv. Space Res., 16(8), 105-118
10. Innocenti, L., Mesland, D. (eds) (1995) EURECA Scientific Results, Adv. Space Res., 16, No. 18, Pergamon, Oxford
11. Weber, P., Greenberg, J.M. (1985) Can Spores Survive in Interstellar Space? Nature 316, 403-407

PRODUCTION OF ORGANIC COMPOUNDS IN INTERSTELLAR SPACE

- Design for Exobiology Experiments on the Earth Orbit-

Kensei Kobayashi, Takashi Kasamatsu and Takeo Kaneko
Department of Chemistry and Biotechnology
Yokohama National University
Hodogaya-ku, Yokohama 240-8501, Japan

Takeshi Saito
Institute for Cosmic Ray Research, University of Tokyo
Midoricho, Tanashi, Tokyo 188, Japan

1. Introduction

For the generation of life, bioorganic compounds are essential. It has been supposed that most of bioorganic compounds essential to the first life were formed in the planets. Various kinds of simulation experiments have been performed to examine how these compounds were abiotically synthesized in primitive planetary environments. In these days, however, the primitive earth atmosphere is thought to be "mildly-reduced," containing carbon monoxide, carbon dioxide, nitrogen and water (Kasting, 1990). If the primitive atmosphere was "mildly-reduced" type, formation of amino acids and other bioorganic compounds essential for origin of life were quite difficult (Schlesinger and Miller, 1983; Chang and Bur-Nun, 1983).

It is known that various kinds of organic compounds exist in space, e.g., in interstellar molecular clouds, comets and carbonaceous chondrites. When comets and meteorites fall to planets, it is possible that they can supply organic compounds to the planets. It is supposed that organic compounds in comets were formed in their precursor bodies, interstellar dust particles (ISDs). ISDs in dense clouds were covered with ice mantles. The composition of ISD ice mantles was not known, but it was estimated that they contain such volatiles as water, carbon monoxide, methane and ammonia from the analytical results of coma of Halley's comet (Yamamoto, 1991). In order to simulate reactions in ISDs, ice mixtures of carbon monoxide (or methane), ammonia and water made in a cryostat were irradiated with high energy protons or ultraviolet light: Organic compounds such as nitriles were detected in the products by

J. Chela-Flores and F. Raulin (eds.),
Exobiology: Matter, Energy, and Information in the Origin and Evolution of Life in the Universe, 213–216.

infrared spectroscopy, but bioorganic compounds could not be detected in them (Moore et al., 1983; Allamandola et al., 1988). Briggs *et al.* (1992) suggested that glycine was formed after ultraviolet irradiation of ice mixture of carbon monoxide, methane, ammonia and water, but the experiments were not conducted quantitatively.

In the present study, formation of organic compounds in simulated ISD ice mantles were quantitatively examined. Then we propose a new exobiology experiment in space station in order to confirm our hypothesis that bioorganic compounds can be abiotically formed from very simple molecules in the actual space environment.

2. Abiotic Synthesis of Amino Acids in Simulated ISDs: Laboratory Experiments

A gas mixture of carbon monoxide (or methane), ammonia and water vapor (1: 1: 1 in volume) was flown against the copper block located in the cryostat evacuated with a turbo molecular pump to make the simulated ISD ices. The ice was irradiated with high energy protons from a van de Graaff accelerator (Tokyo Institute of Technology): Total energy deposited to the ice during irradiation was *ca.* 1.1×10^{22} eV. Blank runs were also performed in the same manner, just without any irradiation.

After irradiation, the temperature of the metal block was naturally raised to a room temperature, and residual matter on the metal block was recovered with water, which was subjected to amino acid analysis before and after acid hydrolysis.

Amino acids such as glycine and alanine were detected in the proton irradiation products from ice mixtures of carbon monoxide (or methane), ammonia and water after acid-hydrolysis, while only trace amounts of amino acids were detected in the products before acid-hydrolysis. This results suggest the possible formation of "amino acid precursors," not free amino acids, in interstellar dust grains by cosmic radiation (Kasamatsu *et al.*, 1997a).

Methanol was also used as a carbon source in ISD-simulation experiments, since (i) it is suggested to be in comets and ISDs, and (ii) ice mixtures of methanol, ammonia and water can be easily and quantitatively made. The mixture including methanol was sealed in a glass tubes, and was irradiated with 35 MeV protons from a Sector Focusing (SF) cyclotron (Institute for Nuclear Study (INS), the University of Tokyo) at *ca.* 0.1 μA, while the temperature of the ice was controlled at 77 K (solid), 298 K (liquid) or 373 K (gas) (See Fig. 1). After irradiation, a part of the products was acid-hydrolyzed and analyzed by high performance liquid chromatography.

The G-value of glycine when methanol was used as a carbon source in gaseous phase was 0.023, which was almost the same as that when carbon monoxide or methane was used in gaseous phase. The G-value when the mixture with methanol was used in solid and liquid phase was 0.00057 and 0.00012, individually (Kasamatsu *et al.*, 1997b).

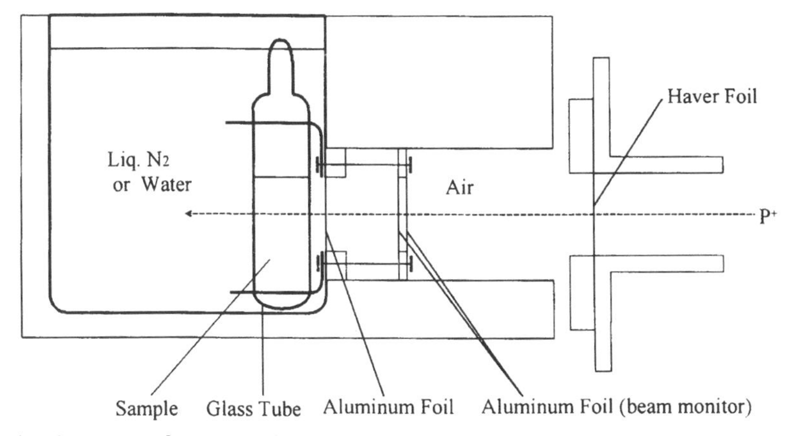

Fig. 1. Apparatus for proton irradiation of the mixture of CH_3OH, NH_3 and H_2O

3. Estimation of Amino Acid Formation Rate in ISDs

Using the present results, we estimate the formation rate of amino acids in ISD environment. Judging from the result of laboratory experiments using methanol, it seems that the G-value of glycine in solid phase is about 1/40 of that in gas phase, when the starting materials with the same composition were used.

Though we do not know the exact composition of ISD ice mantles, we presume that their major components are water, carbon monoxide and ammonia (100: 10: 1 in molar ratio) from the reported composition of coma of Comet Halley (Yamamoto, 1991). Since the G-value of glycine in a gas mixture of the same composition experimentally obtained was 0.00023 (Kasamatsu *et al.*, 1997b), the G-value in the ice mantle of ISDs are estimated as 0.00023/40 (*i. e.,* 6×10^{-6}). Energy flux of high energy particles (E > 100 MeV) in dense clouds is *ca.* 5.4 keV cm^{-2} s^{-1} (Morfill *et al.*, 1976). When we assume that an average residence time of ISDs (major diameter is 0.1 μm) in dense clouds is 10 million years, then we can estimate the concentration of glycine precursors in ISDs as ca. 20 nmol cm^{-3}. Since it is probable that bioorganic compounds like amino acid precursors in ISDs are preserved in comets after aggregation of ISDs in proto solar system, these bioorganic compounds delivered to the primitive earth by comets or their fragments (interplanetary dusts) are quite important sources for building blocks of the first life on earth.

4. Exobiology Experiments in Earth Orbit

Since the idea that bioorganic compounds can be formed in interstellar space obtained through laboratory experiments is striking, it should be confirmed in actual space conditions, such as in a space station in earth orbit. We are designing an apparatus of such exobiology experiments in earth orbit (EEEO).

Candidates for the starting materials of EEEO are (i) a frozen mixture of carbon monoxide (or methane), ammonia and water, and (ii) a frozen mixture of methanol, ammonia and water. The former is a better simulation of ISDs, but it is more difficult to keep frozen in space for years. Here we chose the latter as starting materials since the ice mixture of them can be made more easily.

The experiment is quite simple: The starting mixtures kept frozen are exposed to cosmic radiation in an exposure facility of a space station. If a mixture of water, methanol and ammonia (9:4:3 in molar ratio) of the size of 4 cm$\times$ 4 cm$\times$1 cm is exposed to cosmic radiation. The G-value of glycine in the frozen mixture was 5×10^{-4}, and dose rate of cosmic radiation in earth orbit was reported to be *ca.* 3 mGy d^{-1} (Benton *et al.*, 1991). Then we can expect the formation of 16 pmol of glycine in the mixture after standing there for 3 years: The amount is more than detection limit.

In 2001 an exposure facility of Japanese Experimental Module (JEM) is planned to be launched, which will give a quite promising opportunity to performe EEEO. Further evaluation of the present proposal, such as a low temperature technology to keep the mixture frozen, is in progress.

5. References

Allamandola, L. J., Sandford, S. A. and Valero, G. J. (1988) Photochemical and thermal evolution of interstellar/ precometary ice analogs, Icarus **76**, 225-252.

Bar-Nun, A. and Chang, S. (1983) Photochemical reactions of water and carbon monoxide in earth's primitive atmosphere, J. Geophys. Res. **88**, 6662-6672

Benton, E. V., Frank, A. L., Benton, E. R., Csige, I., Parnell, T. A. and Watts, J. W. Jr. (1991) Radiation exposure of LDEF: Initial results, *NASA-Langley Resarch Center report, NASA CP* **3134** (Part 1), 325-337.

Briggs, R., Ertem, G., Ferris, J. P., Greenberg, J. M., McCain, P. J., Mendoza- Gomez, C. X., and Schutte, W. (1992) Comet Halley as an aggregate of interstellar dust and further evidence for the photochemical formation of organics in the interstellar medium, *Origins Life Evol. Biosphere* **22**, 287-307.

Chyba, C. F., and Sagan, C. (1992) Endogenous production, exogenous delivery and impact-shock synthesis of organic molecules: an inventory for the origins of life, *Nature* **355**, 125-132.

Kasamatsu, T., Kaneko, T., Saito, T., and Kobayashi, K. (1997a) Formation of organic compoundsin simulated interstellar media with high energy particles, *Bull. Chem. Soc. Jpn.* **70**, 1021-1026.

Kasamatsu, T., Kaneko, T., Saito, T., and Kobayashi, K. (1997b) Quantitative estimation of organic formation in interstellar dust environments, *Chikyukagaku* **31**, 181-191.

Kasting, J. F. (1990) Bolide impact and the oxidation state of carbon in the Earth's early atmosphere, *Origins Life Evol. Biosphere* **20**, 199-231.

M. H. Moore, B. Donn, R. Khanna and M. F. A'Hearn (1983) Studies of proton irradiated cometary-type ice mixtures,*Icarus* **54**, 388-405.

Morfill, G. E., Völk, H. J. and Lee, M. A. (1976) Cosmic rays — Astronomy with high energy particles, J. Geophys. Res. **81**, 5841-5852 (1976).

Schlesinger, G. and Miller, S.L. (1983) Prebiotic synthesis in atmospheres containing CH_4, CO, and CO_2: I.Amino acids. J. Mol. Evol. **19**, 376-382.

Yamamoto, T. (1991) Chemical theories on the origin of comets, R. L. Newburn, Jr., M. Neugebauer and

Section 7.
Exobiology on Mars and Europa

THE SEARCH FOR EXTRATERRESTRIAL BIOCHEMISTRY ON MARS AND EUROPA

CHRISTOPHER P. MCKAY
Space Science Division
NASA Ames Research Center
Moffett Field, CA 94035 USA

Abstract.
There is, from a biochemical perspective, only one type of life on Earth. To understand life as a general phenomenon it is essential to consider other types of biochemistry. In this context, fossils are not sufficient, because they do not retain information about the biochemistry of an organism. Based on current knowledge, two worlds in our solar system hold the prospect for finding a biochemical system different from the Earth; Mars and Europa. On Mars, intact organisms may be preserved in the ancient permafrost, while on Europa organisms from a possible ocean may be present near cracks in the icy surface.

1. Introduction

Life on Earth has evolved into a vast array of sizes, shapes, and environments. Despite that diversity, all life on Earth is based on the same biochemistry. All life on Earth originates from the same basic genetic code, the same basic set of proteins, and a handful of other molecules. Lifeforms on Earth are all variations of a single underlying biochemical theme.

In considering life beyond the Earth it is appropriate to ask what parts of terrestrial biochemistry may be expected on other worlds. For example, could life be based on proteins composed of amino acids of the D-form rather than the L-form used on Earth. There is no reason, in principle, that this possibility should be precluded, as laboratory experiments have shown enzyme action in laboratory biochemical systems composed of D-proteins only (Milton et al. 1992).

J. Chela-Flores and F. Raulin (eds.),
Exobiology: Matter, Energy, and Information in the Origin and Evolution of Life in the Universe, 219–227.

The only way to develop an understanding of the biochemistry of life in a general sense is to study several different types of biochemical life forms. Thus, the search for life beyond the Earth is not just a search for fossil evidence of life, but must include a search for the biochemical materials of life; whether dead or alive.

There are several ways in which biochemicals can be differentiated from abiotic organic material. The distinction is important because the solar system and interstellar space are rich in abiotic organic material. Biochemicals are, by definition, organic materials produced by lifeforms. Organic material containing unusually high concentrations of only certain organics would be an indication that some life form formed these products. An example would be soil samples containing a high concentration of the 20 amino acids used by life on Earth. A second example, related to the first, in which biochemicals might differ from abiotically produced organic material is in the prevalence of certain isomers. Examples are the use of L-amino acids and D-sugars by terrestrial life. A third signature of life in organic material on Earth is the presence of an isotopic shift in the carbon 12 to 13 ratio (Schidlowski 1988).

The quintessential requirement for life is liquid water and for this reason a search for life beyond the Earth is, for practical purposes, a search for liquid water (McKay 1991). In our own solar system Mars shows convincing evidence of liquid surface water, albeit billions of years ago. This evidence of fluvial flow is the main motivation for searching for evidence of past life on Mars (McKay 1997). Galileo Orbiter images of Jupiter's moon Europa show clear signs of activity in the surface ice-cover. This observation has led researchers to speculate that there may be an ocean of liquid water underneath a relatively thin ice layer (Carr et al. 1998). In this paper we consider how we might acquire biochemical samples from these two worlds.

2. Mars

There is considerable geological evidence that at some time in the past there was liquid water flowing as stable meandering rivers on the surface of Mars (Carr 1996). This evidence exists primarily in the form of valleys present in the ancient cratered terrain, and the most recent images from Mars Global Surveyor support the theory that these features formed from erosion by liquid water. It is probable that Mars once had a thicker atmosphere and warmer surface conditions than at the present since liquid water is marginally stable under the present thin atmosphere (McKay 1997). From a biological point of view, however, the evidence of liquid water, *per se*, is the motivating factor in the search for life. The specifics of pressure and temperature are secondary as long as conditions allowed for liquid water to

Table 1. Potential Sources of Biochemistry on Mars

- Dormant spores in surface materials.
- Living systems in the subsurface.
- Organisms preserved in salt or amber.
- Frozen organisms in the permafrost.

be present at the surface.

Much effort has been directed toward developing a rationale and a strategy for searching for fossils as evidence of past life on Mars (McKay 1997, Walter and DesMarais 1993). However, as discussed above, a search for biochemical material — dead or alive — must also be considered. Table 1 lists four possible sources of biochemical material on Mars. First, and most unlikely, there may be dormant spores present in the surface materials on Mars. It is clear that the present environmental conditions are too severe for life to grow, but the low temperature and low water activity may allow for hardened spores to survive over a long period of time. The presence of spores in the soil could be a possible explanation for the Viking Labeled Release results (see Zent and McKay 1994) although no simulations to test this idea have been conducted. Another potential source of living material is the deep subsurface. Boston et al. (1992) developed the arguments for the presence of liquid water and a source of chemical energy in the martian subsurface. However, they did not discuss the seemingly impossible ecological problems of spreading life from one martian subsurface niche to another.

Biochemical samples — intact and often viable organisms — can be preserved in dehydrating materials, such as salt and amber. On Earth, there are bacteria present in salt deposits that are hundreds of millions of years old. However, water droplets can move readily through salt and hence it is not possible to confirm that the microorganisms have been present in the salt since its deposition. Preservation in amber provides much more reliable data. Viable bacteria have been recovered from amber that is 25 million years old (Cano and Borucki 1995) and other studies suggest amber up to 90 million years old contains viable microorganisms. Because amber is a product of higher plants it is, however, not likely to be present on Mars.

The most promising source of biochemistry on Mars is probably the permafrost. Organisms frozen in the permafrost would be well preserved. On Earth, Siberian permafrost has shown that microorganisms remain viable after 3.5 million years, at temperatures of $-10°C$ (Gilichinsky et al., 1992). On Mars, the necessary preservation time is 3 to 4 billion years but the temperatures are also much lower, $< -70°C$, so thermal decay would

Table 2. Factors that accelerate oxidation on Mars (McKay 1996)

Process, compared to Earth	Note
No Tectonic Recycling of Organic Sediments	
More Rapid Escape to Space of H_2	Gravity 0.38 $\times$Earth
Less Volcanic Activity	10^{-2} – 10^{-3} $\times$Earth
Smaller Ocean	0.01 – 0.3 $\times$Earth

not limit the long-term survival of life in permafrost. However, low level radioactivity from U, Th, and K in the martian soil would produce about one million rads in 5 million years, and limit the survival of even radiation tolerant microorganisms to a few tens of millions of years. While radioactivity might inactivate microorganisms it would not significantly alter their biochemical makeup. Their remains could therefore be analyzed and compared to terrestrial biochemistry.

There are several factors that may have accelerated the oxidation of early Mars relative to the early Earth (McKay 1996). These factors are listed in Table 2.

If Mars experienced a more rapid oxidation than the Earth, we might expect to find evidence of eukaryotes, and perhaps even multicellular life in the martian fossil record. Presumably O_2 buildup on Mars could have been 1000x faster than on Earth and as a result Mars may have experienced the equivalent of the Cambrian, 3.5 Gyr ago. If this scenario is true, life could have been aerobic and multicellular on Mars before the environment deteriorated. Thus, a search for fossils and biochemical evidence on Mars might include eukaryotes, and even multicellular life (McKay 1996).

3. Europa

Europa, one of the Galilean Moons of Jupiter, may hold liquid water and is therefore one of the most promising objects to search for life in the outer solar system. In addition, it provides an example of a new class of habitable worlds — tidally heated moons around giant planets — that may be directly relevant to the recently discovered extrasolar giant planets.

Theoretical considerations of tidal heating, and direct observations of the surface suggest the possibility that there is an ocean underneath a relatively thin ice cover on Europa (Carr et al., 1998).

If Europa were in a perfectly circular, synchronous orbit it would experience negligible tidal heating from Jupiter. However, Europa, Io, and

Ganymede are in a LaPlace resonance that forces non-zero eccentricities (Cassen et al. 1982). Physically, the energy dissipated by tidal forces on Europa comes from the orbital energy of the satellite and the rotational energy of Jupiter. This tidal heating together with the decay of radioactive elements, and the release of energy from earlier periods represents the heat sources that are warming the interior of Europa (e.g., Squyres et al. 1983).

The low density of Europa leads to the theory that there might be a significant layer of ice near the surface. Earth's moon, which is similar in size to Europa, and Io, both have densities of about 3.5 g cm^{-3} compared to Europa's 3.0 g cm^{-3}. If Europa has differentiated into a rocky core of silicates with Io-like densities, then this would imply an over 100 km thick surface layer of water ice. If the heat flow is large enough, the lower layers of the ice would melt forming an ocean. If instead, Europa's interior is composed of hydrated silicates — the water is distributed within the body — then the surface shell may be relatively thin and completely frozen.

The notion of a subice ocean on Europa has been extensively discussed in the scientific literature for more than 25 years (Lewis 1971, Fanale et al. 1977, Smith et al. 1979, Cassen et al. 1979, Squyres et al. 1983, Ross and Shubert 1987, Ojakangas and Stevenson 1989, McDonald et al., 1998), as well as in the popular literature (Clarke 1982). In addition, some models for the origins of life on Earth suggest a submarine origin at hydrothermal vents (Corliss et al. 1979) and others have invoked impact melting of a frozen ocean (Bada et al. 1994). Within this framework there have been speculations about life in an ocean on Europa (Reynolds et al. 1983).

The dominant source of biological energy on the Earth is sunlight. For Earth as well as Europa, available sunlight exceeds geothermal heating by several orders of magnitude. It is important therefore to consider a possible photosynthetically driven ecology on Europa. Reynolds et al. (1983) considered the possibility that cracks in a thick ice cover would bring ocean water close enough to the surface that photosynthesis could occur. As a terrestrial analog they used algae and diatoms living under sea and lake ice in Antarctica.

In addition to bulk water near the surface, it is possible that liquid water is generated by internal melting driven by the solid state greenhouse effect (Brown and Matson, 1987). Recent modeling by Urquhart and Jakosky (1996) indicated that this effect is much smaller than previous studies had indicated. However, the Urquhart and Jakosky (1996) model used a simple uniform Beer's law parameterization for the penetration of sunlight into the ice. Recent work on sunlight and thermal measurements of thick lake ice on Antarctic dry valley lakes (McKay et al., 1994; Fritsen et al., 1998) shows that the properties of an ice cover are not uniform and that scattering effects cannot be neglected. In the Antarctic, the surface of the lake ice remains frozen while solar heating causes melting 1 meter below the surface and provides a habitat for algae and diatoms.

Life at deep sea vents on the Earth is often described as being independent of the surface biosphere. However, the life forms observed in abundance at these vents (crabs, worms, etc) are part of an ecosystem that is based on chemoautotrophic bacteria consuming H_2S outgassing from the vent and that becomes oxidized with O_2 dissolved in the ambient seawater. This O_2 is originally derived from photosynthesis on the Earth's surface. On Europa, H_2S may be emanating from geothermal vents but there is not likely to be a source of O_2 in the ocean water. However, recent work (Kinkle and Radu, personal communication) indicates that the H_2S reaction with CO_2 can, in the presence of iron, form the basis for a chemoautotrophic ecosystem. CO_2 is likely to be present in the ocean of Europa (Stevens and McKinley 1996).

We can also consider a similar but simpler chemical pathway based on H_2 and CO_2. This combination of gases is used by methanogenic bacteria. Stevens and McKinley (1995) have found an isolated microbial ecosystem deep beneath the surface of the Earth — profoundly independent of the surface biosphere — that is based on this reaction. The H_2 is released by chemical oxidation of ferrous iron in local basaltic rocks. Similar ecosystems have been proposed for Mars (Boston et al. 1992) and are the most relevant existing ecosystem for analogy to Europa. Recent work on the kinetics and rate limits in the consumption of H_2 by methanogens (Kral et al. 1998) shows that organisms can take up H_2 at ppm levels.

If there is an ocean of liquid water on Europa and it contains life then we would like to obtain samples of that life for biochemical studies. One way to acquire samples would be to penetrate the ice cover and directly sample the ocean. However, this would be an expensive and technologically challenging task. An alternative source of biochemistry on Europa might be the dark material associated with cracks in the surface ice. A flyby sampler could excavate some of this material and collect it for return to Earth. This method could be the lowest cost and most expeditious way to acquire samples of alien biochemistry.

Even if there is life on Europa it is unlikely that environments rich in O_2 could have arisen. In order for O_2 to build up there must be a way to lose H_2 to space, or sequester reduced organic material beneath the surface. Neither mechanism is probable in a global ocean, under a thick ice cover. Thus, it is probable that life on Europa, if it is present, would be anaerobic only.

4. Conclusions

To truly expand our understanding of life as a phenomenon in the universe we must find other examples of biochemistry — mere fossils are not enough. By comparing alien and terrestrial biochemistry we can begin to establish what is universal in biochemistry and what is particular to life on Earth. In our own solar system the most promising sites for a near-term search for the biochemically intact remains of life forms are 1) in the subsurface permafrost on Mars, and 2) in the dark material near cracks on the surface ice of Europa.

5. References

Bada, J.L., C. Bigham, and S.L. Miller (1994) Impact melting of frozen oceans on the early Earth: Implications for the origin of life. *Proc. Natl. Acad. Sci.* **91**, 1248-1250.

Boston, P.J., M.V. Ivanov, and C.P. McKay (1992) On the possibility of chemosynthetic ecosystems in subsurface habitats on Mars. *Icarus* **95**, 300-308.

Brown, R.H., and D.L. Matson (1987) Thermal effects of insolation propagation into the regoliths of airless bodies. *Icarus* **72**, 84-94.

Cano R.J., and M.K. Borucki (1995) Revival and identification of bacterial spores in 25- to 40-million-year-old Dominican amber. *Science* **268**, 1060-1064.

Carr, M.H. (1996) *Water on Mars.* Oxford University Press, New York.

Carr, M.H., M.J.S. Belton, C. R. Chapman, M.E. Davies, P. Geissler, R. Greenberg, A.S. McEwen, B.R. Tufts, R. Greeley, R. Sullivan, J.W. Head, R.T. Pappalardo, K.P. Klaasen, T.V. Johnson, J. Kaufman, D. Senske, J. Moore, G. Neukum, G. Schubert, J.A. Burns, P. Thomas, and J. Veverka (1998) Evidence for a subsurface ocean on Europa. *Nature* **391**, 363-365.

Cassen, P., R.T. Reynolds, and S.J. Peale (1979) Is there liquid water on Europa? *Geophys. Res. Lett.* **6**, 731-734.

Cassen, P.M., S.J. Peale, and R.T. Reynolds (1982) Structure and thermal evolution of the Galilean satellites. In *Satellites of Jupiter* (D. Morrison, Ed.), 93-128. Univ. of Arizona Press, Tucson.

Clarke, A.C. (1982) *2010: Odyssey Two*, Ballantine, New York.

Corliss, J.B., J. Dymond, L.I. Gordon, J.M. Edmond, R.P. von Herzen, R.D. Ballard, K. Green, D. Williams, A. Bainbridge, K. Crane, and T.H. van Andel (1979) Submarine thermal springs on the Galapagos Rift. *Science* **203**, 1073-1083.

Fanale, F.P, T.V. Johnson, and D.L. Matson (1977) Io's surface and the histories of the Galilean satellites. In *Planetary Satellites*, (J. Burns, Ed.), 379-405. Univ. of Arizona Press, Tucson.

Fritsen, C.F., E.E. Adams, C.P. McKay, and J.C. Priscu (1998) Liquid water content of permanent ice covers on lakes in the McMurdo dry valleys. In *The McMurdo Dry Valleys of Antarctica: A Cold Desert Ecosystem*, (J.C. Priscu, Ed.), Antarctic Research Series, American Geophysical Union, Washington D.C., **72**, 269-280.

Gilichinsky, D.A., E.A. Vorobyova, L.G. Erokhina, D.G. Fyordorov-Dayvdov, and N.R. Chaikovskaya (1992) Long-term preservation of microbial ecosystems in permafrost. *Adv. Space Res.* **12** (4) 255-263.

Kral, T.A., K.M. Brink, S.L. Miller, and C.P. McKay (1998) Hydrogen consumption by methanogens on the early Earth. *Origins Life Evol. Biosphere*, in press.

Lewis, J.S. (1971) Satellites of the outer planets: Their physical and chemical nature. *Icarus* **15**, 174-185.

McDonald, G.D., M. Bell, and C. Sagan (1998) On the likelihood and detectability of involatile organic material on Europa (in press).

McKay, C.P. (1997) The search for life on Mars. *Origins Life Evol. Biosph.* **27**, 263-289.

McKay, C.P. (1996) Oxygen and the rapid evolution of life on Mars, In *Chemical Evolution: Physics of the Origin and Evolution of Life.* (J. Chela-Flores and F. Raulin, Eds.), pp. 177-184, Kluwer.

McKay, C.P., G.D. Clow, D.T. Andersen, and R.A. Wharton Jr. (1994) Light transmission and reflection in perennially ice-covered Lake Hoare, Antarctica. *J. Geophys. Res.* **99**, 20427-20444.

McKay, C.P., and H. Hartman (1991) Hydrogen peroxide and the origin of oxygenic photosynthesis. *Origins Life* **21**, 157-163.

Milton, R.C. deL., S.C.F. Milton, S.B.H. Kent (1992) Total chemical synthesis of a D-enzyme: The enantiomers of HIV-1 protease show demonstration of reciprocal chiral substrate specificity. *Science* **256**, 1445-1448.

Ojakangas, G.W. and D.J. Stevenson. (1989) Polar wander of an ice shell on Europa. *Icarus* **81**, 242-270.

Reynolds, R.T., S.W. Squyres, D.S. Colburn, and C.P. McKay (1983) On the habitability of Europa. *Icarus* **56**, 246-254.

Ross, M., and G. Schubert (1987) Tidal heating in an internal ocean model of Europa. *Nature* **325**, 133-134.

Schidlowski, M. (1988) A 3,800-million-year isotopic record of life from carbon in sedimentary rocks. *Nature* **333**, 313-318.

Smith, B.A., and Voyager Imaging Team (1979) The Galilean satellites and Jupiter: Voyager 2 imaging science results. *Science* **206**, 927-950.

Squyres, S.W., R.T. Reynolds, P.M. Cassen, and S.J. Peale (1983) Liquid water and active resurfacing on Europa. *Nature* **301**, 225-226.

Stevens, T.O. and J.P. McKinley (1995) Lithoautotrophic microbial ecosystems in deep basalt aquifers. *Science* **270**, 450-454.

Urquhart, M.L. and B.M. Jakosky (1996) Constraints on the solid-state greenhouse effect on the ice Galilean satellites. *J. Geophys. Res.* **101**, 21169 - 21176.

Walter, M.R. and D.J. DesMarais (1993) Preservation of biological information in thermal spring deposits, Developing a strategy for the search for a fossil record on Mars. *Icarus* **101**, 129-143.

Zent, A.P. and C.P. McKay (1994) The chemical reactivity of the martian soil and implications for future missions. *Icarus* **108**, 146-157.

POSSIBLE DEGREE OF EVOLUTION OF SOLAR-SYSTEM MICROORGANISMS.

JULIAN CHELA-FLORES (*)
The Abdus Salam International Centre for Theoretical Physics, Miramare P.O. Box 586; 34100 Trieste, Italy and Instituto de Estudios Avanzados (Universidad Simon Bolivar), Apartado 17606, Parque Central, Caracas 1015A, Venezuela.

Abstract

The present work is concerned with the search for life in the outer Solar System. We devote our attention to the determination of the likely degree of evolution of such biota (Chela-Flores, 1996; 1997a,b,c), and the determination of the Europan environments that should be probed. Archaebacteria-like organisms are the most likely biota to exist on Europa (Delaney *et al.*, 1996). We have argued that evolution should have occurred in Europa and that the experimental test of this conjecture is feasible through a Cryobot/Hydrobot type of space mission (Horvath *et al.*, 1997). Although difficult instrumentation issues are involved in the preparation of a package to search for life in Europa, we have initiated the discussion of what would seem to be a reasonable biological experiment.

1. Introduction

Refuges against evolution seems to be excluded from paleontological evidence (Little *et al.*, 1997). Indeed, Cambrian fauna, such as lamp shells and primitive molluscs were maintained during Silurian times by microorganisms that lived in hydrothermal vents. In the current Cenozoic Era these hot environments have seen the extinction of such fauna. It is then reasonable to assume that microorganisms themselves, in general, would be inexorably subject to evolutionary pressures, independent of their particular habitat. On Earth, amongst single-celled organisms, the eukaryotic cellular plan took a little over one billion years to make its first appearance in the fossil record during the Proterozoic Eon, after prokaryotic communities (stromatolites) were well established in the Archean Eon. It still remains to be confirmed, or rejected, whether the Europan environment may have had liquid water over geologic time, capable of sustaining a primary archaea community. To sum up, two ingredients support the present attempt to identify indicators of the degree of evolution of putative Europan biota:

- The possibility of testing the onset of eukaryogenesis in the outer Solar System.

(*) Research Associate at the School of Theoretical Physics, Dublin Institute for Advanced Studies, 10, Burlington Road, Dublin 4, Ireland.

J. Chela-Flores and F. Raulin (eds.),
Exobiology: Matter, Energy, and Information in the Origin and Evolution of Life in the Universe, 229–234.

• The apparently complete lack of refuges against evolutionary pressures that the Earth biota has been subject to for almost four billion years. Hence, we have pointed out the possibility that the eukaryotic transition may be a general consequence of geological changes during the process of planetary formation, coupled to the effect of natural selection. In other words, the first appearance of eukaryotes on Earth may have occurred independent of the fact that the microorganisms we are familiar with are confined to the biosphere of planet Earth.

2. Is There Life on Europa?

It is generally agreed that in 2 billion years (Gyr) microorganisms oxygenated the whole planet, produced the bulk of the iron ores and fashioned the eukaryotic cell, a transcendental transition referred to as 'eukaryogenesis', which was the first step on the pathway towards intelligent multicellular organisms on this planet. If there are intelligent organisms elsewhere, one possibility is that they may also have gone through the stage eukaryogenesis along their own evolutionary pathway. One very modest first step in the search for the evolution of intelligence that can be implemented inside the Solar System, is to search for extraterrestrial eukaryotes, wherever the search may be likely to succeed with missions especially designed for that purpose. It may be argued that such an endeavour is not technologically impossible, and its implementation could be achieved in a reasonable period of time. One possible site where we could begin the search is under the ice-covered surface of Europa, the icy Galilean satellite of Jupiter. As we have seen in Sec.1 hydrothermal vents on the sea-floor are viable sustainers of life on Earth, independent of energy from the Sun. The possible Europan volcanic activity, due to its interaction with Jupiter and other Galilean moons, renders Europa the number one target in the Solar System for the first identification of a living microorganism that may be unrelated to the terrestrial progenote.

On the other hand, the possibility of detecting biochemicals on the ice surface of Europa has been raised recently (McKay, 1998; Chela-Flores, 1997c). Much detail of the surficial properties has been gained since 1995, when the Galileo mission began the exploration of the Jovian system. A possible mechanism for bringing biochemicals to the surface will be discussed in Sec. 4.3 below. The forthcoming campaign of the NASA Galileo Europa Mission (GEM) during the period 1998-1999 will be restricted to infrared and ultraviolet spectroscopy. Surficial biogenic tests on Europa may have to wait for further missions, in order to allow a thorough search for traces of putative Europan biochemistry or signs of extant life. The possibility of probing for life under the ice, in the presumed Europan ocean will be discussed in Sec. 5 below.

3. Missions for Europa and its Earth analogue, Lake Vostok

3.1. POSSIBLE BIOLOGY EXPERIMENTS

We proceed to discuss two possible biology experiments:

3.1.1. On the ice surface of Europa (to be implemented by means of the shallow Cryobot melter probe [sCmp]),

3.1.2. In the possible Europan ocean (Cryobot/Hydrobot mission, [CHm] cf., Fig 1).

FIGURE 1: A cryobot/hydrobot painting (*) suggesting the search for microorganisms at the bottom of Europa's ocean. The submersible (hydrobot) is seen in the foreground.

3.2. SHALLOW CRYOBOT MELTER PROBE

In analogy with phenomena observed in the dry valley lakes of Antarctica, we have the following situation: a factor in the lack of uniformity in surface brightness and colour of the Europan surface may be the presence of microorganisms, or their biomolecules. In other words, the search for extraterrestrial biochemistry (cf. Sec. 2) on the surface of Europa seems to be a possible relatively straighforward aspect of the Europa campaign. This hypothesis can be tested, for instance, by spectroscopic search of the Europan surface ice. Unfortunately, the forthcoming campaign of GEM does not consider an adequate spectroscopic technique. The surficial biogenic tests that we are now proposing will have to wait for a further orbiting spacecraft provided with microwave spectroscopic facilities, in order to allow a search for biological macromolecules. On the other hand, sCmp seems the right mechanism to test directly for the above-mentioned indicators of life. The motivation for this proposal is given in the following section.

3.3. THE GROWTH OF THE ALGAL MATS IN THE DRY VALLEY LAKES

An important lesson may be drawn from the growth of the algal mats in the dry valley lakes. A phenomenon involving large biomasses of benthic algal mats has been observed: Pieces of these mats escape through about five meters of lake ice that cover perennially these lakes (Wharton, Jr.*et al.,* 1983). Buoyant forces are capable of detaching algal mats formed in the benthic regions. This leads to pieces of the mats to rise to the bottom of the ice cover and to freeze into the ice. Through the combined

---

(*) Courtesy Jet Propulsion Laboratory. Based on U.S. government- sponsored research under contract NAS7-1260.

action of ablation of ice from the upper surface and the formation of new ice at the top of the liquid water, some pieces of these mats reach the upper lake surface in about one decade. The geophysicists and glaciologists responsible for this discovery estimate that such mat escape is an important mechanism for the distribution of these algae to the exterior of the lakes.

Granted the main hypothesis of this work, namely that a certain degree of evolution may have taken place in Europa, starting from chemoautolithotrophic microorganisms at hydrothermal vents at the bottom of the Europa Ocean, more advanced cells, not necessarily analogues of archaebacteria, as previously assumed, may have evolved, as we know that hydrothermal vents are not refuges against evolution (Chela-Flores, 1997c). We have argued that analogues of diatoms, for instance, are possible. For this reason algal mats consisting of either prokaryotes, or eukaryotes may have been lifted up to the visible surface, or at least the accessible subsurface of Europa. One possibility of interpreting the lack of uniformity in surface brightness and colour of the Europan surface, particularly in the anti-Jovian hemisphere, suggests that the impurities in the surface ice may contain some microorganisms, or at least some biochemicals. This hypothesis, although speculative, is based on the experience gained in the perennially ice-covered lakes of the McMurdo Dry Valleys. In those environments there are communities of prokaryotes co-existing with eukaryotes.

4. The Cryobot/Hydrobot mission

A combined approach is needed in the CHm biology experiment. The Cryobot is based on an advanced type of a Philberth probe. Within that space there would be an "in-situ chemistry laboratory" (ISiCL). Its instrumentation would aim to determine whether the ocean exhibited one set of requirements for "life we would recognise". We believe that the detection of life and the evolutionary stage it has reached should be one of the primary goals of a Europa exploration program (Horvath *et al.,* 1997).

4.1. INSTRUMENTATION FOR *IN SITU* ANALYSIS

We wish first of all to define in some detail some of the minimum ISiCL equipment that is needed in the biology experiment. The optical system that would be proposed for the Hydrobot depends on the chemical composition of the putative microorganisms. The technique relies on the material having the ability to induce luminescence by the application of various dyes. The dyes are detected with a microscope. If microscopic fluorescence is taken as a possible means for probing for life, some advantages and some challenges are immediately evident.

First and foremost, in contrast with the Martian search for life, the typical resolution needed for fluorescence micrographs of chromosomes is 10 μm (Lodish, H.*et al.,* 1995). Such resolution is well within the scope of a light microscope. Martian research is linked to the electron microscope, since the nodules in the Allan Hills meteorite, which are currently under discussion, have been suggested to be nanobacteria, 50-75 nm long and 10-20 nm in diameter (Benoit and Taunton, 1997).

The difficulties of going beyond a light microscope in the Hydrobot are evident. We believe that since only a *light* microscope is needed, viewing a 'robotic biologist' inside the Hydrobot is feasible with simple optics and sampling arms.

4.2. ABUNDANCE AND DISTRIBUTION OF SINGLE-CELLED ADVANCED MICROORGANISMS IN THE EARTH OCEANS AND IN LAKE VOSTOK

In one litre of seawater one may find as many as ten million diatoms, which may be considered the primary foodstuff of marine fauna (Hoover, 1979). Some species of Pennales are known to dwell under the permanently ice-covered lakes of the Antarctic dry valleys (Oro *et al.*, 1992), one of the Earth environments that may resemble in some aspects the Europan ocean.

On the other hand, Lake Vostok is the most important Earth analogue of the Europan environment (Ellis-Evans and Wynn-Williams, 1996). A main objective of the preliminary plans for the Europan campaign is Lake Vostok. Once the planetary protection protocols are duly taken into account, forthcoming knowledge of the microflora that populates Lake Vostok will be of great value for testing the instrumentation requirements for the Europa campaign.

4.3. APPLICATION OF BANDING TECHNIQUES WITH UNCERTAINTY ON CHEMICAL DETAIL OF THE GENOME OF A EUROPAN MICROORGANISM

Quinacrine fluorescent dye inserts itself between base pairs in the DNA helix producing the so-called Q-bands, which for the purpose of the robotic mission, would probably suffice.

The more involved use of Giemsa stain in order to produce the more permanent R-bands, as suggested in our previous work (Chela-Flores, 1997c; Alberts *et al.,* 1983) is probably an unnecessary complication.

Adjacent areas stain differently; bands give a clear indication of slightly different modes of DNA compaction, which is the hallmark that the organism has reached a higher degree of evolution. Diverse forms of heterochromatin are the clearest hallmarks of eukaryoticity (Chela-Flores, 1998).

In general, it may be argued that gene activity is correlated with light-staining bands (for instance, genes that are transcriptionally active are light-staining (Watson, *et al.,* 1987)). The question is not so much what is the chemical detail of the genome, as the degree of compaction it may have reached.

5. Discussion and Conclusions

We have discussed an analogy with phenomena observed in the dry valley lakes of Antarctica: a factor in the lack of uniformity in surface brightness and colour of the Europan surface may be the presence of microorganisms, or their biomolecules.

This hypothesis can be tested, for instance, by spectroscopic search of the Europan surface ice. The proposed Europa Ocean Discovery mission (Edwards *et al.*, 1997) does not plan to search directly for surface biochemistry or biology, in the sense of Sec. 3.

We conclude that the above-mentioned orbital, as well as the Hydrobot/Cryobot missions are both necessary. The Discovery-type of mission, however, should also include the appropriate equipment for microwave spectroscopy.

REFERENCES

Alberts, B., Bray, D., Lewis, J., Raff, M., Roberts, K. and Watson, J.D. (1983). *Molecular Biology of the Cell.* Garland Publishing: New York. p.399.

Benoit, P.H. and Taunton, A.E. (1997). *The challenge of remote exploration for extraterrestrial fossil life.* In: Proceedings of Optical Science, Engineering, and Instrumentation SD 97 Symposium: Instruments, Methods and Missions for Investigation of Extraterrestrial Microorganisms. The International Society for Optical Engineering, Bellingham, WA, USA. pp.98-108.

Chela-Flores, J. (1996). *Habitability of Europa: possible degree of evolution of Europan biota.* (Abstracts), 5th. Capistrano Conf. San Juan Capistrano Research Institute, 31872 Camino Capistrano, San Juan Capistrano, CA 92675, USA, 12-14 November. p. 21.

Chela-Flores, J. (1997a). A Search for Extraterrestrial Eukaryotes: Biological and Planetary Science Aspects. In: *Astronomical and Biochemical Origins and the Search for Life in the Universe.* Eds. C.B. Cosmovici, S. Bowyer and D. Werthimer. Editrice Compositore: Bologna. pp. 525-532.

Chela-Flores, J. (1997b). A Search for Extraterrestrial Eukaryotes: Physical and Biochemical Aspects of Exobiology. *Origins Life Evol. Biosphere* (in press).

Chela-Flores, J. (1997c). *Testing for evolutionary trends of Europan biota.* Proc. Optical Science, Engineering, and Instrumentation **3111**. SD97 Symposium (San Diego, California, 29 July-1 August, 1997): *Instruments, Methods and Missions for Investigation of Extraterrestrial Microorganisms.* The International Society for Optical Engineering, Bellingham, WA, USA. pp. 62-271.

Chela-Flores, J. (1998). First steps in eukaryogenesis: Origin and evolution of chromosome structure. *Origins Life Evol. Biosphere* **28,** 215-225.

Delaney, J., Baross, J., Lilley, M. and Kelly, D. (1996). *Hydrothermal systems and life in our solar system.* Europa Ocean Conference, (Abstracts), 5th. Capistrano Conf. San Juan Capistrano Research Institute, 31872 Camino Capistrano, San Juan Capistrano, CA 92675, USA, 12-14 November. p. 26.

Edwards, B.C., Chyba, C.F., Abshire, J.B., Burns, J.A., Geissler, P., Konopliv, A.S., Malin, M.C., Ostro, S.J., Rhodes, C., Rudiger, C.,. Shao, X-.M., Smith, D.E., Squyres, S.W., Thomas, P.C., Uphoff, C.W., Walberg, G.D., Werner, C.L., Yoder, C.F., Zuber, M.T. (1997). *The Europa Ocean Discovery mission.* Proc. Optical Science, Engineering, and Instrumentation **3111**. SD97 Symposium (San Diego, California, 29 July-1 August, 1997): *Instruments, Methods and Missions for Investigation of Extraterrestrial Microorganisms.* The International Society for Optical Engineering, Bellingham, WA, USA. pp. 249-261.

Ellis-Evans, J.C. and Wynn-Williams, D. (1996). A great lake under the ice, *Nature* **381**, 644-646.

Hoover, R. (1979). *Those Marvelous Myriad Diatoms.* Nat. Geographic Magazine, June. pp.870-878.

Horvath, J., Carsey, F., Cutts, J., Jones, J., Johnson, E., Landry, B., Lane, L., Lynch, G., Chela-Flores, J., Jeng, T.-W. and Bradley, A. (1997). *Searching for ice and ocean biogenic activity on Europa and Earth.* Proc. Optical Science, Engineering, and Instrumentation **3111**. SD97 Symposium (San Diego, California, 29 July-1 August, 1997): *Instruments, Methods and Missions for Investigation of Extraterrestrial Microorganisms.* The International Society for Optical Engineering, Bellingham, WA, USA. pp. 490-500.

Little, C.T.S., Herrington, R.J., Maslennikov, V.V., Morris, N.J. and Zaykov, V.V. (1997). *Silurian hydrothermal-vent community from the southern Urals, Russia.* Nature **385,** 146-148.

Lodish, H., Baltimore, D., Berk, A., Zipursky, S.L., Matsudaira, P. and Darnnell, J. (1995). *Molecular Cell Biology*. 3rd ed. Scientific American Books. W.H. Freeman: New York. p.354.

McKay, C.P. (1998). *The search for extraterrestrial biochemistry.* In: Chela-Flores, J. and Raulin, F. (eds.). (1998). Exobiology: Matter, Energy, and Information in the Origin and Evolution of Life in the Universe. Kluwer Academic Publishers, Dordrecht, The Netherlands. (This volume, pp. 219-227.)

Oro, J. Squyres, S.W., Reynolds, R.T., and Mills, T.M. (1992). *Europa: Prospects for an ocean and exobiological implications.* In: Exobiology in Solar System Exploration. G.C. Carle, D.E. Schwartz, and J.L. Huntington, (eds.). pp. 103-125.

Watson, J.D., Hopkins, N.H., Roberts, J.W., Steitz, J.A. & Weiner, A.M. (1987). *Molecular Biology of the Gene.* 4th. ed. Menlo Park, California: The Benjamin/Cummings Publishing Co. pp. 685-686.

Wharton, Jr., R.A., Parker, B.C. and Simmons, Jr., G.M. (1983). *Distribution, species composition and morphology of algal mats in Antarctic dry valley lakes,* Phycologia **22**, pp. 355-365.

IS THERE AN ALTERNATIVE PATH IN EUKARYOGENESIS?

An Astrobiological View on Making the Nucleated Cell.

JOSEPH SECKBACH[1], THOMAS E. JENSEN[2], KOICHIRO MATSUNO[3], HAKOBU NAKAMURA[4], MAUD M. WALSH[5] and JULIAN CHELA- FLORES[6]

[1]Hebrew University of Jerusalem, 91905 Israel[1]; [2]Department of Biological Sci. Lehman College, Bronx. NY 10468. USA. [3]Dept. Bioengineering. Nagaoka University. of Techno. Nagaoka 940-21, Japan. [4]Biol. Institute, Faculty of Science. Konan University, Kobe 658, Japan. [5]Inst. Environ. Studies, Louisiana State University, Baton Rouge, LA. 70803, [6]International. Center for Theoretical Physics, Trieste, Italy.

Abstract

The transition from prokaryotic to eukaryotic cells ('Eukaryogenesis') is still a biological mystery. The present paper revisits the question of the origin of the eukaryotic cell and suggests that the biochemical, ultrastructural aspects and the renewed efforts in space missions in Solar System exploration will present us with an opportunity for answering the question: Is there an alternative path in Eukaryogenesis?

1. Introduction

1.1. ORIGIN OF EUKARYOTES BY INTRA-CELLULAR NATURAL SELECTION

Natural selection, a mechanism for evolution of organisms first proposed by Charles Darwin and now generally accepted by biologists, can be applied to the evolution of cells as well. This concept of cellular evolution through natural selection provides an alternative to the endosymbiotic origin of the organelles in eukaryotes as the key to Eukaryogenesis.

Prokaryotic cells have been shown to possess a wide array of inclusion bodies. One group of prokaryotes, the Cyanobacteria, have been shown to collectively possess some 35 different inclusion bodies of various morphologies (Jensen 1985) and it is likely that inclusions were common in other prokaryotic groups going back in time before eukaryotic cells existed. These inclusions in prokaryotes probably came and went for 3.5 billion years. If an inclusion would allow more efficient cellular reproduction, then by natural selection this cell line and the inclusion was retained. At some point in time a nucleus was formed in a prokaryotic cell, probably as an endomembrane event such as suggested

---

[1] John P. Laborde **Chair** for Sea Grant and Visiting Professor at the Dept. Biol. Sci., Louisiana State University (508 Life Sciences Building) Baton Rouge. LA. 70803. USA

J. Chela-Flores and F. Raulin (eds.),
Exobiology: Matter, Energy, and Information in the Origin and Evolution of Life in the Universe, 235–240.

by Jensen (1989, 1994) or by Nakamura (1975, 1994). If a complex structures like the nucleus could arise by an endomembrane event then most likely other inclusions, such as mitochondria, chloroplasts, and Golgi apparatus, could originate in the same way. One possibility is that Eukaryogenesis resulted from membrane differentiation to compartmentalize metabolic pathways (Nakamura 1994). Cyanobacteria contain complete systems for photosynthetic and respiratory metabolisms, and one cyanobacterial species, *Chlorogloea fritshii*, has been proposed as a candidate for the ancestral eukaryote (Nakamura, 1975, 1994).

2. Would the Search for Eukaryotes in the Solar System Contribute to Answer the Question: "Is There an Alternative Path in Eukaryogenesis?"

Archaebacteria-like organisms (ALO) are the most likely biota to exist on Europa, a satellite of Jupiter (Delaney et al.,1996; Horvath et al.,1997). Furthermore, an experimental test of this conjecture is feasible through a lander space craft capable of penetrating the Europan surface ice-layer by melting through it (Horvath et al., 1997). Miniaturization of the traditional fluorescent microscope may make it possible to distinguish through visual examination between prokaryotes and eukaryotes.

On Earth the eukaryotic cellular plan amongst single-celled organisms took a little over one billion years to make its first appearance in the fossil record during the Proterozoic Eon, after prokaryotic communities (stromatolites) were well established in the Archean Eon. It still remains to be confirmed, or rejected, whether the Europan environment may have had liquid water over geologic time capable of sustaining a primary Archaean community. Recent reports about possibilities of ancient life on Mars (McKay et al. 1996; cf. Harvey & McSween, 1996; Anders,1996) suggest that such environments are not unique in our solar system.

3. Is Eukaryogenesis a Unique Consequence of Evolution on Earth ?

3.1. DISTINCTIVE CHARACTER OF THE EUKARYOTIC CELLULAR PLAN

Two main reasons favor the appearance and distinctive character of the eukaryotic cellular plan. Firstly, a critical step in the diversification of single celled organisms may have been the loss of the ability to manufacture cell-wall biomolecules (peptidoglycan) of the prokaryotes. Without the constraint that these biomolecules impose on cell shape, both Archaea and Eubacteria have been able to diversify beyond the Domain Monera.

Secondly, Earth-bound eukaryotes are not habitually extremophiles, but their diversification may share a common thread with Archaea. Eukaryotes, in spite of not being able to exploit fully all the extreme environments may, nevertheless, invade to a certain extent those niches sometimes at the disposal of the other extremophiles (Roberts 1998).

As discussed above, the identification of primitive eukaryotes is not straightforward. Another difficulty is that the simplest criterion to recognize them, a membrane-bounded set of chromosomes or organelles, does not help to identify unambiguously eukaryotes. Indeed, there are prokaryotic organisms that do have membrane-bounded nucleoids; *Gemmata oscuriglobus* (Fuerst and Webb, 1991) and *Nitrosolobus multiformis* (Jensen, 1994) contain an invagination of the plasma membrane segregating its nucleoid by a double membrane.

On the other hand there are some eukaryotes that lack some vital organelles like mitochondria (*Staurojoenina* and *Mixotricha paradoxa*, protists symbiotic in termites, or in the giant Ameba *Pelomyxa palustris* [see Margulis and Olendzenski, 1992] and the parasite flagellate *Giardia*). Protozoa and Animalia as well as fungi do not possess plastids. Additionally, prokaryotic symbionts of the protozoan *Euplotidium itoi* contain proteins typical of eukaryotes (Rosati et al., 1993). Further complicating identification is the fact that silicification experiments of certain microorganisms (Westall, 1995) suggest that the identification of eukaryotes at the morphological, or molecular levels, particularly during the process of fossilization, may lead to some confusion.

3.2. BIOTA TYPICAL OF HOT SPRINGS AND SEA-FLOOR ON EARTH

Possible analogs for life in the oceans of Europa may be found in some extreme environments on Earth, where organisms from all kingdoms may be found. Several Archaea and Bacteria thrive in extreme high levels of temperatures. The ubiquitous unicellular algae Cyanidiaceae are the only eukaryotes growing in acidic (pH of 2- 4) thermophilic (up to 57°C) hot-springs (Seckbach, 1972, 1994). These primitive rhodophytans can thrive in pure carbon-dioxide (Seckbach et al., 1970). Similarly, close to the warm vents near mid-ocean ridges there are dense invertebrate communities, in which the chemosynthetic bacteria live in symbiosis with the metazoan components of the ecosystem. Bivalves and gastropods live in symbiosis with their bacteria in their gill tissues. Vestimentiferans, polychaetous annelids have a very broad distribution throughout the ocean floor (Rosati et al., 1993).

In the bottom of the Gulf of Mexico there is an underwater lake of brine, so dense that it remains in a depression of the sea-floor. A vast salt deposit produces the gradual seeping through of gases that autotrophic bacteria can turn into foodstuff, and these bacteria become the first stage in the food chain of clams, mussels and tubeworms.

3.3. ARE HOT-SPRINGS REFUGES AGAINST EVOLUTION?

Because hot springs are so foreign to our everyday experience and also far removed from catastrophes that may exterminate other ecosystems, the hot spring environment has been assumed to be a sort of refuge against evolutionary pressures and a background in which new species evolve.

This 'refuge' concept has prevailed even amongst exobiologists who have speculated on the possible origin of life at the bottom of the putative Europan ocean. Previous authors have only considered that Archaea may have also evolved in the benthic regions heavily dependent on bacterial chemosynthesis. However, Little et al. (1997) have identified fossils from the earliest hydrothermal-vent community found so far, dating from the late Silurian Period, over 400 Myr BP (i.e., 0.4 billion years ago). This particular community has its own case of species extinction, as the fossils have been identified and correspond to lamp shells (inarticulate brachiopods) and snail-like organisms (monoplacophorans).

We cannot maintain with certainty today that the analogous conditions that may exist elsewhere in the Solar System, particularly in Europa, will induce the appearance of ALO. This is purely a question of an experimental or observational nature. However, given the common origin of all the Solar System, the favorable conditions in Europa may have been conducive to the first steps in evolution, including the first appearance of Archaea and eventually of further domains. This source of life itself (i.e, hydrothermal vents at the bottom of the Europa ocean) cannot be considered a permanent refuge for Archaea, but the transition into a complex cell, possibly a eukaryote has first of all to be ruled out by experimental tests. The effort is worth making, as Eukaryogenesis has been the most momentous step in the evolution towards intelligence.

Contemporary parallel evolution may depend on a totally primary production based on chemosynthetic bacteria. This possibility cannot yet be excluded from occurring in outer solar- system satellites, where frozen water, or water in hydrated silicates, is known to exist. We have discussed elsewhere (Chela-Flores, 1997) one kind of experiment for detecting the degree of evolution of microorganisms in the Europan ocean. This effort would provide the testing ground to the question "Is there an alternative path in Eukaryogenesis?".

3.4 BIOCHEMICAL CONSIDERATIONS

At present we lack experience with parallel evolution on Earth vs. extraterrestrial environments, so it is essential that present and future efforts concerning Europa should not be confined to the possibility of designing equipment capable of recognizing exclusively analogues of the lowest domains of Earth biota (Archaea and Bacteria). It will be important to look not only for cellular structure, but also for chemicals that may indicate the presence of precursors of biomolecules. Over geologic time, comets may have contributed significantly to the organic inventory of Europa, as it has been recently shown that the Hale-Bopp comet contains in its nucleus such chemicals: carbon monoxide, hydrogen cyanide, formaldehyde, and many other compounds rich in carbon (Cruikshank, 1997).

An evolutionary significant event that could have happened on the primitive Earth [and on other planets] would be a process of enhancing energy concentrations locally in addition to that of synthesizing prebiotic molecules. Once the concentration energy was be released, it could have driven further molecular association and dissociation. Experimentally, it has been shown that thermal heterocomplex molecules from amino acids that could have been ubiquitous on the primitive Earth [and similar life-carrying solar-system planets], have carried with themselves excitations whose lifetimes are sufficiently large, so as to materialize microscopic heat engines (Matsuno, 1995, 1996). It seems plausible that from this chemical evolution scenario, further biological evolution may have led to Prokaryotes and then to early eukaryotes.

4. Conclusions

Evidence from modern organisms suggests that there is no need to utilize an endosymbiotic scenario to explain the origin of the eukaryotic cell and the inclusions found it. The eukaryotic cell with its inclusions may have come into existence by intracellular natural selection of endomembrane structures in the same manner that birds, fish and other organisms came into existence by natural selection.

Our present knowledge about the diversity of life in extreme environments on Earth suggests that exploration of the solar system, particularly the Jovian satellite Europa, may shed some light on basic questions in microbiology, particularly the issue of Eukaryogenesis.

5. Acknowledgments

The senior author (JS) appreciates the very kind hospitality of **Dr. Jack R. Van Lopik**, Exec. Dir. of the Sea Grant and his faithful assistant Mrs. **Brenda Henning** and thanks **Dr. Russell L. Chapman**, Exec. Dir. of CCEER, for usage of the Biol. Dept. All are members of the Louisiana S. U. at Baton Rouge, where this Ms was completed.

6. References

Anders, E. (1996). Science **274,** 2119-2121.

Chela-Flores, J. (1997). Origins Life Evolution. Biosphere (in press).

Cruikshank, D.P.(1997) "Stardust Memories", Science, **275**, 1895-1896.

Delaney, J., Baross, J., Lilley, M. and Kelly, D. (1996). Europa Ocean Conference, (Abstracts), 5th. Capistrano Conf. San Juan Capistrano Research Institute, 31872 Camino Capistrano, San Juan Capistrano, CA 92675, USA, 12-14 November, p. 26.

Fuerst, J.A. and Webb, R.J. (1991) Proc. Natl. Acad. Sci. USA, **88**, 8184-8188.

Harvey, R.P. and McSween, Jr., H.P. (1996). Nature **382**, 49-51 .

Horvath, J., Carsey, F., Cutts, J., Jones, J., Johnson, E., Landry, B., Lane, L., Lynch, G., Chela-Flores, J., Jeng, T.-W. and Bradley, A. (1997). Proc. Optical Science, Engineering, and Instrumentation 3111. Instruments, Methods and Missions for Investigation of Extraterrestrial Microorganisms. San Diego, California, USA. pp. 490-500.

Jensen, T.E. (1985) Arch. Hydrobiol.pl. **71**, 33-73.

Jensen, T.E. (1989) Cytobios **60,** 47-61.

Jensen, T.E. (1994) in Seckbach, J., (ed.), *Evolutionary Pathways and Enigmatic Algae: Cyanidium Caldarium (Rhodophyta) and Related Cells.* Kluwer Academic Publishers: Dordrecht, The Netherlands, pp. 53-66.

Little, C.T.S., Herrington, R.J., Maslennikov, V.V., Morris, N.J. and Zaykov,V.V. (1997) Nature **385**, 146-148.

Margulis, L. (1992) in: Margulis, L and Orlendzenski, L. (eds.) *Environmental Evoltion.* The MIT Press, Cambridge. pp. 149-199.

McKay, D.S., Gibson, E., Thomas-Keprta, K.L., Hogatolla, V., Romanek, C., Clemett, S., Chillier, S., Maechling, C.R. and Zare, R.N. (1996) Science **273**, 924-930.

Matsuno, K. (1995) J. Theor. Biol. **173**, 137-145.

Matsuno, K. (1996) J. Theor. Biol. **180**, 105-109. and in: Chela-Flores and Raulin F. (eds.) *Chemical Evolution: Physics of the Origin and Evolutionof Life.* Kluwer Academic Publishers. pp. 231-238.

Nakamura, H. (1975) Viva Origino **4**: 43-61.

Nakamura, H. (1994) in Seckbach, J., (ed.), *Evolutionary Pathways and Enigmatic Algae: Cyanidium Caldarium (Rhodophyta) and Related Cells.* Kluwer Academic Publishers: Dordrecht, The Netherlands, pp. 3-18.

Roberts, D.McL.(1998) in: Seckbach, J.(ed.) *Enigmatic Microorganisms and Life in Extreme Environments.*

Kluwer Academic Publishers. Dordrecht, The Netherlands (in preparation).

Rosati, G. , Lenzi, P. and Franco, U. (1993) Micron, **24**, 465-471.

Seckbach, J. (1972) Microbios, **5**, 133-142.

Seckbach, J. (1994) in: Seckbach, J., (ed.), *Evolutionary Pathways and Enigmatic Algae: Cyanidium Caldarium (Rhodophyta) and Related Cells*. Kluwer Academic Publishers: Dordrecht. pp. 99-112.

Seckbach, J., Baker, F. A. and Shugarman, P.M. (1970) Nature **227**: 744-745.

Trowell, S., Wild, J., Hovrath, J., Jones, J., Johnson, E. and Cutts J., Europa Ocean Conference, (Book of Abstracts), Capistrano Conference No. 5, San Juan Capistrano Research Institute 12-14 November, p. 76, 1996.

Westall, F. , Boni, L. and Guerzoni, E. (1995) Paleontology **38**, 495-528.

DID EARTHLIFE COME FROM MARS?

PAUL DAVIES
Centre for Basic Research
P.O. Box 389
Burnside
South Australia 5066

1. Traffic of material between planets

In 1871 Lord Kelvin delivered an address to the British Association in Edinburgh, in which he offered the following observation [1]: "When two great masses come into collision in space, it is certain that a large part of each is melted; but it seems also quite certain that in many cases a large quantity of debris must be shot forth in all directions, much of which may have experienced no greater violence than individual pieces of rock experience in a landslip or in blasting by gunpowder." Kelvin went on to speculate that if one of the cosmic bodies should be a planet bearing life, then "we must regard it as probable in the highest degree that there are countless seed-bearing meteoric stones moving through space." In this manner, Kelvin thought, life may have come to Earth from elsewhere in the universe.

The possibility that rocks could be blasted off a planet by a cosmic collision and convey life to another planet was until recently discounted. Partly this was due to a general reluctance to contemplate catastrophic astronomical events, but also because it was thought that any collision powerful enough to eject material into space would inevitably destroy any living organisms.

In recent years this scepticism has been replaced by an acceptance that material can be transferred between the planets of the solar system relatively unscathed. The discovery of martian meteorites on Earth suggests that from time to time Mars will suffer an impact by an asteroid or a comet with enough violence to propel debris into solar orbit. Some fraction of this debris will eventually be swept up by the Earth. The known martian meteorites have not suffered very severe shocking, leaving open the possibility that any embedded microorganisms could survive ejection from the martian surface. The claim by McKay et. al. [2] that one such meteorite, ALH 84001, contains fossilised bacteria invites the conjecture that a similar meteorite might convey live microbes from Mars to Earth, perhaps in the form of dormant spores (assuming for now that there has been microbial life on Mars).

The theory that life can propagate between planets through outer space is called panspermia, and was popularised by Svante Arrhenius many years ago [3]. However, Arrhenius postulated "naked panspermia", involving exposed individual bacteria travelling through space unprotected. This theory, which was revived in modern form by Hoyle and Wickramasinghe [4], must be distinguished from "rocky panspermia", in which microbes transit between planets cocooned within rocks.

J. Chela-Flores and F. Raulin (eds.),
Exobiology: Matter, Energy, and Information in the Origin and Evolution of Life in the Universe, 241–244.

2. Hazards of interplanetary travel

For a live organism to journey successfully between Mars and Earth inside an ejected rock, it must overcome several hazards: ejection from Mars, radiation, cold, entry into Earth's atmosphere and a potentially hostile planetary environment.

A large cosmic impact with Mars would result in the vaporisation of the impactor and much of the surface rocks at ground zero. The ensuing shock wave would compress the surrounding rock, raising the temperature to lethal values. However, as pointed out by Melosh [5] surface rocks on the periphery of the impact site cannot be compressed because they are subjected only to atmospheric pressure from above. These surface rocks will be accelerated vertically without severe shocking as the compressed lower strata rebound, and excavate the impact crater. If the impact is large enough, the unshocked surface rocks could reach escape velocity. Any microorganisms within the rocks would be protected from atmospheric frictional heating by the rocky material. (The incoming impactor will in any case evacuate a tunnel of air as it passes through the atmosphere.) In spite of the enormous impulse, organisms of microbial size could withstand the g forces involved. Melosh has estimated that a major impact would project about a billion tonnes of material into space in which microbes would remain alive, embedded in rocks of average diameter one metre.

Once in space the major hazard is radiation. The deadliest form of radiation is ultra-violet from the sun, but this is easily screened out by a thin layer of rock. Solar flares produce penetrating bursts of protons intermittently, but these would not be a major hazard from organisms protected by a metre or more of rock. High energy cosmic rays are more penetrating, but again, a metre of rock would provide a good shield. The instrinsic radioactivity of the rock itself might present the greatest hazard, but based on estimates of the resilience of microorganisms to radiation [6] it seems probable that some terrestrial-type bacteria could withstand several million years in space under these conditions. This is easily long enough to reach Earth. Computer models by Gladman et. al. [7] suggest that some 2.5% of the total Mars ejecta would have hit Earth after 10 million years, some of it much sooner.

The cold of outer space does not present a problem. Indeed, there is evidence that freeze-drying bacteria positively aids their resistance to radiation [7]. The heat of re-entry is more hazardous. Most one-metre rocks hitting Earth's atmosphere either explode of vaporise. A shallow incoming trajectory would be the most favourable. The interior of the rock need not be unduly heated by the short period of frictional braking, which would melt only the surface. If a rock eventually fragments in mid-air, it could spill its microbial cargo into the atmosphere, from where the organisms could drift slowly to the surface. Alternatively, those fragments of rock that reach the surface at terminal velocity would not hit the ground or ocean with any great violence.

In the past the bombardment of the planets was more severe. At the end of the period of late heavy bombardment, 3.8 billion years ago, there would have been a prolific traffic of material between Mars and Earth. Melosh has shown [5] that the fragment size of the ejecta scales with the mass of the impactor, so that much larger (and hence more protective) rocks would have been flung into space at that time.

3. When Mars and Earth were twins

The analysis of the hazards of interplanetary travel suggest that bacteria could survive the journey from Mars to Earth. However, would conditions on arrival prove congenial? There is strong evidence that at the end of the period of late heavy bombardment Mars was warm and wet, and similar to Earth. The presence of volcanoes on Mars implies the existence there of ancient hydrothermal systems similar to those along mid-ocean ridges on Earth, that are known to play host to rich ecosystems. The discovery of a deep subterranean biosphere on Earth [8] suggests a similar biosphere may have existed on Mars, inhabited by deep-living chemotrophic organisms - primary producers needing no light or organic matter, that make biomass directly from the chemical and thermal energy coming up from the crust. If a martian chemotroph landed in the sea on Earth, possibly as a dormant spore, it may be conveyed to a hydrothermal system by ocean currents, and revived. Thereafter it would multiply and spread.

It is conceivable that life on Earth started this way. There are several reasons to favour Mars over Earth as the original site for biogenesis. Mars had all the ingredients necessary for life to get started - liquid water, organics from comets, minerals such as iron and sulphur, reducing gases like hydrogen and methane, clays and other materials to offer catalytic surfaces, and geothermal heat sources. In addition, it offered several advantages over Earth. A very large impact of the sort that happened during the heavy bombardment would have created a rock vapour atmosphere that boiled the oceans and sterilised the surface of the Earth [9]. It would have also sent a heat pulse deep into the surface strata, killing any subsurface microbes to a depth of about a kilometre. Being a smaller planet, Mars was less of a target for such large impacts. It also escaped the really huge impact that created Earth's moon and drove off most of the volatile material. Mars' smaller size implies a faster rate of surface cooling, enabling microbes to inhabit a deeper comfort zone in the rocks, and thus be better protected from the impacts. Other refugia from bombardment exist: large impacts would displace rocks into orbit, where they would have been spared the rigours of the rock vapour atmosphere. From this relative haven, life could eventually return to re-colonise the planet when some of the ejected rocks fell back to the surface. Because of the lower escape velocity on Mars, it would be easier for microbes to survive this temporary displacement than in the case of Earth. Chris McKay has speculated [10] that Mars may have had a head start over Earth because of the lower level of bombardment, enabling life to evolve to a relatively advanced stage by the time Earth became habitable.

4. Did Earthlife go to Mars?

Although Mars' lower gravity favours the transfer of material from Mars to Earth, some rocks will inevitably go the other way. There is an excellent chance that these rocks will inoculate Mars with terrestrial microbes. Earth bacteria that reached Mars 3.8 billion years ago would have found conditions very much like home. Colonisation is a therefore a distinct possibility. For this reason alone I think it is very likely that there was life on Mars between 3.8 and 3.6 billion years ago. It may still be there, deep beneath the surface.

We may thus identify several scenarios for early life in the solar system:

- Life began on Mars and spread to Earth.
- Life began on Earth and spread to Mars.
- Life began on another body and colonised Earth and perhaps Mars too.
- Life originated independently on both Earth and Mars.
- Life is unique to Earth; Mars is sterile.

If the cross-contamination of Earth and Mars was intense enough, the two planets may have formed a coupled ecosystem, with obvious consequences for early evolution. If that was so, it may be impossible for us to determine which planet, if either, was the site of the ultimate origin of life. Although the first four scenarios all predict that life existed on Mars at some stage, the philosophical consequences of an independent origin of martian life are dramatically different from the case of simple panspermia.

The possibility of the exchange of live organisms between Earth and Mars raises interesting questions about the quarantining of spacecraft and planetary protection in the event of a Mars sample return mission, assuming there is extant life on Mars [11]. It is usually argued that martian organisms would present no threat to life on Earth because pathogens must be closely adapted to their hosts. But if life on Mars and Earth had a common origin we cannot be sure that martian microbes might not be similar to terrestrial pathogens. On the other hand, if cross-contamination is ongoing, many martian microbes may have reached Earth during the course of evolutionary history, without any dire consequences. (Unless, of course, one attributes episodes of mass extinction to diseases caused by the transfer of martian germs during enhanced bombardment.)

Finally, it will happen that some fraction of rocky ejecta will be flung out of the solar system altogether by Jupiter. However, the probability of a terrestrial rock hitting an earthlike planet in another star system, or vice versa, is infinitesimal. Thus, unlike in the scenario of Arrhenius, any straightforward version of rocky panspermia would not result in life spreading across the galaxy. (For a more complicated version involving the possibility of the interstellar transfer of life, see McKay [12].)

References

1. Quoted in Arrhenius, S.: *Worlds in the Making*, Harper, London (1908), p. 219.
2. McKay, D.S. et. al.: *Science* 273 (1996), 924.
3. Arrhenius, S.: *Worlds in the Making*, Harper, London (1908).
4. Hoyle, F., and Wickramasinghe, C.:*Our Place in the Cosmos*, Dent, London (1993).
5. Melosh, H.J.: *Geology* 13 (1985) 144; *Nature* 332 (1988), 687; *The Planetary Report* 14, No. 4 (1994), 16.
6. Horneck,G, Bucker, H., and Reitz, G.: *Adv. Space Res.* 14 (1994), 1041; Weber, P., and Greenberg, M.: *Nature* 403. See also the papers by Horneck and Mileikowsky in this volume.
7. Gladman, B. et. al.: *Science* 271 (1996), 1387.
8. Gold, T.: *Proc. Natl. Acad. Sci. USA* 89 (1992), 6045.
9. Sleep, N., Zahnle, K., Kasting, J. and Morowitz, H.:*Nature* 342 (1989), 139.
10. McKay, C.: *Origins of Life and Evolution of the Biosphere* 27 (1997), 263.
11. Nealson, K. et. al. *Mars Sample Return: issues and recommendations*, National Academy Press, Washington (Sagan, C.:*The Planetary Report* 14, No. 4 (1994), 3.
12. McKay, C.: *Mercury* 25, No. 6 (1996), 15.

ULTRASTRUCTURE IN THE CARBONATE GLOBULES OF MARTIAN METEORITE ALH84001

Frances Westall[1], Pietro Gobbi[2], Dane Gerneke[3] and Giovanni Mazzotti[2]
1. University of Bologna, DIPROVAL, via S. Giacomo 7, 40126 Bologna, Italy. frances@geomin.unibo.it
2. University of Bologna, Istituto di Anatomia Umana, via Irnerio 48, 40126 Bologna, Italy.
3. University of Cape Town, Electron Microscope Unit, Rondebosch, 7700 Cape, South Africa

Introduction

The complex history of meteorite ALH84001 of probable Martian origin has been well documented, starting with two shock events which strongly influenced the rock (Treimann, 1995; McKay *et al.*, 1996). Carbonate was deposited in pores and in pre-existing fractures in the rock between these two events (*op. cit.*). The carbonate deposits on the fracture surfaces are typically globular in shape (Fig. 1), forming flattened discs. They are characteristically rimmed by Mg-rich carbonate or alternating rims of Mg- and Fe-rich carbonate (Figs. 1, 2), and sometimes some apatite (McKay *et al.*, 1996). The centers of the globules are Fe-rich.

The origin of the carbonate is hotly disputed. Scott et al. (1997a,b) believe that they were formed by high temperature shock melting of the rock on the surface of Mars, with subsequent injection into the fractures, whereas others propose either a high or relatively low temperature hydrothermal origin (Romanek *et al.*, 1994; Mittlefehldt, 1994; Harvey and McSween, 1996; Gleason *et al.*, 1997). Gleason et al. (1997) suggest that variable composition of CO_2 rich hydrothermal fluids infiltrating the rock while on Mars would explain the disequilibrium demonstrated by the complex zoning of the carbonate globules. They calculate that the fluid reactions took place at relatively low temperatures (<300°C) over a period of about 1 MY. Lastly, a low temperature biological (bacterial) origin was suggested by McKay *et al.* (1996) on the basis of a number of lines of reasoning, including the presence of PAHs, nano-sized magnetite and bacteriomorph structures. Critical to this theory is the temperature of formation of the carbonate since temperatures much above 110°C are not conducive to the presence nor the preservation of organic life. As noted above, estimates of the temperature of formation of the carbonate range from >650°C (Harvey and McSween, 1996; Bradley *et al.*, 1996; Scott *et al.* 1997), <300°C (Gleason *et al.*, 1997), between 125-205°C (Valley *et al.*, 1997a), to <100°C (Valley *et al.*, 1997b).

Apart from the discussion concerning the temperature of formation of the carbonate globules, the purported bacteriomorph structures described by McKay *et al.* (1996) have come under closer scrutiny. The bacteriomorphs of McKay *et al.* (*op.cit.*) are small, oval to elongated structures, sometimes segmented, ranging from about 100 nm to up to 1 μm in length. The authors compared them to very small terrestrial bacteria (nanobacteria, cf. Folk, 1993). Bradley *et al.* (1997a) reinterpreted aligned particles in the carbonate as magnetite crystals, formed at high temperatures, epitaxially oriented along the crystal axis. Further studies (Bradley *et al.*, 1997b) supported their

J. Chela-Flores and F. Raulin (eds.),
Exobiology: Matter, Energy, and Information in the Origin and Evolution of Life in the Universe, 245–250.

suggestion that the elongated, parallel bacteriomorphs could be magnetite whiskers or substrate lamellae with the segmented aspect enhanced by Au coating. While some of the oriented structures could be related to weathering and incipient clay formation on the carbonates, McKay et al. (1997) argued that not all the bacteriomorph structures could be explained crystallographically. Complicating the interpretation of the globules and the microstructures contained within them is the fact that the carbonates show evidence of having been partially dissolved and McKay et al. (1996), in fact, gave an alternative interpretation for the bacteriomorph structures, relating them to the effects of dissolution.

In order to provide additional information about the nature and the ultrastructure of the carbonate globules, we studied some globules on a fracture surface of ALH84001 using high resolution SEM. This paper describes our preliminary results.

High resolution scanning electron microscope studies of a fragment of ALH84001

A 0.045 g cube, including a fracture surface containing a few carbonate globules, was examined using a JEOL 849 FEI-SEM at the University of Bologna and a Leo S440 (with Kevex-Sigma EDX, backscatter and cathodoluminescence detectors) at the University of Cape Town. The sample was subdivided and a 600-300 µm surface containing two zoned globules, measuring about 70 µm in diameter, and a number of smaller, semi-circular "blebs" of Mg carbonate <10 µm in diameter was investigated.

This surface was initially studied without coating and subsequently with a C/Pt coating (2.5 nm thick). The carbonate globules and the smaller "blebs" of carbonate are readily located using the backscatter detector (Fig. 1). The rim of the globules is complexly zoned containing an undulating layer of Fe-carbonate of various thickness (Fig. 2). The Fe-carbonate globules consist of both blocky and crushed grains with the blocky grains exhibiting fine, crystallographically-determined lamination (Figs. 3, 5). Crush zones < 1 µm thick occurring within some of the blocky grains may possibly be crystallographically related (Fig. 5). Some of the surfaces of the blocky carbonate show corrosion features (Fig. 5) (due to dissolution, McKay et al. [1996] or to devolatisation, Scott et al. [1997b]). This is most common on the laminated surfaces exposed parallel to the main fracture. A number of the blocky Fe-carbonate grains

---

Figure 1. Backscatter image of a fracture surface of ALH84001 containing two carbonate globules (arrows). The Mg-carbonate appears as a dark rim around the globules and as dark spots on the fracture surface. Scale 50 µm.

Figure 2. Secondary electron image of the edge of one globule showing blocky Fe-carbonate (Fe), amorphous silica coating (Si), and Mg/Fe carbonate layered rim (vertical wall at the top of the photograph). Scale 10 µm.

Figure 3. Blocky Fe-carbonate with inclusions. Note the elongated inclusions parallel to the crystalline lamellae (black arrows). The other inclusions are subangular to subround and exhibit no particular orientation. Some inclusions are banded (white arrows and inset, [scale 50 nm]). Amorphous silica (Si) coats the inclusions on the upper fracture surface of the blocky carbonate. Scale 200 nm.

Figure 4. Amorphous silica layer with a reticulated texture resembling "burst bubbles" coating a fracture surface in the blocky Fe-carbonate. Overlying blocky Fe-carbonate arrowed. Scale 2 µm.

Figure 5. Blocky carbonate surface showing etched laminae parallel to the crystallographic axes, coated by amorphous silica. An internal crush zone (possibly crystallographically oriented) is arrowed. Scale 2 µm.

Figure 6. Detail of the porous, blobby, amorphous silica. Scale 200 nm. Inset shows that the blobs are in contact with one another and are cemented by smooth silica (scale 50 nm).

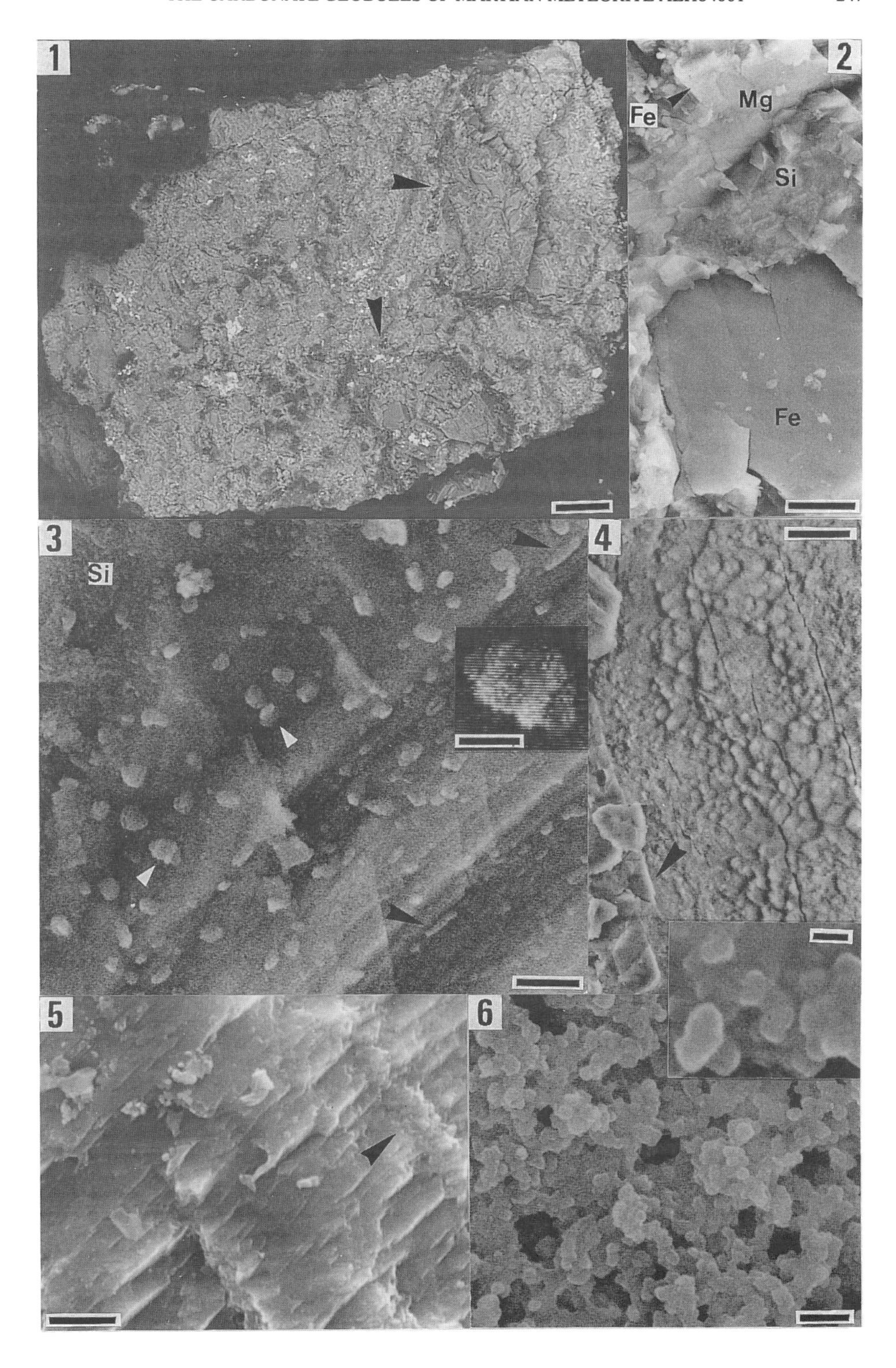
1
2
Fe
Mg
Si
Fe
3
Si
4
5
6

contain "floating" inclusions (i.e. inclusions not in contact with each other) (Fig. 4). Rounded, subrounded to subangular inclusions range from 20-100 nm in size and elongated inclusions are up to 130 nm in length and <10 nm in width. Some of the inclusions are banded (Fig. 4), others are cleaved so that they resemble halves of a shell. The more elongated inclusions (ca. 130 nm in length) are aligned parallel to the crystallographic axis (similar to the structures described by Bradley et al., 1997b), whereas the other inclusions appear to be disorientated. They are not particle surface phenomena as they occur in freshly cleaved material. These inclusions are also common in the Fe bands embedded within the Fe rim. They are much rarer in the Mg-carbonate. The composition of the inclusions is unknown, as they are too small to be measured using the EDX system.

Fracture surfaces in the carbonate globules parallel to the main fracture surface, as well as the orthopyroxene fracture surface, are coated with a thin (ca. 100 nm) film of amorphous, blobby-textured silica (opal) (Figs. 4, 6). The corroded Fe-carbonate surfaces are also coated. The silica consists of round to oval blobs, 50-100 nm in diameter, cemented in a smooth silica matrix (Fig. 6). The blobs are in contact with each other, in contrast to the "floating" inclusions in the blocky carbonate grains. At low magnifications the silica deposit exhibits a reticulate pattern, resembling a plane of "burst bubbles" (Fig. 4). Its reticulate, porous nature is confirmed at higher magnifications (Fig. 6). This blobby coating also coats the subangular to subrounded inclusions floating in the blocky carbonate (Fig. 3).

Chromite crystals included in the Fe-carbonate are angular to subangular fragments of less than 1 μm to a few micrometers in size. They also contain irregularly-shaped inclusions, <50 to 200 nm in size. Some of the chromite grains have been fractured and dislocated by shock fracturing. A few of the larger chromite grains contain round cores, about 1 μm in diameter, of a different composition (lower atomic number; it gives a similar backscatter signal to the Fe-carbonate and exhibits similar crystalline cleavage).

Compared to the grains in the carbonate globules, the surrounding orthopyroxene exhibits an amorphous, ill-defined morphology from which the inclusions typical of the Fe-carbonate are absent. However, chromite grains in the orthopyroxene also display inclusions.

Discussion

Owing to the preliminary nature of our study, the following discussion is a first, tentative interpretation of our observations. Our data are not sufficient to determine whether the carbonate globules were deposited as a result of shock meting (as proposed by Scott et al., 1997a, b) or to deposition from hydrothermal fluids flowing through the fractures (Mittlefehldt, 1994; Harvey and McSween, 1996; Romanek et al., 1994).

The nature and origin of the submicron "floating" inclusions in the Fe-carbonate is enigmatic. The correspondence of the elongated inclusions with the structural lamellae of the carbonate implies a crystalline origin and we concur with the suggestion of Bradley et al. (1997a,b) that they could represent epitaxially oriented magnetite whiskers or the protruding edges of the carbonate lamellae. The other subrounded to subangular "floating" inclusions obviously preceded the formation of the carbonate. One explanation for their presence is that they may represent ground up particles of the orthopyroxene host rock, produced during the first impact shock, which were suspended in the fluid which deposited the carbonate, and locked into the crystal matrix. The presence of broken chromite grains containing identical inclusions in both the orthopyroxene and the carbonate globules is in favour of this hypothesis. Alternatively the inclusions may have an external origin and were introduced by a carbonate-rich fluid of external origin. This can be tested once the composition of the inclusions is known. The nano-sized bacteriomorph structures described by McKay et al. in their 1996 paper are not similar to our "floating" inclusions, but they also suggested an alternative, mineralogical interpretation, related to the dissolution of the carbonate, for some of the nano-sized structures.

The cores of different composition (possibly carbonate) within some of the chromite grains in the blocky Fe-carbonate may represent alteration of the chromite fragments by the carbonate-bearing fluid which deposited the globules. Would alteration explain the clear, round borders of the cores though?

The amorphous, blobby, porous silica deposited in fractures running through both the carbonate and the orthopyroxene postdates the deposition of the carbonates (as noted by Valley et al., 1997) and an event which caused the carbonate corrosion features. The bubbly, porous texture of the silica deposit suggests that it was deposited from a volatile-rich fluid (equilibrium temperatures of the silica indicate 125-205°C, Valley et al., 1997). This deposit is very similar texturally to modern, hydrothermal sinters (Cady and Farmer, 1996) and fossiliferous, Early Archean sediments flushed by silica-rich, hydrothermal fluids (Westall, unpub. data).

Our observations support previous suggestions (based on REE analysis, Mittlefehldt, 1994; Mittlefehldt and Lindstrom, 1994) that the rock was influenced by an hydrothermal event. Although Gleason et al. (1997) suggest that the carbonates, themselves, derived from hydrothermal fluids (on the evidence of the indications of thermal disequilibria and changing composition of the fluids depositing the carbonates), they believe that the hydrothermal event documented by the REEs predated the formation of the carbonate globules. Whether or not this was the case for the REEs, our data show that there was another hydrothermal event which postdated later fracturing, crushing and corrosion of the carbonate.

Conclusions

All the nanoscale structures described here appear to be of mineralogical origin, or can be explained by mineralogical phenomena. To date, we have seen no structures which could be attributed exclusively to fossil life but we stress that our investigation is still at a preliminary stage. There are similarities between the amorphous silica deposit in the late stage fractures and hydrothermally deposited silica in modern and Early Archean sediments deposited near hydrothermal vents.

The sequence of events revealed by our analysis of the carbonate globules is thus: (1) shock fracturing and possible contemporaneous grinding of the orthopyroxene, possibly producing the anhedral fragments of chromite included in the carbonate, as well as the submicron-sized subangular to subrounded "floating" inclusions which obviously precede the carbonate; (2) deposition of the carbonate; (3) dissolution/devolatisation of the carbonate/fracturing of the carbonate (and chromite inclusions); (4) deposition of amorphous silica by hydrothermal fluids along the second generation fracture planes, possibly associated with the second generation fracturing and corrosion of the carbonate.

Acknowledgements

We gratefully acknowledge M.J. de Wit (from the NRC) for funding for SEM time, A. Fantazzini and W. Williams for photographic help.

References

Bradley, J.P., Harvey, R.P. and McSween, H.Y. Jr. (1996) Magnetite whiskers and platelets in ALH84001 Martian meteorite: Evidence of vapour growth phase, *Geochim. Cosmochim. Acta*, **60**, 5149-5155.

Bradley, J.P., Harvey, R.P., and McSween, H.Y., Jr, (1997) *Nature*, **390**, 454.

Cady, S. and J. Farmer, (1996) Fossilization processes in siliceous thermal springs: trends in preservation along thermal gradients, in G.R. Bock and J.A. Goode (Eds.), *Evolution of hydrothermal ecosystems on Earth (and Mars)*, Wiley, Chichester , pp. 150-173.

Folk, R.L., SEM imaging of bacteria and nannobacteria in carbonate sediments and rocks, J. *Sediment. Petrology.*, **63**, 990-999, 1993.

Gleason, J.D., Kring, D.A., Hill, D.H., and Boynton, W.V. (1997) Petrography and bulk chemistry of Martian orthopyroxeneite ALH84001: implications for the origin of secondary carbonates. *Geochim. Cosmochim. Acta* , **61**, 3503-3512.

Harvey, R.P. and McSween, H.Y. Jr. (1996) A possible high temperature origin for the carbonates in the martian meteorite AL84001, *Nature*, **382**, 49-51.

McKay, D.S., Gibson, E.K. Jr., Thomas-Keprta, K.L., Vali, H., Romanek, C.S., Clemett, S.J., Chillier, X.D.F., Maechling, C.R. and Zare, R.N., (1996) Search for past life on Mars: Possible relic biogenic activity in martian meteorite ALH84001, *Science*, **273**, 924-930.

McKay, D.S., Gibson, E.K. Jr., Thomas-Keprta, K.L., Vali, H., (1997), *Nature*, **390**, 455

Mittlefehldt , D.W. (1994) ALH84001, a cumulative orthopyroxene member of the martian meteorite clan, *Meteoritics*, **29**, 214-221.

Mittlefehldt, D.W. and Lindstrom, M.M. (1994) Geochemical evidence for mixing of three components in martian meteorite ALH84001, *Meteoritics*, **29**, 504 (abst.).

Romanek, C.S., Grady, M.M., Wright, I.P., Mittelfehldt, P.W., Socki, R.A., Pillinger, C.T. and Gibson, E.K. (1994), Record of rock-fluid interaction on Mars from meteorite ALH84001, *Nature*, **372**, 655-657.

Scott, E.R.D., Yamaguchi, A., and Krot, A.N. (1997) Petrological evidence for shock melting of carbonates in the martian meteorite ALH84001, *Nature*, **387**, 366-379.

Treimann, A.H. (1995) A petrographic history of martian meteorite ALH84001: Two shocks and an ancient age, *Meteoritics*, **30**, 294- 302.

Valley, J.W., Eiler, J.M., Graham, C.M., Gibson, E.K. Jr., Romanek, C.S. and Stolper, E.M. (1997) Low temperature carbonate concretions in the martian meteorite ALH84001: Evidence from stable isotopes and mineralogy, *Science*, **275**, 1633-1638.

Valley, J.W., Eiler, J.M., Graham, C.M., Gibson Jr., E.K., and Romanek, C.S. (1997b) Ion microprobe analysis of oxygen and carbon isotope rations in the ALH84001 meteorite. *Lunar Planet. Sci. XXVIII*

PRODUCTION AND DETECTION OF ORGANIC COMPOUNDS ON MARS

KENSEI KOBAYASHI and TAKEO KANEKO
Department of Chemistry and Biotechnology, Yokohama National University, Hodogaya-ku, Yokohama 240-8501, Japan

YUKISHIGE KAWASAKI
Mitsubishi Kasei Institute of Life Science
Minamiooya, Machida, Tokyo 194, Japan

TAKESHI SAITO
Institute for Cosmic Ray Research, University of Tokyo
Midoricho, Tanashi, Tokyo 188, Japan

1. Introduction

In Viking mission in 1976, no organisms nor organic compounds were detected in Mars surface regolith (Soffen, 1976). On the other hand, McKay et al. (1996) detected polycyclic aromatic hydrocarbons (PAHs) in Martian meteorite ALH 84001 by two-step laser mass spectrometry (L^2MS). From the presence of PAHs in it, together with other observations, they concluded that it was plausible that there were past life of Mars. Since life on Mars is still controversial, we need further analyses of Martian samples in order to obtain solid evidences for organisms or bioorganic compounds on Mars.

NASA is now performing and planning several Mars missions, including sample return projects. Independent from them, we have been discussing on Mars exobiology mission by international collaboration, called "Return-to-Mars-Together (RTMT)" since 1994 (Kobayashi et al., 1996; Saito, 1996). Not like NASA's Mars missions, the RTMT proposal is now centering around *in situ* analyses of microorganisms and organic compounds on Mars. Here we discuss what kind or organic compounds (including organisms) existed in the past Mars or exist in the present Mars, and what kind of techniques we can apply to detect them.

2. Possible Organic Compounds and Organisms on Mars

In our earth, life was supposed to be born about 3.8 billion years ago, using organic

J. Chela-Flores and F. Raulin (eds.),
Exobiology: Matter, Energy, and Information in the Origin and Evolution of Life in the Universe, 251–254.

compounds formed from the primitive earth atmosphere and organic compounds delivered by extraterrestrial bodies such as comets. If the early Mars had dense atmosphere which was closed to the primitive earth's - a mixture of carbon dioxide, carbon monoxide, nitrogen and water - and liquid water on its surface, the same kinds of chemical evolution processes could be occurred to form bioorganic compounds. When a gas mixture of CO_2, CO, N_2 and H_2O at various mixing ratios, simulating the primitive earth atmosphere, was irradiated with high energy protons (major components of cosmic ray) , amino acid precursors (compounds which give amino acids after hydrolysis) and uracil was detected in the products (Kobayashi and Tsuji, 1997; Kobayashi et al., 1997, 1998).

Comets and meteorites should have hit upon not only earth but also Mars to give their organic compounds to the planets (Kasamatsu et al., 1997). Thus it is quite plausible that Mars had a wide variety of abiotically-formed organic compounds - both endogenous and exogenous. In the primordial "sea" of Mars, the first organisms may have been born using these organic compounds. If life was born on Mars, organic compounds biotically produced were added to Mars besides the abiogenous organic compounds. If life was not born on Mars, however, abiogenous organic compounds can be reserved without biotic degradation. These compounds would give us valuable insights for chemical evolution occurred on the earth before the origin of life.

Water on the surface of the early Mars has been lost in the history of Mars, and a part of water has been reserved in underground frozen soils of Mars. Thus it is quite plausible that Mars organisms - dead or alive - and/or organic compounds can be detected there. It is quite important to know how deep the frozen layer exists in the Mars.

3. Detection Methods for Organisms on Mars

Biological experiments in Viking Mission were designed based on incubation techniques. It is claimed, however, that only a part of terrestrial microorganisms can be detected with incubation techniques. A more comprehensive technique is now required to detect organisms on Mars since we have no information on Mars organisms, if they live there.

We proposed a fluorescence microscopic technique as a new detection methods for organisms on Mars in the RTMT project (Kawasaki and Tsuji, 1996). A conceptual figures of the proposed method is shown in Fig. 1. A soil sample was mixed with fluorescent reagents such as 6-carboxyfluorescein diacetate acetoxymethyl ester (CFDA-AM) and 1-anilinonaphtalene 8-sulfonate (ANS): CFDA-AM is an esterase substrate which gives fluorescence after hydrolyzed by active enzymes; ANS gives fluorescence when adsorbed to cell membranes. The fluorescence images are digitally detected with CCD cameras, and are transmitted to Earth.

This is a promising method by which we can detect all of the terrestrial organisms examined. The present method with ANS can be applied to the detection of not only

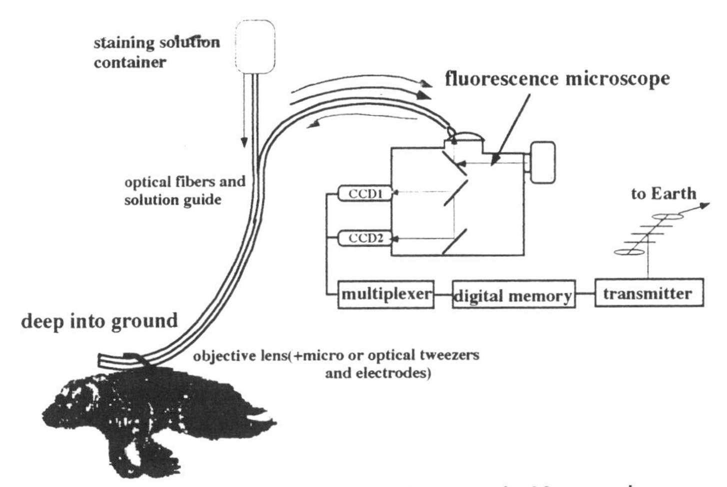

Fig. 1 Conceptual figure of *in situ* detection system for Mars organisms

living cells but also dead cells.

4. Detection Methods for Organic Compounds on Mars

In Viking Mission were used a pyrolysis-gas chromatograph/mass spectrometer (PY-GC/MS) (Bieman, 1974). The analytical results did not showed any evidence of the presence of organic compounds in surface regolith of Mars. It does not mean that there is no organic compounds in Mars, since (i) the examined surface sample contained superoxides and peroxides which might destroy any organic compounds there, and (ii) PY-GC/MS can be detected only a part of organic compounds such as aromatic hydrocarbons. Organic compounds on Mars may be present in the place without the oxidants: Ices in polar caps and underground frozen soils are two possible candidates. Here we discuss several detection methods for the organic compounds on such places on Mars in the future Mars mission.

After surface water of Mars was lost, the organisms in Mars may have died-out or survived, and organic compounds on Mars -either the biogenic and the abiogenic- may have been denatured to some extent. If these compounds were quite highly denatured in Mars environments, polyaromatic hydrocarbons (PAHs) are possible targets of analysis, like in the case of ALH 84001. On the other hand, if they are quite well-preserved, there may be bioorganic compounds such as biogenic protein (or abiogenic proteinoids). Such a wide range of organic compounds from living enzymes to simple PAHs are among our targets.

As previously mentioned the fluorescence microscopic technique can be applied to detect life-related compounds like proteins, nucleic acids and cell membranes when we use appropriate fluorescent reagents. In other cases, some mass spectrometric analysis and amino acid analysis after acid-hydrolysis seem to be a favorable choice, since amino acids can be detected in hydrolysates of a wide range of biogenic and abiogenic samples.

PAHs are detected in ALH84001 with a two-step laser ionization/desorption mass spectrometer (McKay et al., 1996), but this technique can hardly applied to in situ analysis. Here we propose that our fluorescence microscopy can be applied to detection of PAHs. When a fluorescent image and spectra of PAHs mixed with simulated Mars sand, each PAH can be very sensitively detected, even a single object with 100 nm diameter can be detected. It is also possible to discriminate some PAHs with their spectra. The present detection limit of PAHs is as small as 100 molecules if they are aggregated.

5. Conclusion

In the future Mars mission, we propose fluorescence microscopy to detect organisms. There are many candidates for methods to detect organic compounds on Mars, which includes PY-GC/MS, laser desorption mass spectrometers such as MALDI-MS and an amino acid detection system after acid-hydrolysis. Fluorescence microscopy can be included in the list if PAHs are possible organic compounds on Mars.

It must be emphasized that we do not have comprehensive techniques to analyze organic compounds, and that we must know the types of organic compounds before analytical techniques are chosen. Studies on stability of organic compounds in Mars environment must be examined now.

6. References

Biemann, K. (1974) The results on the Viking gas chromatograph-mass spectrometer experiment, *Origins Life* **5**, 417-430.

Kawasaki, Y., and and Tsuji, T. (1996) Searching for extraterrestrial life - developement of detection system of microorganisms-, *Viva Origino*, **24**, 293-304.

Kasamatsu, T., Kaneko, T., Saito, T., and Kobayashi, K. (1997), Formation of organic compounds in simulated interstellar media with high energy particles, *Bull. Chem. Soc. Jpn.*, **70**, 1021-1026.

Kobayashi, K., and Tsuji, T. (1997) Abiotic synthesis of uracil from carbon monoxide, nitrogen and water by proton irradiation, *Chem. Lett.*, **1997**, 903-904.

Kobayashi, K., Kasamatsu, T., Sato, T., Ishikawa, Y., and Saito, T. (1996) Strategy for the detection of bioorganic compounds on Mars, in J. Chela-Flores and F. Raulin (eds.). *Physics of the Origin and Evolution of Life*, Kluwer Academic, Dordrecht, pp. 381-388.

Kobayashi, K., Sato, T., Kajishima, S., Kaneko, T., Ishikawa, Y., and Saito, T. (1997) Possible complex organic compounds on Mars, Adv. Space Res., 19, 1067 -1076.

Kobayashi, K., Kaneko, T., Saito, T., and Oshima, T. (1998) Amino acid formation in gas mixtures by particle irradiation, *Origins Life Evol. Biosphere*, in press.

McKay, D. S., Gibson Jr., E. K., Thomas-Keprta, K. L., Vali, H., Romanek, C. S., Clemett, S. J., Chillier, X. D. F., Maechling, C. R. and Zare, R.N. (1996) Search for past life on Mars: Possible relic biogenic activity in Martian meteorite ALH 84001, Science **273**, 923-930.

Saito, T. (1996) Return-to-Mars-Together, *Viva Origino* **24**, 243-255.

Soffen, G. A. (1976) Scientific results of the Viking Mission, *Science* **194**, 1274-1276.

PREBIOTIC SYNTHESIS BY LIGHTNING IN MARTIAN VOLCANIC PLUMES

RAFAEL NAVARRO-GONZALEZ and VLADIMIR A. BASIUK
Laboratorio de Química de Plasmas y Estudios Planetarios, Instituto de Ciencias Nucleares, Universidad Nacional Autónoma de México, Ciudad Universitaria, México D.F. 04510, MEXICO

Abstract. A thermochemical-hydrodynamic model of lightning in explosive volcanic eruption columns has been used to assess the potential significance of volcanic lightning in the production of key molecules needed for chemical evolution and the origins of life in early Mars. Martian magmatic gases were assumed to be composed of 40% H_2O, 15% S, 5% N, and 40% C in various oxidation states ranging from CH_4 to CO_2. It is predicted that NO is the prime nitrogen containing species formed from mildly reducing (O/C = 0.75) to oxidizing (O/C = 2) conditions. NH_3 and HCN are abundant products under reducing conditions (O/C < 0.25). HSCN is only formed in reducing conditions, and HCHO is more efficiently produced at O/C ≤ 1.5. The chemical composition of gases emitted by primitive Martian volcanoes is not known but presumably they had a O/C ratio ≤ 1. Therefore, volcanic lightning may have played an important role in the process of chemical evolution in early Mars.

1. Introduction

At the end of the heavy bombardment process, conditions on Mars may have been favorable for the emergence of life. According to the orthodox theory advanced by Oparin (1924, 1936) and Haldane (1928), and examined experimentally by Miller (1953, 1955), the atmosphere played an important role in the formation of raw materials (such as hydrogen cyanide and formaldehyde) necessary for the synthesis of the building-blocks of life (*e.g.*, amino acids, purines, pyrimidines and sugars). The early Martian atmosphere is believed to have been dominated by carbon dioxide, molecular nitrogen and water vapor (McKay and Stober, 1989). Under these conditions, the high-temperature core of the lightning channel does not produce hydrogen cyanide (Stribling and Miller, 1987) but instead nitric oxide (Chameides and Walker, 1981). NO appears not to be directly involved in prebiotic synthesis except when reduced to ammonia (Summers and Chang, 1993). Therefore the problem arises as to the origin of the basic ingredients for life. The delivery of intact prebiotic material from the interstellar medium to the terrestrial planets *via* comets is an attractive solution to the problem (Chyba *et al.*, 1990). Nevertheless, there could be a number of unidentified microenvironments in early Mars that were equally favorable for prebiotic synthesis. Volcanic eruptions could perhaps played a role in chemical evolution since they were exceptionally common, globally distributed (Mouginis-Mark *et al.*, 1992)

J. Chela-Flores and F. Raulin (eds.),
Exobiology: Matter, Energy, and Information in the Origin and Evolution of Life in the Universe, 255–260.

and highly explosive (Wilson and Head, 1983; Gregg and Williams, 1996) during the early history of Mars. In particular, the explosive eruption column of a volcano is the most interesting locale due to (1) the presence of reduced magmatic gases; (2) the production of volcanic lightning activity; and (3) the fast escape of the nascent molecules from the high-temperature zone at sonic or supersonic speeds (Basiuk and Navarro-González, 1996; Navarro-González *et al.*, 1996). The aim of this paper is to explore the significance of this microenvironment using a thermochemical-hydrodynamic model of the lightning channel.

2. Theoretical Modeling of Volcanic Lightning

2.1 COMPOSITION OF VOLCANIC GASES

The chemical composition of the volatiles outgassed by primitive volcanoes is not known. As a first approximation, it was assumed that the Martian magmatic gases had a mole or atomic fraction similar to the volatiles found in CI chondrites (Dreibus and Wänke, 1989; Chyba, 1991): 0.40 water, 0.40 atomic carbon, 0.15 atomic sulfur, and 0.05 atomic nitrogen. Nitrogen and sulfur were considered to be in elemental form (*e.g.*, N_2 and S_2, respectively) whereas carbon was studied in various oxidation states ranging from the highly reduced (CH_4) to the highly oxidized (CO_2) states. In the simplest case, the simulations were performed assuming that the magmatic gases were not contaminated by volatiles present outside the volcanic vent environment prior to the occurrence of volcanic lightning. In later cases, the effects of dilution with ground water and the atmosphere were investigated. The Martian atmosphere was assumed to be composed by 80% carbon dioxide and 20% molecular nitrogen (Kasting, 1993).

2.2 SIMULATION OF VOLCANIC LIGHTNING

Lightning in volcanic eruptions differs from thunderstorm lightning basically in the chemical composition of the species subjected to the high temperature in and around the lightning channel. Most notably is the presence of ash and a high water vapor content. The sudden deposition of the electric energy along the lightning channel produces a thermally equilibrated plasma at temperatures in excess of 10,000 K. As the lightning channel expands and cools by mixing with surrounding material, the chemical species in the hot channel maintain a state of chemical equilibrium as long as their relaxation times are shorter than the cooling rate of the hot channel. Eventually, a temperature is reached at which the rates of reactions become too slow to keep chemical equilibrium and the mixing ratios of the chemical species freeze-out at this point. This freeze-out temperature (T_f) is likely to be characteristic for each chemical species. For the case of nitric oxide, T_f has been found to be about 2660 K for lightning in the contemporaneous Earth's atmosphere (Borucki and Chameides, 1984).

For simplicity it was assumed that T_f was similar to all the species formed by volcanic lightning. The equilibrium mixing ratios of H, H_2, OH, H_2O, H_2O_2, O, O_2, O_3, C_{gas}, CO, CO_2, CHO, CH, CH_2, CH_3, CH_4, C_2H_6, C_2H_5, C_2H_4, C_2H_3, C_2H_2, C_2H, C_2,

HCHO, HCO_2H, CH_3OH, N, N_2, N_2O, NO, NO_2, CN, HCN, HOCN, NH, NH_2, NH_3, N_2H_4, CH_3NH_2, CH_3CN, CH_3CHO, C_2H_5CHO, C_2H_5CN, CH_2CHCN, HC_3N, C_3H_8, C_3H_6, C_3H_4, $C_2H_5NH_2$, C_2N_2, C_2H_5OH, S, S_2, S_3, S_4, S_5, S_6, S_7, S_8, CS_2, SH, H_2S, SO, SO_2, SO_3, CH_3SH, COS, C_2H_5SH, and HSCN were calculated at 2660 K using thermochemical data from the NIST Database: Structures and Properties. The presence of ash in the volcanic lightning channel was not included in the model at this stage.

The product energy yields (PX) for species X formed by volcanic lightning were calculated according to the following expression (Borucki and Chameides, 1984): $P_X = a\, f^{\circ}_X (T_f)$; where a is the number of molecules heated to, or above T_f, per unit discharge energy and equals 3.2×10^{18} molecule J^{-1}, and $f^{\circ}_X (T_f)$ is the mixing ratio of species X at T_f.

3. Results and Discussion

The products formed during the turbulent cooling of the hot volcanic lightning channel are: H_2, SO_2, H_2S, O_2, NO, COS, CS_2, HCN, NH_3, CH_2O, C_2H_2, and HSCN. Their product energy yields vary considerably depending on the initial oxidation state of the carbon atom present in the magmatic gases. Molecular hydrogen is the most abundant product with an energy yield of 3.5×10^{18} molecule J^{-1} at O/C = 0. As the oxidation state of the carbon increases, its yield rapidly decreases by a factor of 0.1 at O/C = 1 and 0.02 at O/C = 2. Hydrogen sulfide and sulfur dioxide are the next most important products under reducing ($P_{H_2S} = 2.1 \times 10^{17}$ molecule J^{-1}) and oxidizing ($P_{SO_2} = 3.9 \times 10^{17}$ molecule J^{-1}) conditions, respectively; their yields become analogous under slightly neutral conditions ($P_X \cong 4 \times 10^{16}$ molecule J^{-1} at O/C $\cong$ 0.9). Molecular oxygen forms from mildly reducing (O/C $\cong$ 0.25) to oxidizing conditions with a maximum yield of $P_{O_2} = 8.5 \times 10^{15}$ molecule J^{-1} at O/C = 2. Carbon disulfide is an abundant product under reducing conditions ($P_{CS_2} = 2.6 \times 10^{15}$ molecule J^{-1} at O/C = 0) and becomes irrelevant at O/C > 1.25. Carbonyl sulfide is formed with a fairly insensitive yield ($P_{COS} \cong 10^{15}$ molecule J^{-1}) from reducing to neutral conditions and thereafter steadily decreases over an order of magnitude.

The predicted profiles for the nitrogen-containing species are shown in Figure 1 as a function of the initial oxidation state of the carbon atom present in magmatic gases ranging from the most reduced (O/C = 0 for methane) to the most oxidized (O/C = 2 for carbon dioxide) states. Nitric oxide is produced in low yield under reducing conditions ($P_{NO} = 1.4 \times 10^{11}$ molecule J^{-1} at O/C = 0) but rapidly becomes the prime nitrogen-containing species formed from mildly reducing (O/C $\cong$ 0.75) to oxidizing (O/C = 2) conditions, where it reaches a maximum yield of 1.9×10^{15} molecule J^{-1}. NH_3 is an important product under reducing conditions with a maximum yield of $P_{NH_3} = 3.2 \times 10^{13}$ molecule J^{-1}; as the oxidation state of carbon increases, P_{NH_3} markedly falls to 1.1×10^{12} molecule J^{-1} at O/C = 1 and then steadily to 1.1×10^{11} molecule J^{-1} at O/C = 2. Hydrogen cyanide is efficiently formed at O/C $\leq$ 0.75 and becomes the dominant

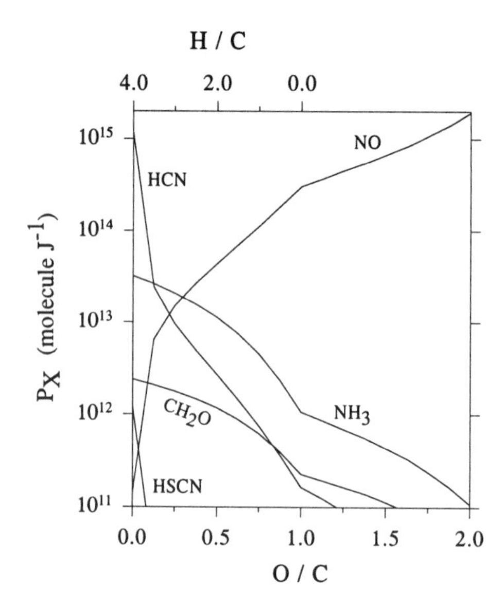

Figure 1. Product energy yields, Px, for selected compounds formed by volcanic lightning at various initial oxidation states of carbon ranging from CH_4 to CO to CO_2.

nitrogen containing species under extremely reducing conditions where it reaches a maximum yield of $P_{HCN} = 1.2 \times 10^{15}$ molecule J^{-1}. HCN, NO and NH_3 are formed in comparable yields at O/C $\cong$ 0.25. Hydrogen thiocyanide is a trace product formed only under reducing conditions with the highest energy yield of 1.2×10^{12} molecule J^{-1} at O/C = 0.

Figure 1 also contains the predicted profile for formaldehyde. CH_2O is efficiently formed under reducing conditions (O/C $\leq$ 1) with an energy yield of about 10^{12} molecule J^{-1}, and then steadily falls over an order of magnitude in neutral or mildly oxidizing conditions. Acetylene (not shown in Fig 1) is the only other organic molecule formed under extremely reducing conditions where its maximum energy yield is about 2.8×10^{13} molecule J^{-1}.

An important aspect to consider is the instant mixing of magmatic gases with the background atmosphere and/or with underground or ground water prior to the occurrence of volcanic lightning. These effects were investigated using a mildly reducing volcanic mixture with a O/C = 0.5. The surrounding atmosphere was assumed to be composed of 80% CO_2 and 20% N_2. It was found that the production efficiencies of HCN, NH_3 and CH_2O are relatively insensitive to atmospheric or water dilution below 40% and thereafter rapidly diminish. In contrast, that of HSCN drastically drops at about 25% dilution. NO becomes the prime product when the magmatic gases are highly diluted with the atmosphere ($P_{NO} = 3.1 \times 10^{16}$ molecule J^{-1}) or water ($P_{NO} = 4.2 \times 10^{14}$ molecule J^{-1}).

The chemical composition of magmatic volatiles outgassed by primitive Martian volcanoes is not known. Even for the Earth this is a difficult problem to unravel due to the extensive metamorphosis that Archean basalts have experienced through geologic time. The carbon species present in contemporaneous basalts is dominated by CO_2. This is consistent with thermochemical considerations using magmas at 1200°C with an oxygen fugacity, fO_2, that corresponds approximately to the fayalite-magnetite-quartz buffer ($10^{-12} \geq fO_2 \leq 10^{-8}$ atm) (Heald *et al.*, 1963). It is possible that the oxidation state of the primordial terrestrial mantle was more reduced and was gradually oxidized by recycling water from the surface followed by outgassing of hydrogen (Kasting, *et al.*, 1993). On the Moon, the two carbon gases that are released upon acid hydrolysis are CO and CO_2. The O/C ratio in the Apollo 15 basalts (Gibson *et al.,* 1975) is approximately 1.5 and is consistent with that expected during crystallization of lunar basalts, assuming T = 1123°C and $fO_2 = 10^{-13.1}$ to $10^{-13.7}$ atm (Wellman, 1970). The Martian basalts have a greater Fe / (Mg + Fe) ratio than terrestrial and lunar basalts and consequently Martian magmatic gases could have been significantly more reduced with a O/C $\leq$1. Therefore, volcanic lightning may have played an important role in the production of HCN, NH_3, CH_2O and HSCN in early Martian plumes.

4. Acknowledgments

This work was supported by grants from the National Council of Science and Technology of Mexico (CONACyT-4282-E9406) and the National Autonomous University of Mexico (DGAPA-IN102796).

5. References

Basiuk, V. A., and R. Navarro-González 1996. Possible Role of Volcanic Ash-Gas Clouds in the Earth's Prebiotic Chemistry. *Origins Life Evol. Biosphere* **26**, 173-194.

Borucki, W. J., W. L. Chameides 1984. Lightning: Estimates of the Rates of Energy Dissipation and Nitrogen Fixation. *Rev. Geophys. Space Phys.* **22**, 363-372.

Chameides, W. L., and J. C. G. Walker 1981. Rates of Fixation by Lightning of Carbon and Nitrogen in Possible Primitive Atmospheres. *Origins of Life* **11**, 291-302.

Chyba, C. F. 1991. Terrestrial Mantle Siderophiles and Lunar Impact Record. *Icarus* **92**, 217-233.

Chyba, C. F., P. J. Thomas, L. Brookshaw, and C. Sagan 1990. Cometary Delivery of Organic Molecules to the Early Earth. *Science* **249**, 366-373.

Dreibus, G., and H Wänke 1989. Supply and Loss of Volatiles Constituents during the Accretion of Terrestrial Planets. In *Origin and Evolution of Planetary and Satellite Atmospheres* (S. K. Atreya, J. B. Pollack, and M. S. Matthews, eds.), pp. 268-305. Univ. of Arizona Press, Tucson.

Gibson, E.K., Jr., S. Chang, K. Lennon, G.W. Moore, and G.W. Pearce 1975. Sulfur abundances and distributions in mare basalts and their source magmas. *Proc. Lunar Sci. Conf.* **6**, 1827-1301.

Gregg, T. K. and S. N. Williams 1996. Explosive Mafic Volcanoes on Mars and Earth: Deep Magma Sources and Rapid Rise Rate. *Icarus* **122**, 397-405.

Haldane, J. B. S. 1928. The Origin of Life. *Ration. Ann.* **148**, 3-10.

Heald, E.F., J.J. Naughton, and I.L., Jr. Barnes 1963. The Chemistry of Volcanic Gases 2. Use of Equilibrium Calculations in the Interpretation of Volcanic Gases. *J. Geophys. Res.* **68**, 545-

557.

Kasting, J. F. 1993. Earth's Early Atmosphere. *Science* **259**, 920-926.

Kasting, J.F., D.H. Eggler, and S.P. Raeburn 1993. Mantle Redox Evolution and the Oxidation State of Archean Atmosphere. *J. Geol.* **101**, 245-257.

McKay, C. P., and C. R. Stoker 1989. The Early Environment and its Evolution on Mars: Implications for Life. *Rev. Geophys.* **27**, 189-214.

Miller, S. L. 1953. A Production of Amino Acids under Possible Primitive Earth Conditions. *Science* **117**, 528-529.

Miller, S. L. 1955. Production of Some Organic Compounds under Possible Primitive Earth Conditions. *J. Am.. Chem.. Soc.* **77**, 2351-2361.

Mouginis-Mark, P. J., L. Wilson, and M. T. Zuber 1992. The Physical Volcanology of Mars. In *Mars* (H. H. Kieffer, B. M. Jakosky, C. W. Snyder, and M. S. Matthews, eds.), pp. 424-452. Univ. of Arizona Press, Tucson.

Navarro-González, R., V. A. Basiuk, and M. Rosenbaum 1996. Lightning Associated to Archean Volcanic Ash-Gas Clouds. In *Chemical Evolution: Physics of the Origin of Life* (J. Chela-Flores, and F. Raulin, eds.), pp. 123-142, Kluwer Academic Publ., Dordrecht.

Oparin, A. I. 1924. *Proiskhozhdenie Zhizni*. Isd Moskovskii Rabotchii, Moscow.

Oparin, A. I. 1936. *The Origin of Life*. Macmillan, New York.

Stribling, R., and S. L. Miller 1987. Energy Yields for Hydrogen Cyanide and Formaldehyde Synthesis: The HCN and Amino Acid Concentrations in the Primitive Ocean. *Origins of Life* **17**, 261-273.

Summers, D. P., and S. Chang 1993. Prebiotic Ammonia from Reduction of Nitrite by Iron(II) on the Early Earth. *Nature* **365**, 630-633.

Wellman, T.R. 1970. Gaseous species in equilibrium with the Apollo 11 holocrystaline rocks during their crystallization. *Nature* **225**, 716-717.

Wilson, L., and J. W. Head 1983. A comparison of volcanic eruption processes on Earth, Moon, Mars, Io and Venus. *Nature* **302**, 663-669.

EARLIEST MICROBES ON EARTH AND POSSIBLE OCCURRENCE OF STROMATOLITES ON MARS

V.C. TEWARI
Wadia Institute of Himalayan Geology, Dehra Dun, 248001, India.

Abstract

The carbon isotope ratios in 3.8 Gyr old Isua metasediments in Greenland and convincing evidence of life in the form of cyanobacterial stromatolites and microfossils present in Western Australia and South Africa clearly indicate that photosynthetic life was present about 3.5 Gyr ago on the planet Earth. Three major types of microbes Archaea, Bacteria and Eucarya in evolutionary order have been recognised as the earliest microbes on Earth. The first life was a simple micro organism (the RNA world).

Recent discovery of possible filamentous bacteria from Martian meteorite ALH 84001 and presence of polycyclic aromatic hydrocarbons (PAHs) by NASA scientists is quite significant. There is vast scope for future exobiological research on Martian biota and possible occurrence of fossil microbialites (stromatolites) on Mars. Stable isotopic fractionation study of Martian microbial carbonate sediments and the latest evidence for presence of liquid water on the surface of Mars will help in understanding the comparative early history of Origins of Life and evolution of biosphere on Earth and Mars.

1. Introduction

The earliest microbes and microbially laminated structures (stromatolites) on planet Earth are recorded from Early Archean Apex Basalt of northwestern Western Australia (~ 3465 million years old; Schopf, 1993). A diverse assemblage of filamentous prokaryotic microbial fossils (cyanobacteria) have been described which suggest that the earliest photosynthesizing oxygen producing cyanobacteria had evolved on Earth around 3500 million years. It is quite interesting in evolutionary history of the cyanobacteria that some Archean genus ***(Archaeoscillatoriopsis grandis)*** is morphologically comparable to Mesoproterozoic genus ***Oscillatoriopsis media*** reported from Deoban Cherts of the Lesser Himalaya, India (Tewari, 1989).

J. Chela-Flores and F. Raulin (eds.),
Exobiology: Matter, Energy, and Information in the Origin and Evolution of Life in the Universe, 261–265.

A. grandis is also similar to the modern cyanobacterium ***Oscillatoria chalybea*** (Schopf, 1993 p. 646). These cyanobacteria were mainly responsible for development of stromatolitic buildups, the first megascopic evidence of life on Earth (Tewari, 1984, 1989, 1993).

The discovery of microfossils from the Martian meteorite ALH 84001 and presence of polycyclic aromatic hydrocarbons (Mc Kay et al, 1996; Mc Kay, 1997) is quite significant in search for early life and stromatolites on Mars. Schidlowski (1988) on the basis of carbon isotope ratios from 3.8 Gyr old Isua Complex (Fig. 1) has indicated the presence of phototrophic life on Earth. The possible occurrence of stromatolites on Mars has been discussed in recent years by various stromatophiles (Schidlowski, 1992; Mc Kay, et al, 1992; Knoll and Walter, 1996; Tewari, 1997 a, b, c).

2. Exobiology on Mars

A comparative geological and exopalaeobiological study may reveal that the early records of life (4.0 - 3.5 Gyr) would have better preserved on Mars since Earth has suffered plate tectonics and orogenic events during early Archaean. The extensive search for fossilised evidence of life like Cyanobacteria, stromatolites and eukaryotic biota (J. Chela Flores, 1997, Tewari, 1997a, b, c) from Mars will provide direct evidence of exopalaeontology.

2.1 SEARCH FOR EXTRA TERRESTRIAL STROMATOLITES (SETS)

The earliest stromatolites on Earth were formed by prokaryotic microbial communities in Archean sediments. Hence, a search for extra terrestrial stromatolites on Mars or any other planet should be a targeted mission of search for evidence of extinct life. Water rich permafrost polar region of Mars is such a possible locality for extensive search. Figure 2 shows the completely preserved 'Cone' of a Mesoproterozoic stromatolite taxa ***Conophyton garganicus*** and its axial zone with microstructure from Gangolihat Dolomite of the Lesser Himalaya, India (Tewari, 1981). Figure 3. indicates the flow chart of the search for fossil evidence of past life (microfossils, stromatolites and thrombolites etc.) on Mars (reproduced from Mc Kay, 1992) for a future MRSR mission.

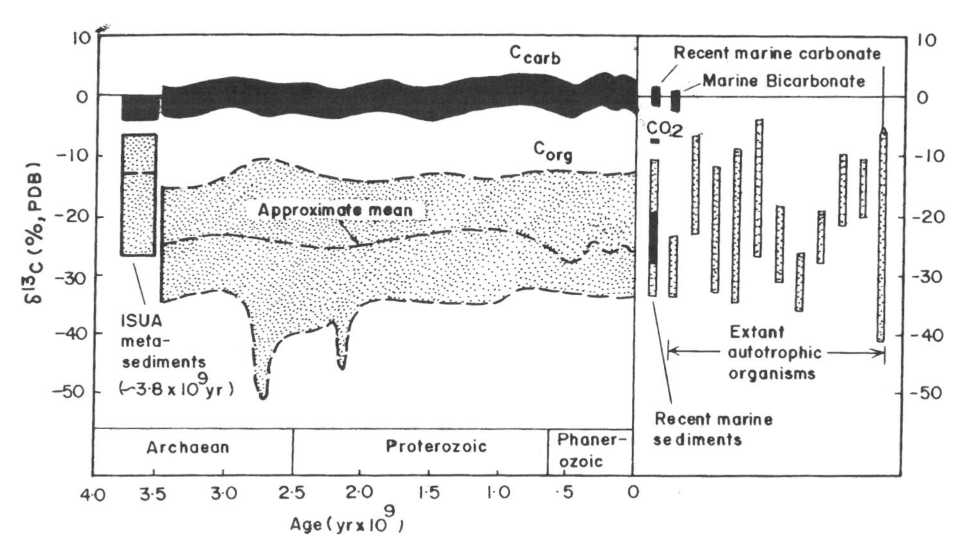

Figure 1. Carbon isotopic values for carbonate and organic carbon sediments from Isua complex on Earth (3.8 billion years, after Schidlowski, 1988). Similar carbon isotope values in ancient Martian organic material may indicate biological origins.

Figure 2. ***Conophyton garganicus*** **(cone) and its microstructure of the axial zone, a Mesoproterozoic stromatolite from Gangolihat Dolomite, Lesser Himalaya, India**

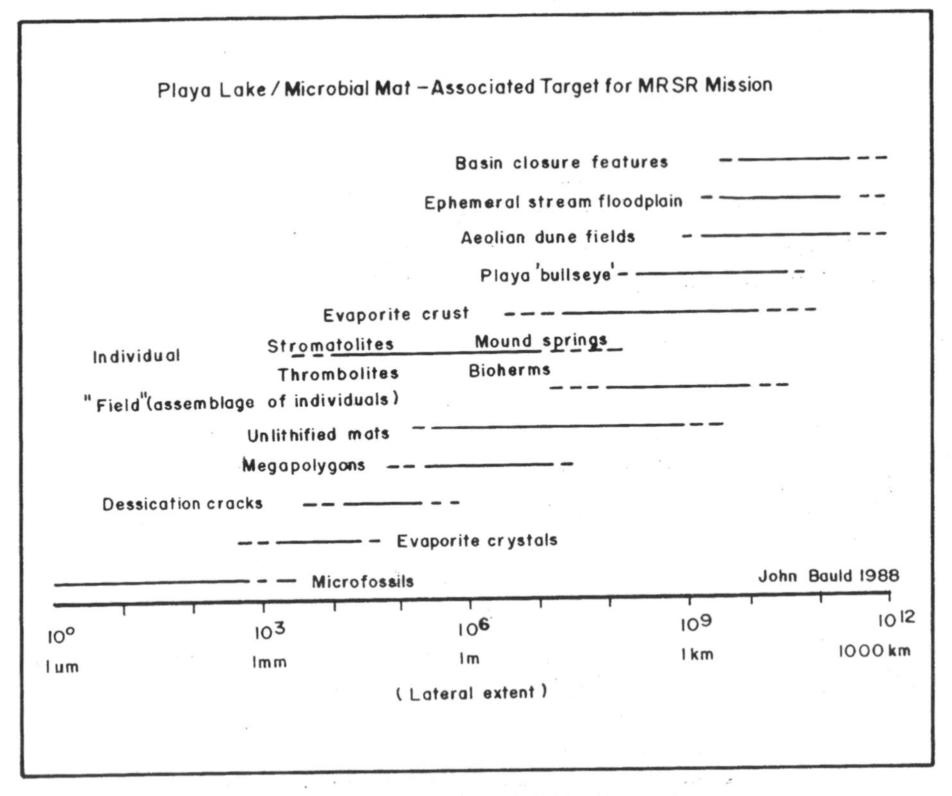

Figure 3. Search for fossil evidence of past life on Mars; a future MRSR Mission (reproduced from C.P. Mc Kay, 1992).

3. Conclusion

Stromatolites on Earth were formed by interaction of benthic microbial communities and carbonate sediments in shallow water environments during the early history of evolution of life. The recent evidences for presence of liquid water on the surface of Mars is an indicator for extraterrestrial search for life on another planet.

Acknowledgements

I am grateful to Professors Julian Chela Flores, I.C.T.P., Trieste, Italy, Manfred Schidlowsli, M.P.I., Germany, C.P. McKay, NASA, USA, Ms Francis Westall, Italy, and M. Chadha, India for discussions. I am thankful to Director, W.I.H.G., Dehra Dun for providing facilities. Mrs. Prabha Kharbanda is thanked for ably typing the manuscript.

References

Chela-Flores, J. (1997). A search for extraterrestrial eukaryotes : biological and planetary science aspects. In: ***Astronomical and Biochemical origins and the search for Life in the Universe***. Eds. C.B. Cosmovici, et al. Editrice Compositore Bologna pp. 525-532.

Knoll, A. H. and Walter, M.R. (1996). The limits of palaeontological knowledge : finding the gold among the dross, In: ***Evolution of hydrothermal ecosystems on Earth (and Mars?),*** Wiley, Chichester (Ciba Foundation symposium 202 pp. 198-213.

Mc Kay, C.P. (1992). Mars : A reassessment of its interest to biology, In: ***Exobiology in Solar System Exploration***. Eds. G. Carle *et al.* The proceedings of a symposium of Exobiology Program of NASA pp. 67-81.

Mc Kay, C.P. (1997). The search for life on Mars. ***Origins of Life and Evolution of the Biosphere*** *27*, 263-289.

Mc Kay, D.S., Gibson, E.K. *et al* (1996). Search for past life on Mars : possible relic biogenic activity in Martian meteorite ALH 84001. ***Science***, *273*, 924-930.

Schidlowski, M. (1988). A 3,800 million year old isotopic record of life from carbon in sedimentary rocks. ***Nature,*** *333*, 313-318.

Schidlowski, M. (1992). Stable carbon isotopes : possible clues to early life on Mars. ***Adv. Space Res***. *12* (4), *4* (101) *4* (110).

Schopf, J.W. (1993) Microfossils of the Early Archean Apex Chert : New evidence of the antiquity of life. ***Science,*** *260*, 640-646.

Tewari, V.C. (1981) The systematic study of Precambrian stromatolites from the Gangolihat Dolomite, Kumaon Himalaya. ***Him. Geol***. *11*, 119-146.

Tewari, V.C. (1984) Stromatolites and Precambrian - Lower Cambrian biostratigraphy of the Lesser Himalaya, India, ***Geophytology Spl. Publ.*** (1983), 71-97.

Tewari, V.C. (1989). Upper Proterozoic-Lower Cambrian stromatolites and Indian stratigraphy. ***Him. Geol.*** *13*, 143-180.

Tewari, V.C. (1993). Precambrian and Lower Cambrian stromatolites of the Lesser Himalaya, India. ***Geophytology***, *23(1),* 19-39

Tewari, V.C. (1997a) Possibility of life on Mars ***Ashmika,*** 4-5 (in Hindi), WIHG, India.

Tewari, V.C. (1997b). Earliest microbes on Earth and possible occurrence of stromatolites on Mars ***Trieste Conference on Chemical Evolution V*** Exobiology: Matter, Energy and Information in the origin and Evolution of Life in the ***Universe (abstract),*** p. 39.

Tewari, V.C. (1997c). Vendotaenids : Earliest megascopic multicellular algae on Earth (in press).

Section 8.
The Interstellar Medium, Comets and Chemical Evolution

COMETARY CONTRIBUTIONS TO PLANETARY ATMOSPHERES

TOBIAS OWEN
University of Hawaii Institute for Astronomy
2680 Woodlawn Drive, Honolulu, Hawaii 96822 USA

AND

AKIVA BAR-NUN
Department of Geophysics & Planetary Sciences
Tel-Aviv University, Ramat-Aviv, Tel-Aviv, Israel

Extended Abstract.

Scientists have speculated for many years about the possible contributions comets may have made to the volatile inventories of the inner planets. Interest in such speculation has been enhanced by the realization that comets have the potential to deliver organic materials to the early Earth, possibly to assist in the origin of life (Oro 1961). While it has been possible to argue for the probability that comets must have struck the Earth (Wetherill 1975; Sill and Wilkening 1978; Ip and Fernandez 1988) it has proved difficult to marshal specific chemical evidence that would establish beyond doubt that a cometary component was present among the Earth's volatiles.

Instead, more attention has been paid to a possible meteoritic contribution, in part because the chondritic meteorites were known to contain a so-called "planetary component" of noble gases whose relative abundances appeared to match those in Earth's atmosphere. There was also a widespread assumption that these meteorites were representative of the rocky materials making up the Earth (e.g., Turekian and Clark 1975; Anders and Owen 1977).

However, there was always a problem with the meteorites as a source of volatiles because of the high value of Xe/Kr in the "planetary component," about 10 times the value in the Earth's atmosphere. It was generally assumed that the "missing xenon" on Earth was buried in shales or ice, but careful searches have failed to find it (Wacker and Anders 1984; Bernatowycz et al. 1985). Furthermore, the relative abundances of the isotopes

J. Chela-Flores and F. Raulin (eds.),
Exobiology: Matter, Energy, and Information in the Origin and Evolution of Life in the Universe, 269–273.

in atmospheric xenon are distinctly different from those found in the meteorites. Yet krypton/argon and the krypton isotopes are closely meteoritic. Thus unless a very special fractionation and outgassing sequence is invoked (e.g., Pepin 1991), it is difficult to see how the meteorites could have supplied the heavy noble gases. Perhaps they were delivered by comets.

Unfortunately, we have no detections of noble gases in comets. Lacking direct data, it is useful to attempt laboratory experiments that could simulate the formation of comets in the outer solar nebula. A. Bar-Nun has carried out a long series of such experiments, finding a large temperature-dependent fractionation of noble gases that are trapped in amorphous ice (Bar-Nun et al. 1985; 1988; Laufer et al. 1987; Owen et al. 1991). In the case of the noble gases, ice forming at about 50 K traps Ar, Kr, and Xe in relative abundances closely resembling those in the Earth's atmosphere (Owen et al. 1992).

The model for comet formation should then include the following steps: Icy grains from a collapsing fragment of an interstellar cloud sublime as they reach the midplane of the forming solar nebula and recondense on the rapidly cooling carbon-silicate grain cores (Lunine et al. 1991). In the Uranus- Neptune region of the solar nebula, this re-condensation takes place at a temperature near 50 K. Ambient gas from the solar nebula is trapped in the condensing ice, producing grains that accrete to form a comet nucleus. If this is correct, the gases contained in the comet ices should reveal the abundance pattern produced by this process. We have argued that the relative abundances of CO^+ and N_2^+ observed in comet tails as well as the generally low level of nitrogen in comets are consistent with the trapping of solar nebula gas by the ice that formed the comets (Owen and Bar-Nun 1995a, b; Notesco and Bar-Nun 1996). That being the case, the noble gases should follow this same course and when we finally have the sensitivity to detect them in comets, we expect to find the pattern of abundances observed in the amorphous ice produced in the laboratory experiments.

Thus we are still missing two critical links between comets and the Earth: We don't **know** that comets actually carry the noble gases in the abundance pattern we find in Earth's atmosphere and the laboratory work does not yet have the necessary precision to test for fractionation of xenon isotopes, so we can't be sure that cometary xenon remsembles the xenon in our atmosphere. We must therefore seek other tests of the model.

One possibility is to look for isotopic signatures among the other volatile elements (besides the noble gases) that are specific to comets. Unfortunately, this approach also founders on limited data. The most thoroughly investigated element is carbon, and despite some early observations with large error bars, it appears that $^{12}C/^{13}C$ in comets has the same value found in the Sun, the meteorites and the Earth (Krankowsky 1991; Jewitt et al. 1997). The same is true for $^{15}N/^{14}N$ and $^{32}S/^{34}S$, at least in the case of Comet Hale-Bopp (Jewitt et al. 1997).

However, the story is very different for D/H. The most precise determination of D/H in cometary H_2O is that by the mass spectrometers on the Giotto spacecraft, leading to a value of $(D/H)_{H_2O} = 3.1 \pm 0.4 \times 10^{-4}$ for Comet P/Halley (Eberhardt et al. 1995; Balsiger et al. 1995). Subsequent Earth-based observations of the sub-mm HDO line at 464.925 GHz have yielded $(D/H)_{H_2O} = 2.9 \pm 1.0 \times 10^{-4}$ for Comet C/1996 B2 (Hyakutake) (Bockelee- Morvan et al. 1998) and $3.2 \pm 1.2 \times 10^{-4}$ for comet C/1995 O1 (Hale-Bopp) (Meier et al. 1998). Despite the large error bars on the ground-based determinations, it appears safe to conclude that the value of D/H in cometary water is distinctly higher than the value in Earth's oceans, where $D/H = 1.6 \times 10^{-4}$. If these three comets are truly representative, it is not possible to make the oceans simply out of melted comets—some other source of water with a low level of D/H is needed.

One possibility is offered by the rocks that make up most of the mass of the planet. Water vapor in the inner solar nebula could have assumed a value of $D/H = 0.8 \times 10^{-4}$ through exchange of deuterium with solar nebula H_2 (Lécluse and Robert 1994). If this water vapor were then adsorbed on rocky grains that ultimately accreted to form Earth's rocks, it would have been incorporated in the planet, where it would be available to mix with cometary water to form the oceans (Owen 1997). In fact, the high value of O/C at the Earth's crust had already suggested that another source of H_2O in addition to comets was required for the Earth (Owen and Bar-Nun 1995a).

Atmospheric neon provides another line of evidence for an initial internal reservoir of volatiles on Earth. In our atmosphere, $^{20}Ne/^{22}Ne = 9.8$, whereas the solar wind value of this ratio is 13.7 (Anders and Grevesse 1989). The neon trapped in mantle rocks shows values approaching the solar one (Craig and Lupton 1987; Honda et al. 1991), suggesting that the atmospheric neon underwent a significant fractionation process. Evidently nitrogen and carbon escaped this process, even though they have lower atomic weights than neon. The laboratory work on trapping of gases in amorphous ice shows that neon does not get trapped unless temperatures are well below 25 K (Laufer et al. 1987). Hence we do not expect comets to contain neon, and the low upper limit of $Ne/O < 0.005\times$ solar in Comet Hale-Bopp (Krasnopolsky et al. 1997) is certainly consistent with this expectation. Thus we could account for the fractionated neon in the atmosphere today as a relic of very early history on the Earth, a history that was obliterated for carbon and nitrogen (but not neon!) by the subsequent delivery of these elements (in appropriate compounds) by comets (Owen et al. 1992; Owen 1997).

So again cometary bombardment of the Earth appears to solve an atmospheric problem (the neon isotope fractionation) and this adds support to a model of cometary delivery of volatiles. Until we detect and measure noble gases in comets, however, we have only the laboratory experiments on low temperature ice to buttress the argument. Fortunately there are three missions to comets planned for the next decade that may provide the data we need: CONTOUR, DS-4 and ROSETTA. All three will carry mass spectrometers and sample systems capable of investigating the gases carried in cometary ices. Meanwhile, we can expect additional indirect searches for cometary contributions to planetary atmospheres from studies of Mars and Jupiter, both of which also show signs of cometary delivery of volatiles (Owen 1997; Owen et al. 1997).

References

Anders. E. and Grevesse, N. (1989) Abundances of the elements: Meteoritic and solar, *Geochim. Cosmochim. Acta* **53**, 197-215.

Anders, E. and Owen, T. (1977) Mars and Earth: Origin and abundances of volatiles, *Science* **198**, 317-332.

Balsiger, H., Altwegg, K., and Geiss, J. (1995) D/H in Comet P/Halley, J. *Geophys. Res.* **100**, 5827-5834.

Bar-Nun, A., Herman, G., Laufer, D., and Rapport, M L. (1985) Trapping and release of gases by water ice and implications for icy bodies, *Icarus* **63**, 317-332.

Bar-Nun, A., Kleinfeld, I., and Kochavi, E. (1988) Trapping of gas mixtures by amorphous water ice, *Phys. Rev. G* **38**, 7749-7754.

Bernatowicz, T. J., Kennedy, B M., and Podosek, F.A. (1985) Xe in glacial ice and the atmospheric inventory of noble gases, *Geochim. Cosmochim. Acta* **49**, 2561-2564.

Bockelée-Morvan, D., and 11 co-authors. (1998) Deuterated Water in comet C/1996 B2 (Hyakutake) and its implications for the structure of the primitive solar nebula, *Icarus*, in press.

Craig, H., and Lupton, J.E. (1976) Primordial neon, helium, and hydrogen in oceanic basalts, *Earth Planet. Science Lett.* **31**, 369-385.

Eberhardt, P., Reber, M., Krankoswsky, D., and Hodges, R.R. (1995), The D/H and $^{18}O/^{16}O$ ratios in water from comet P/Halley, *Astron. Astrophys.* **302**, 301-316.

Honda, M., McDougall, I. Patterson, D., Doulgeris, A., and Clague, D.A. (1991) Possible solar noble-gas component in Hawaiian basalts, *Nature* **349**, 149-151.

Ip, W.H., and Fernandez, J.A. (1988) Exchange of condensed matter among the outer and terrestrial protoplanets and the effect on surface impact and atmospheric accretion, *Icarus* **74**, 47-61.

Jewitt, D.C., Matthews, H.E., Owen, T.C., and Meier, R. (1997) Measurements of ^{12}C ^{13}C, ^{14}N ^{15}N, and ^{32}S ^{34}S ratios in Comet Hale-Bopp (C/1995 O1), *Science* **278**, 90-93.

Krankowsky, D. (1991). The composition of comets, in R. L. Newburn, M. Neugebauer, J. Rah (eds.), *Comets in the Post Halley Era*, Kluwer Academic Publishers, Dordrecht, pp. 855-878.

Krasnopolsky, V.A., Mumma, M.J., Abbott, M., Flynn, B.C., Meech, K.J., Yeomans, D.K., Feldman, P.D., Cosmovici, C.B. (1997) Detection of soft X-rays and a sensitive search for noble gases in comet Hale-Bopp (C/1995 O1), *Science* **277**, 1488-1491.

Laufer, D., Kochavi, E., and Bar-Nun, A. (1987) Structure and dynamics of amorphous water ice, *Phys. Rev. B* **36**, 9219-9227.

Lécluse, C. and Robert, F. (1994), Hydrogen isotope exchange reaction rates: Origin of water in the inner solar system, *Geochim. Cosmochin. Acta* **58**, 2927-2939.

Lunine, J. I., Engel, S., Rizk, B., and Horanyi, M. (1991) Sublimation and reformation of icy grains in the primitive solar nebula, *Icarus* **94**, 333-334.

Meier, R. Owen, T.C., Matthews, H.E., Jewitt, D.C., Bockelée-Morvan, D., Biver, N., Crovisier, J., and Gautier, D. (1998) A Determination of the HDO/H_2O Ratio in Comet C/1995 O1 (Hale-Bopp), *Science* **279**, 842-844.

Notesco, G. and Bar-Nun, A. (1996) Enrichment of CO over N_2 by their trapping in amorphous ice and implications of Comet P/Halley, *Icarus* **122**, 118-121.

Oro, J. (1961) Comets and formation of biochemical compounds on the primitive earth, *Nature* **190**, 389-390.

Owen, T. (1997) From planetesimals to planets: Contributions of icy planetesimals to planetary atmospheres, in Y. J. Pendleton and A. G. M. Tielens (eds.), *From Stardust to Planetesimals*, Astron. Soc. Pacific, San Francisco, pp. 435-452.

Owen, T.C., Atreya, S.K., Mahaffy, P.R., Neimann, H.B., and Wong, M H. (1997) On the origin of Jupiter's atmosphere and the volatiles on the Medicean stars, in C. Barbieri, J. H. Rahe, T. V. Johnson and A. M. Sohas (eds.), *The Three Galileos: The Man, The Spacecraft, The Telescope*, Kluwer, Dordrecht, pp. 289-297.

Owen, T. and Bar-Nun, A. (1995a) Comets, impacts and atmospheres, *Icarus* **116**, 215-226.

Owen, T. and Bar-Nun, A. (1995b) Comets, impacts and atmospheres II: Isotopes and noble gases, in K. Farley (ed), Conference Proceedings No. 341: *Volatiles in the Earth and Solar System*, AIP, New York, pp. 123-138.

Owen, T., Bar-Nun, A., and Kleinfeld. I. Noble gases in terrestrial planets: Evidence for cometary impacts?, in R. L. Newburn, M. Neugebauer, J. Rahe (eds.), *Comets in the Post Halley Era*, Kluwer Academic Publishers, Dordrecht, pp. 429-438.

Owen, T., Bar-Nun, A., and Kleinfeld, I. (1992) Possible cometary origin of heavy noble gases in the atmospheres of Venus, Earth and Mars, *Nature* **358**, 43-46.

Pepin, R.O. (1991) On the origin and early evolution of terrestrial planet atmospheres and meteoritic volatiles, *Icarus* **92**, 2-79.

Sill, G.T. and Wilkening, L. (1978) Ice clathrate as a possible source of the atmospheres of terrestrial planets, *Icarus* **33**, 13-27.

Turekian, K.K. and Clark, S.P., Jr. (1975) The non-homogenous accumulation model for terrestrial planet formation and the consequences for the atmosphere of Venus, J. *Atmos. Sci.* **32**, 1257-1261.

Wacker, J.F. and Anders, E. (1984) Trapping of xenon in ice and implications for the origin of the Earth's noble gases, *Geochim. Cosmochim. Acta* **48**, 2372-2380.

Wetherill, G.W. (1975) Late heavy bombardment of the moon and terrestrial planets, *Proc. Lunar Sci. Conf.* **VI**, pp. 1539-1561.

COMETS AS A SOURCE OF LIFE'S ORIGINS

J. MAYO GREENBERG AND AIGEN LI
Laboratory Astrophysics, University of Leiden, Postbus 9504, 2300 RA Leiden, The Netherlands
e-mail: greenber@strw.LeidenUniv.nl; agli@strw.LeidenUniv.nl

Abstract. The proposition that comet dust provided the necessary ingredients for prebiotic chemical evolution on the earth is supported in five ways: (1) comet dust contains prebiotic organics; (2) comet dust is chiral; (3) comet dust could have been brought "safely" to the earth's surface as a consequence of its fluffy character and of fragmentation processes occurring in the earth's vicinity; (4) the morphological as well as chemical structure of comet dust particles of 3–5 μm provide the thermodynamic basis for directed chemical organization; (5) each 3 km size comet deposited 10^{25} such seeds of comet dust on the prebiotic earth.

1. Introduction

While it has been correctly argued that comets are not the greatest contributors of organics to the earth's surface (Chyba & Sagan 1992) we believe it is *incorrect* to use this fact to imply that comets are not the most likely contributor to the origins of life. We use the plural of the word origins because it is more likely than not that the way comets delivered prebiotic organics to the earth, and undoubtedly also to Mars, the beginning of life had a multitude of chances both in time and space. The problem of demonstrating this devolves itself into showing how the chemical composition, the chirality and the morphology of comet organics and comet dust combine to provide appropriate packages of complex organics which when landing in water provide highly concentrated, albeit small, volumes of prebiotic chemical evolution sites.

The story starts with the chemical evolution of tenth micron interstellar dust grains in the clouds of gas and dust in space, how they become chiral, how they coagulate in the protosolar nebula into comets and how the fragile

J. Chela-Flores and F. Raulin (eds.),
Exobiology: Matter, Energy, and Information in the Origin and Evolution of Life in the Universe, 275–285.

fluffy nature of comet nuclei provides the physical basis for "gentle" delivery of comet organics to the earth.

2. Interstellar and precometary dust

The basic model of interstellar dust consists of three populations of particles (Li & Greenberg 1997). The major mass is in tenth micron particles consisting of silicate cores with organic refractory (complex organic molecules) mantles. Additionally there are very small carbonaceous particles/large molecules. In molecular clouds the large particles accrete additional mantles of frozen molecules and in the dense clouds there is also accretion of the very small particles along with the "ices". This is schematically shown in Fig. 1. The nature of the organic mantle material varies depending on whether the dust is in a low density diffuse cloud or a molecular cloud (Tielens et al. 1996; Greenberg & Li 1996a; Schutte et al. 1998). There are significant variations in the relative proportions of C, N, O and H in the complex organics in different regions. In diffuse clouds the organic mantle is strongly depleted in oxygen and hydrogen, whereas in molecular clouds complex organic molecules are present with more abundant fractions of oxygen and hydrogen (Greenberg et al. 1995). Furthermore, the ratio of the mass of organic mantles to the silicate core is highly variable. In the unified model for diffuse cloud dust of Li & Greenberg (1997) this ratio is $V_{OR}/V_{si} = 0.95$, whereas matching the silicate polarization in the Orion B-N object requires $V_{OR}/V_{si} \approx 2$ (Greenberg & Li 1996a). It is of interest to note that the mass spectra of comet Halley dust - as obtained by Kissel & Krueger (1987) and presumably representing the ultimate molecular cloud collapse phase - gave about equal masses of organics and silicates in the dust which implies a volume ratio of about 2. At the other extreme is the region towards the galactic centre which appears to have a very low ratio $V_{OR}/V_{si} = 0.23$ (Tielens et al. 1996). We shall assume that the organic refractory mantles in the final stages of cloud contraction are most closely represented by the properties obtained for Halley dust; i.e. $M_{OR}/M_{si} = 1$ and with an atomic distribution as given in Table 1 for comet dust organics .

3. Chemical composition of the comet nucleus

The constituents of comet nuclei as aggregated interstellar dust may be divided into two basic categories: refractories and volatiles. The refractories consist of the silicate cores, the organic mantles and the very small carbonaceous/large molecule particles. The volatiles consist of the ices accreted on the dust in the final stage of collapse of the presolar cloud. By combining the relative proportion of the organics as represented by the Halley dust

TABLE 1. Stoichiometric distribution of the elements in laboratory organics compared with the comet Halley mass spectra normalized to carbon.

	Lab Organics			Halley		
	Volatile[†]	Refractory[†]	Total	PICCA(gas)	Dust	Total[‡]
C	1.0	1.0	1.0	1.0	1.0	1.0
O	1.2	0.6	0.9	0.8	0.5	0.6
N	0.05	> 0.01	>0.03	0.04	0.04	0.04
H	1.70	1.3	1.5	1.5	1.0	1.2

[†] Division between volatile and refractory is here taken at a sublimation temperature less than or greater than $\sim$ 350K respectively.
[‡] Assuming equal amounts of dust and gas.

with the silicates of interstellar dust and incorporating the general distribution of volatiles as shown in Table 2 from observations of comet coma it has been possible to deduce a generic composition for comet nuclei as shown in Table 3. For details of this procedure see Greenberg (1998).

It is often stated that the interstellar dust ice mantles contain molecular species which are a reasonable facsimile of comet coma species. In fact it is suggested that if the dust molecules could be observed up to the time of comet nucleus formation the construction of a nucleus model would be straightforward. The reason for this is that some of the coma molecules are difficult to associate with solar nebula chemistry using partially evaporated interstellar dust; e.g. HNC, CH_4 (Mumma et al. 1993). Even though the dust mantle data is limited generally to prestellar molecular clouds or post (massive) star formation regions we believe it is instructive to demonstrate the extent to which the comet nucleus molecular pattern is defined as compared with using coma molecules as a starting point. As in the case of coma molecules we note that there is quite a spread in abundances in dust ices. We have chosen high mass embedded protostellar sources rather than molecular clouds as more representative of precometary dust. This is arguable but recall that this is only for comparison purposes. Using the relative values for the more abundant molecules which we have abstracted from Table 4 are $H_2O:CO:CH_3OH:CO_2:H_2CO:CH_4:NH_3:O_3 = 100:5:7:5:5:2:5:1$. The resulting molecular constituent abundances in the comet nucleus are listed in Table 3, column (b). We see, as expected, a small but significant

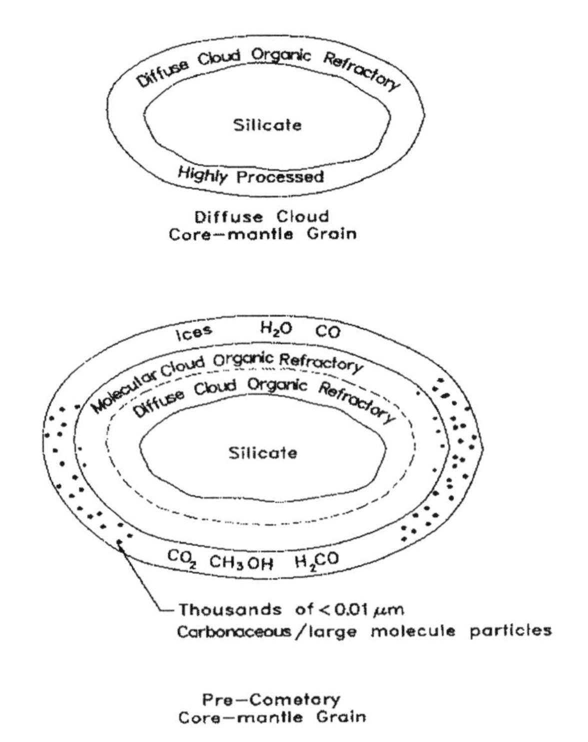

Figure 1. A schematic description of the morphological and chemical structure of core-mantle interstellar dust grains in diffuse cloud regions and in the latest stage of the collapse of an interstellar cloud. The ices are both accreted and created along with the molecular cloud organics. The very small particle/large molecule components of the interstellar dust accrete along with the ices in the dense cloud.

reduction in the total mass fraction of water down to 0.26 but for the rest the changes are not very large. It is evident that a nominal water fraction of about 0.30 is impossible to escape so long as we constrain the comet in terms of the core-mantle dust model and solar system abundances.

4. Chirality in interstellar dust and in comets

Bonner and his colleagues (Bonner 1991; Bonner & Rubinstein 1990) have suggested that the circularly UV polarized radiation from neutron stars could selectively enhance the chirality of interstellar dust organic mantles. Since a neutron star emits left and right circularly polarized radiation equally at opposite poles, if a cloud of dust were to pass by the star symmetrically (equal amounts “above and below”) the net chirality in the cloud would be zero; i.e., equal amounts of left handed and right handed excesses. Such symmetrical passage is unlikely so that most clouds of dust would have

TABLE 2. Molecular abundances in the coma of comets*.

Molecule	C/Hyakutake at 1 AU	Others at 1 AU
H_2O	100	100
CO	5-30	2-20
CO_2	≤ 7	3-6
CH_4	0.7	≤ 0.5-2
C_2H_2	0.3-0.9	
C_2H_6	0.4	
CH_3OH	2	1-7
H_2CO	0.2-1	0.05-4
NH_3	0.5	0.4-0.9
N_2		0.02
HCN	0.15	0.1-0.2
HNC	0.01	
CH_3CN	0.01	
HC_3N	≤ 0.02	
H_2S	0.6	0.3
OCS	0.3	≤ 0.5
S_2	0.005	0.02-0.2
SO_2		≤ 0.001

* – taken from Bockélee-Morvan (1997), in: *Molecules in Astrophysics: Probes and Processes* (E.F. van Dishoeck, ed.), IAU Symp. 178, Kluwer, p.222

a net enantiomeric excess.

The question to be answered is whether random passages of clouds past neutron stars will produce a reasonable expectation of a presolar cloud having a significant enantiomeric excess. The answer to the question resolves itself into two parts: (1) what is the total number of active neutron stars in the galaxy? (2) what is the circularly polarized ultraviolet flux from these sources?

We can attempt to answer these questions on the basis of observations (Greenberg et al. 1994; Hermsen et al. 1994). The crab pulsar is the youngest and is *known* to be about 1040 years old. The oldest pulsar detected is approximately 20 million years old. If we assume that type II supernova explosions occur about every 60 years to produce these pulsars we estimate about $2 \times 10^7/60 = 3 \times 10^5$ pulsars currently active in the Milky Way with periods extending from 30 msec for the crab to about 0.40 sec for PSR B0656 + 14 (see Hermsen et al. 1994 and references therein). The

TABLE 3. Distribution by mass fraction of the major chemical constituents of a comet nucleus: (**a**) as derived from comet volatiles, (**b**) as derived from dust ice mantles.

Materials	Mass Fraction (a)	Mass Fraction (b)
Sil.	0.26	0.26
Carb.	0.086	0.092
Organ.Refr.	0.23	0.23
H_2O	0.31	0.26
CO	0.024	0.02
CO_2	0.030	0.03
CH_3OH	0.017	0.03
H_2CO	0.005	0.02
(other)	0.04	0.05

TABLE 4. Molecules observed in interstellar ice mantles*. Abundances refer to observations of background sources (B), if available. Otherwise, the composition towards high-mass embedded protostellar sources (hmE) is listed.

Molecule	Abundance (%)	Comments
H_2O	100	B
CO (apolar)	10-40	B
CO (polar)	1-10	B, hmE
CH_3OH	$[\leq 4]$-10	hmE
CO_2	$[\leq 0.4]$-10	hmE, tentative
H_2CO	$[\leq 1]$-10	hmE, tentative
H_2	≥ 1	hmE
CH_4	~ 2	hmE, tentative
NH_3	≤ 10	B, hmE
O_3	≤ 2	hmE
XCN	?	hmE
OCS/XCS	?	hmE

* – Taken from Schutte (1996), in: The Cosmic Dust Connection (J.M. Greenberg, ed.), Kluwer, p.1

relationship between the age and the period is based on the spin down rate assuming a constant magnetic field. The total luminosity L is proportional to ω^4. We note that, according to the newest information, the ratio of the X-ray luminosity to the spin down power appears to be roughly constant and comparable to that of the crab pulsar. We shall take the crab as our standard pulsar. If we assume that the X-ray luminosity is an extension of the visible-ultraviolet luminosity and if we assume that the spectral distribution follows $\nu F(\nu)$ = constant as is observed for the crab pulsar at a distance of 2 kpc then $\nu F(\nu) = 1015\,\mathrm{Hz}\times10^{-29}\mathrm{watts/m^2} \approx 10^{-11}\,\mathrm{erg\,cm^{-2}\,s^{-1}}$ which, at 1 pc, corresponds to about $4 \times 10^{-5}\,\mathrm{erg\,cm^{-2}\,s^{-1}}$ or 2.5×10^6 UV photon $\mathrm{cm^{-2}s^{-1}}$ in the decade 3-30 ev; i.e. 10 ev photons.

We picture, in an ideal situation, an interstellar grain (within a cloud) travelling perpendicular to the spin axis of a neutron star at a distance of closest approach R. Along the neutron star spin axis the synchrotron radiation is circularly polarized while at an angle, the light is elliptically polarized. For a dust grain passing above the neutron star the right handed mirror image dust molecules would be selectively destroyed leaving an excess of left handed molecules, and vice versa below. Given an ultraviolet flux of Φ_0 at 1 pc, the net circularly polarized fluence along the path indicated is

$$F = \int_{-\infty}^{+\infty} R\,\Phi_0\,dy/[V_d\,(R^2 + y^2)^{3/2}] = 2\,\Phi_0\,(R/pc)^{-2}\,(R/cm)(V_d/cm\,s^{-1})^{-1} \quad (1)$$

The distance from the crab at which the total fluence, F, acting on a typical grain (relative speed $10\,\mathrm{km\,s^{-1}}$) corresponds to the laboratory flux acting for about 50 hrs is given roughly by $R \approx 40$ pc. This normalizes the astrophysical phenomena to the laboratory efficiency (Greenberg et al. 1994). Note that we had used only a single wavelength and a single amino acid. This result should be subjected to a more complete experimental study in which also the circularly polarized UV processed ice mantle is considered to direct the presence of an e.e.

The neutron stars with fluxes at 1 pc equal to that of the crab nebula at 40 pc are all those with $(\omega/\omega_C)^4 \geq (40)^2$ which leads to all those with $T < 6.3\,T_C = 210$ ms. Thus all pulsars with ages $\leq$ 1.7 million years – about 3×10^4 - qualify as candidates to produce a very *high* enantiomeric excess in a cloud passing within 1 pc. With a mean random cloud speed of $10\,\mathrm{km\,s^{-1}}$ the time between cloud-pulsar collisions (for this number of "effective" pulsars) is $\tau_{C-P} = 3 \times 10^8$ years.

Should a cloud "collide" with a neutron star it will generally only survive an additional few tens of millions of years (well before the next possible encounter) and thus the dust which it contains will either have dissipated, or partaken in the collapse leading to star formation (and possible comet formation), before the next encounter. Thus the probability of finding an individual cloud which has survived a neutron star encounter is $\approx 3 \times 10^7/3\times10^8 = 0.1$. If we were to relax the enantiomeric excess requirement by a factor of 5, the fraction of clouds with 10% enantiomeric excess increases by a factor of 10; i.e., all presolar clouds would satisfy this requirement. A more precise evaluation of the probability of a presolar cloud having an excess of left (or right) handed molecules can be made but in any case it appears highly probable that a *significant* fraction of solar systems started off with a *significant* enantiomeric excess.

5. Preservation of chirality in comets and in the parent bodies of meteorites

There exists incontrovertible evidence for the preservation of complex molecular interstellar dust ingredients in comets (Kissel & Kruger 1987; Mumma et al. 1993). The evidence that the H_2O of interstellar dust was not heated during the formation is also well documented and such volatile molecules as CH_4 and even CO appear to have been preserved in comets in the kinds of relative abundances deduced from observations of interstellar dust. The model of comets (Greenberg & Hage 1990) as aggregated interstellar dust leads also to the fact that comets are very fluffy bodies whose overall density is about $0.3\,\mathrm{g\,cm^{-3}}$. This is consistent with a density of $0.28\,\mathrm{g\,cm^{-3}}$ deduced from non-gravitational motions of comets (Rickman 1989).

There is no question that comets have provided a major fraction of the organics to the earth's surface and that half or more of the oceans were brought by comets in the first 1000 million years of the earth's existence (Greenberg 1986; Chyba 1990; Chyba & Sagan 1992; Chyba et al. 1990). Note that the complex organics alone in the interstellar dust model (Greenberg 1982, 1993) count for at least half the mass of the (ice) water (Greenberg 1998) so that one comet of about 1 km radius would deliver about 1% of the earth's total current biomass in the form of these organics.

The preservation of the presolar dust chirality in comets appears assured on the basis of the fact that there was never liquid water in the nucleus as well as the fact that the temperature probably never rose above about 80 K (Haruyama et al. 1993; Sirono 1998). Further confirmation of this as well as the presence of a high enantiomeric excess in the comet nuclei may be inferred from the recent work on the Murchison meteorite (Engel & Macko 1997). They have found an unambiguously non theoretical value of $L/D > 2$ for some of the amino acids in the meteorite organics and since comet nuclei material has undoubtedly undergone less heating and metamorphosis than the asteroidal/meteorite material we suggest that the comet nuclei have an even higher degree of homochirality.

The low mean density (Greenberg & Hage 1990; Li & Greenberg 1998; Sekanina 1996) of $0.3\,g\,cm^{-3}$ makes it possible for a (dense) atmosphere (Kasting 1993) to have substantially cushioned the comet nucleus fall. Secondly, the low density loose aggregation leads to the creation within the nucleus of expansion shocks rather than compression shocks (Zeldovich & Raizer 1967). This causes the nucleus to fragment and ablate into smaller parts which are better cushioned by the atmosphere. Thirdly, it has been found that cometary nuclei with radii $\geq 0.1 - 0.3$ km will split tidally (Rickman & Greenberg 1998) even before impacting the earth; i.e. at distances $\leq 5 - 8$ earth radii. Consequently, streams of fragments and debris form in Earth-crossing orbits more often than large nuclei cause impacts. Finally, the morphological structure and the physical properties of the core-mantle materials in the individual dust components within the fragments provide a further basis for preserving the non-volatile organics. The heat within the comet fragment is concentrated in the ice mantles of the submicron dust grains because amorphous ice has an extremely low sound speed. The dust ice mantles act like a heat shield for the organics and furthermore, by evaporation, lead not only to cooling of the structure but the gas expansion produces further fragmentation. The final fragments with densities as low as or lower than $0.1\,g\,cm^{-3}$, like comet dust after all volatiles are removed (Greenberg & Hage 1990), may float gently to the earth.

6. Comet dust as seeds for life's origins

Krueger & Kissel (1989) have suggested how small comet dust grains ($3 - 5\,\mu m$ in size) made up of highly porous aggregates of silicate-core organic refractory mantle particles as pictured by Greenberg & Hage (1990) and schematically shown in Fig. 1 as diffuse cloud dust fulfill the requirements for chemical thermodynamics to start molecular organization. Each comet dust particle sitting within a water bath (the primitive ocean) satisfied the conditions for non-equilibrium thermodynamics as in Nicolis & Prigogine (1977). Now, as we picture the additional advantage of homochirality, there appears to be an even greater probability that each comet dust grain provided a seed for life's origins. Furthermore, even if each comet of 3 km radius deposited only 0.1% of its mass as comet dust in the appropriate size range there would have been 10^{25} such seeds (Greenberg & Li 1996b).

7. Conclusions

The finding of enantiomeric excesses of $L/D \geq 2$ in amino acids in the Murchison meteorite is indeed an exciting result (Engel & Macko 1997). It seems to confirm the *quantitative* prediction of the likelihood of 10%

or greater excess in the interstellar dust cloud (Greenberg 1996) which collapsed to form the sun and planets which follows from the Bonner-Rubinstein hypothesis of neutron stars being the cause of chirality in the interstellar dust (Bonner & Rubinstein 1990; Bonner 1991) molecules in the Greenberg dust model.

If interstellar dust chirality is preserved in meteorites it must also be preserved in comets and consequently one of the prime analyses of comet material in the ROSETTA mission should be the search for chirality.

Since the small comet dust particles provide a remarkably suitable environment for directed prebiotic chemical evolution (Kruger & Kissel 1989; Nicolis & Prigogine 1977) the discovery of left handedness in relics (meteorites) of the early solar system provides a strong basis for believing that comets were the main source of the origin of life on earth and probably on Mars as well.

8. Acknowledgements

We are grateful for the support by NASA grant NGR 33-018-148 and by a grant from the Netherlands Organization for Space Research (SRON). One of us (AL) wishes to thank Leiden University for an AIO fellowship and the World Laboratory for a scholarship.

References

Bockélee-Morvan, D., in: *Molecules in Astrophysics: Probes and Processes* (E.F. van Dishoeck, ed.), IAU Symp. 178, Kluwer, p.222

Bonner, W.A. & Rubinstein, F., in: *Prebiological Self-organization of Matter* (C. Ponnamperuma & F.R. Eirich, eds.), p.35, 1990.

Bonner, W.A., The origin and amplification of biomolecular chirality, *Origins of Life & Evolution of the Biosphere* 11, 59, 1991.

Chyba, C.F., Impact delivery and erosion of planetary oceans in the early inner Solar System, *Nature* 343, 129-133, 1990.

Chyba, C.F., Thomas, P.J., Brookshaw, L. & Sagan, C., Cometary delivery of organic molecules to the early earth, *Science* 249, 366-373, 1990.

Chyba, C.F. & Sagan, C., Endogenous production, exogenous delivery and impact-shock synthesis of organic molecules: an inventory for the origins of life, *Nature* 355, 125-132, 1992.

Engel, M.H. & Macko, S.A., Isotopic evidence for extraterrestrial non-racemic amino acids in the Murchison meteorite, *Nature* 389, 265-268, 1997.

Greenberg, J.M., What are comets made of - a model based on interstellar dust, in: *Comets* (L. Wilkening, ed.), University of Arizona Press, p.131-163, 1982.

Greenberg, J.M., The chemical composition of comets and possible contribution to planet composition and evolution, in: *The Galaxy and the Solar System* (R. Smoluchowski, J.N. Bahcall & M.S. Matthews, eds.), University of Arizona Press, p.103-115, 1986.

Greenberg, J.M. & Hage, J.I., From interstellar dust to comets: a unification of observational constraints, *Astrophys. J.* 361, 260-274, 1990.

Greenberg, J.M., Physical and chemical composition of comets - from interstellar space to the earth, in: *The Chemistry of Life's Origins* (J.M. Greenberg, C.X. Mendoza-Gmez

& V. Pirronello, eds.), Dordrecht, Kluwer, p.195-207, 1993.
Greenberg, J.M., Kouchi, A., Niessen, W., Irth, H., van Paradijs, J., de Groot, M. & Hermsen, W., Interstellar dust, chirality, comets and the origins of life: life from dead stars? *J. Biol. Phys.* 20, 61-70, 1994.
Greenberg, J.M., Li, A., Mendoza-Gómez, C.X., Schutte, W.A., Gerakines, P.A. & de Groot, M., Approaching the interstellar grain organic refractory component, *Astrophys. J.*, 455, L177-L180, 1995.
Greenberg, J.M., Chirality in interstellar dust and comets: life from dead stars?, in: *Physical Origin of Homochirality in Life, AIP Proc. 379* (D.B. Cline, ed.), p.185-210, 1996.
Greenberg, J.M. & Li, A., What are the ture astronomical silicates? *Astron. Astrophys.* 309, 258-264, 1996a.
Greenberg, J.M. & Li, A., From cosmic formation of chiral bioorganics in interstellar dust to comet seeds of life's origins: are 10^{25} chances enough?, in: *Chemical Evolution: Physics of the Origin and Evolution of Life* (J. Chela-Flores & F. Raulin, eds.), Kluwer, p.51-71, 1996b.
Greenberg, J.M., Making a comet nucleus, *Astron. Astrophys.* 330, 375-380, 1998.
Hermsen, W., Kuiper, L., Diehl, R., et al., Gamma-ray pulsar studies with COMPTEL, *Astrophys. J. Suppl.* 92, 559, 1994.
Haruyama, J., Yamamoto, T., Mizutani, H. & Greenberg, J.M, *Journal of Geophysical Research* 98, No. E8, p.15079, 1993.
Kasting, J.F., Early evolution of the atmosphere and ocean, in: *The Chemistry of Life's Origins*, (J.M. Greenberg, C.X. Mendoza-Gómez & V. Pirronello, eds.), Dordrecht, Kluwer, p.149-176, 1993.
Kissel, J. & Krueger, F.R., The organic component in dust from comet Halley as measured by the PUMA mass spectrometer on board Vega 1, *Nature* 326, 755-760, 1987.
Krueger, F.R. & Kissel, J., Biogenesis of cometary grains - thermodynamic aspects of self-organization, *Origins of Life & Evolution of the Biosphere* 19, 87-93, 1989.
Li, A. & Greenberg, J.M., A unified model of interstellar dust, *Astron. Astrophys.* 323, 566-584, 1997.
Li, A. & Greenberg, J.M., From interstellar dust to comets: infrared emission from comet Hale-Bopp (C/1995 O1), *Astrophys. J. (Letters)*, in press, 1998.
Mumma, M.J., Stern, S.A. & Weissman, P.R., Comets and the origin of the solar system: Reading the Rosetta stone, in: *Planets and Protostars III* (E.H. Levy, J.I. Lunine & M.S. Matthews, eds.), University of Arizona Press, Tucson, p.1177-1252, 1993.
Nicolis, G. & Prigogine, I., Self-organization in Nonequilibrium Systems, Wiley, 1977.
Rickman, H., The nucleus of comet Halley: surface structure, mean density, gas and dust production, *Adv. Space Res.* 9, No.3, 59-71, 1989.
Rickman, H. & Greenberg, J.M., in: *Hights in Astronomy*, in press, 1998.
Sekanina, Z., Tidal breakup of the nucleus of comet Shoemaker-Levy 9, in: *The Collision of Comet Shoemaker-Levy 9 and Jupiter*, (K.S. Noll, et al., eds.), Cambridge University Press, p.55-80, 1996.
Schutte, W.A., in: *The Cosmic Dust Connection* (J.M. Greenberg, ed.), Kluwer, p.1, 1996.
Schutte, W.A., et al., *Astron. Astrophys.*, to be submitted, 1998.
Sirono, S., *PhD Thesis*, Hokkaido University, 1998.
Tielens, A.G.G.M., Wooden, D.H., Allamandola, L.A., Bregman, J. & Witteborn, F.C., *Astrophys. J.* 461, 210, 1996.
Zeldovich, Ya.B. & Raizer, Yu.P., *Physics of Shock Waves and High-temperature Hydrodynamic Phenomena*, Academic Press, New York, 1967.

FORMATION OF MOLECULAR HYDROGEN: THE MOTHER OF ALL MOLECULES

An Experimental Investigation

V. PIRRONELLO[1], O. BIHAM[2], C. LIU[3], L. SHEN[3], G. VIDALI[3]
[1]*Istituto di Fisica, Universita' di Catania*
Viale A. Doria, 6 – 95125 Catania, Sicily, Italy
e-mail: vpirrone@cdc.unict.it
[2]*Racah Institute, The Hebrew University, Jerusalem, Israel*
[3]*Physics Department, Syracuse University, Syracuse, NY 13442, USA*

1. Introduction

The problem of the formation of molecular hydrogen is certainly the most important one in astrophysical environments, and in interstellar clouds in particular, because H_2 is the most abundant molecule in space and, as such, it is the most frequent collisional partner of all other particles in the gas phase. This relevant peculiarity makes of H_2 what has been called (Duley and Williams, 1984) the "seminal molecule", the *mother* of all other molecular species; it, in fact once ionized by UV photons in diffuse clouds and by cosmic rays in dense, initiates ion-molecule reactions (the only ones that can efficiently take place in the cold interstellar medium because characterized by low energy activation barriers) chains that have been succesfully proposed to synthesize most of observed species in space. Furthermore molecular hydrogen, together with the other species, provides a very efficient mechanism for cooling clouds. Molecules, in fact, once are collisionally excited, deexcite in cascade through rotovibrational transitions emitting photons to which the cloud itself is transparent. This cooling mechanism helps the gravitational collapse of interstellar clouds, increases the formation rate of stars and has great impact in the structure and evolution of galaxies.
In this note we report on measurements of molecular hydrogen synthesis for the first time performed in conditions and on materials close to interstellar reality (Pirronello et al., 1997a,b).

2. The Mechanism of H_2 Formation

The problem of the formation of molecular hydrogen arises from the fact that such simple molecule is not efficiently formed by the direct reaction of two hydrogen atoms colliding in the gas phase. This is a consequence of the fact that the protomolecule, just formed in a high vibrational level, has to release quickly (roughly in a time comparable with its vibrational period) an energy excess of about four and a half electronvolt (or at

J. Chela-Flores and F. Raulin (eds.),
Exobiology: Matter, Energy, and Information in the Origin and Evolution of Life in the Universe, 287–290.

least a good fraction of it) to became stable. In the gas phase such a proto-molecule is isolated and the only way to achieve this goal is through the emission of a photon; this process is however quite slow because involves roto-vibrational forbidden transitions and the usual unavoidable result is that the two hydrogen atoms restart wondering in the cloud. This mechanism, that is important at the early stages of the universe for the formation of primordial molecular hydrogen, is inside already developed galaxies unable to explain the observed abundances of H_2. It was then proposed already decades ago (Gould and Salpeter, 1963; Hollenbach and Salpeter, 1970; Hollenbach, Werner and Salpeter, 1971; Williams, 1968) that a three body reactions (with the third one taking the excess energy) would have been the way to overcome the problem. In the interstellar medium, due to the low densities involved (n_H of the order of 10^2 cm^{-3}, in the so called "diffuse clouds"), the third body cannot be another atom, not even H that is the most abundant, but it must be a dust grain.
The series of processes that have to occur are: collision and sticking of two H atoms with a grain of interstellar dust, mobility of both of them and, upon encounter, recombination with formation of H_2 and in most of cases release in the gas phase.

3. Experimental

The experimental set up and methodology has been already described in detail elsewhere (see e.g. Pirronello et al., 1997a,b), here there is just a short summary.
The measurements have been performed in a cylindrical UHV (Ultra High Vacuum) chamber (in which pressure can be maintained as low as 10^{-10} torr during the whole experiment). In the centre of the chamber, on a substrate that can be cooled by liquid helium flow to about 5 or 6 K, it can be mounted the sample, whose surface is chosen to be the dust grain analogue where to measure H_2 recombination. From two triple differentially pumped lines, aiming to the same spot on the target, atomic hydrogen and deuterium (obtained from H_2 and D_2 previously dissociated in a cavity by a radio frequency input) are admitted and encounter for the first time on the target surface.
Detection of molecules synthesized on the surface is obtained by means of a rotatable quadrupole mass spectrometer tuned on mass 3 to detect HD, a possibility given by our choice to use isotopically labelled hydrogen to get rid of background problems.
The efficiency of the process "r", defined by the ratio of the HD signal "I" (detected after placing the mass spectrometer probe in front of the surface and in between the two lines) divided by the signal relative to the incoming beam "I_{in}" (previously measured placing the quadrupole probe in front of H and D lines), has been obtained as a function of the surface temperature during irradiation. Such an efficiency "r" has been measured for each run both during irradiation of the surface with H and D atoms

$$r_{irr} = I_{irr}/I_{in}$$

and after a TPD (Temperature Programmed Desorption), a procedure of linear warming of the target performed after each irradiation to avoid neglecting contribution to molecular hydrogen formation coming from species that, especially at the lowest temperatures, had remained stuck on the surface

$$r_{TPD} = I_{TPD}/I_{in}$$

The total efficiency at each irradiation temperature is then given by

$$r_{tot} = r_{irr} + r_{TPD}$$

4. Results and Discussion

Among the rich collections of results obtained, we will here report on some of the most interesting relative to "olivine", a magnesium and iron silicate. A natural, hence polycrystalline, stone mechanically polished until shiny, whose surface is certainly a better analogue than any model surface to interstellar silicate (that by the way are believed to be amorphous in structure, due to their wide Full Width at Half a Maximum in the 9.7 μn spectral feature). In Table 1 it is shown the efficiency of molecular hydrogen recombination as a function of the irradiation temperature, in the temperature range of interest in interstellar clouds.

TABLE 1. Efficiency of Molecular Hydrogen Recombination

T(K)	r_{irr}	r_{TPD}
10	.04	.08
12	.05	.06
14	.06	.05
16	.06	.04
18	.07	.03

Table 1 shows how recombination efficiency of molecular hydrogen varies with the surface temperature during irradiation with atoms. Measured values are up to one order of magnitude lower than theoretical estimates by Hollenbach and Salpeter (1971), obtained under some simplifying assumptions for a monocrystalline water ice surface.

Besides these quantitative results, that by the way can be straightforwardly applied in interstellar cloud chemical models, some other interesting experimental evidences were collected that shed light on the type of processes that take place on the sample during irradiation with atomic hydrogen. These experimental evidences are relative to several measurements performed after irradiation with H and D atoms at the lowest temperatures (5K-6K), when warming up the substrate to perform a Temperature Programmed Desorption. The only difference between each run was the irradiation time and hence the coverage (number H and D atoms deposited on the olivine exposed surface. During TPD the quadrupole mass spectrometer was tuned on mass 3 and the outcome of each of these runs was a peak of HD release versus temperature.

It is useful to notice that the amount of HD detected during irradiation at the lowest irradiation temperatures was always very small. During TPD starting from 5-6K no signal was detected until a temperature between 9 and 10K was reached by the olivine layer. Furthermore, going from the lowest coverage values up to higher coverage ones, the HD peak collected during TPD showed a peculiar behaviour: the position of its maximum migrated from higher to lower temperatures. This behaviour is characteristic of second order desorption. Such an expression means only that the species that desorb were not originally adsorbed on the surface but, on the contrary, two different partners were adsorbed and when they became mobile they migrated on the surface, encountered each other and only then desorbed. These evidences imply that H atoms are adsorbed on the surface as atoms and that mobility is not assured by tunneling at all substrate temperatures but some sort of thermal activation is needed; in fact, until a certain temperature is reached adatoms remain immobile in their adsorption sites and hence no recombination and molecular hydrogen formation occur. Even if results (given in Table 1) are not that different from Hollenbach and Salpeter (1971) calculations, from a qualitative point of view the onset temperature for mobility of H adatoms and the second order desorption are closer to Smoluchowski (1981, 1983) calculations.

In a second order desorption , the desorption yield is proportional to the square of the coverage, hence at least in the low coverage regime (the important one in interstellar environments) the famous expression of Hollenbach, Werner and Salpeter (1971) for the production rate (per cubic centimetre, per second) of molecular hydrogen should be substituted by the other one (Pirronello et al., 1997b)

$$R(cm^{-3}s^{-1}) = \tfrac{1}{2}\,(n_H\, v_H\, S_H\, A\, t_H)^2\, n_g\, \alpha$$

where

n_H is the number density of atomic H in the gas phase
v_H is the average speed of H atoms in the gas phase
S_H is the sticking coefficient of H atoms on grains
A is the exposed surface of the grain
t_H is the residence time of H adatoms on the grain
n_g is the number density of grains in clouds
α is mobility times recombination probability

These expression should be used in chemical models of interstellar clouds, both for diffuse and dense stages, because it is conceivable that low coverage of hydrogen atoms on grains will be encountered; in dense clouds the grain surface will be covered by ice and results relative to this other substrate should be obtained.

5. REFERENCES

Duley, W.W. and Williams, D.A. (1984) *Interstellar Chemistry*, Academic Press, London
Gould, R.J. and Salpeter, E.E. (1963) Astrophysical J. **138**, 393
Hollenbach, D.J. and Salpeter, E.E. (1971), Astrophysical J. **163**, 155
Hollenbach, D.J., Werner, M.W., and Salpeter, E.E. (1971), Astrophysical J. **163**, 165
Pirronello, V., Biham, O., Liu, C., Shen, L., and Vidali, G. (1997b) Astrophysical J. **483**, L131
Pirronello, V., Liu, C., Shen, L., and Vidali, G. (1997a) Astrophysical J. **475**, L69
Smoluchowski, R (1981), Astrophys. Space Science **75**, 353
Smoluchowski, R (1983), J. Physical Chemistry **87**, 4229
Williams, D.A. (1968), Astrophysical J. **151**, 935

PRE-BIOTIC CHEMISTRY INDUCED BY ION IRRADIATION OF ICES

G. STRAZZULLA
Osservatorio Astrofisico di Catania
Città Universitaria
I-95125 Catania, Italy

1. Introduction

In the last years many experimental results have been obtained on physical-chemical effects induced by fast ions bombarding frozen gases (CO, CO_2, CH_3OH, CH_4, NH_3, H_2O, etc.) and mixtures simulating ice targets in space (frosts on interstellar grains, planets, satellites, comets, etc.). In particular it has been demonstrated the formation of new, also complex, species and of refractory organic residues. These new species could be of interest to exobiology. It is in fact possible their synthesis by cosmic ion irradiation of astrophysical objects (e.g., icy satellites, comets, interstellar ices). Thus a complex chemistry can be driven on those objects and the new species could have been delivered on primitive Earth so contributing to the development of the early life.

Noteworthy fluxes of energetic (keV-MeV) particles impinge on solid surfaces of astrophysical objects such as interstellar dust and planetary objects, made of refractory (carbonaceous and/or silicates) materials and/or ices. Ion irradiation produces a number of effects whose study has been based on laboratory simulations of relevant targets bombarded with fast charged particles under physical conditions more or less similar to the astrophysical ones. In particular ion irradiation of simple ices at low temperatures (10-150K) produces new species both less and more, even much more, complexes than the irradiated ones (e.g., Johnson 1990, Strazzulla and Johnson 1991, Strazzulla 1997).

Here I give a brief summary of the results obtained so far, with a view to their relevance for exobiology.

J. Chela-Flores and F. Raulin (eds.),
Exobiology: Matter, Energy, and Information in the Origin and Evolution of Life in the Universe, 291–294.

2. Experimental

Many results obtained by "in situ" IR spectroscopy in the 2.2-25 microns range have been reported by different groups. The used experimental apparatus essentially consists of a scattering chamber with IR-transparent windows, facing an IR spectrophotometer. The vacuum is better than 10^{-7} mbar. Frosts are accreted onto a substratum put in contact with a cold finger (10-300 K), by admitting gas (mixtures) into the chamber, through appropriate valves. Ion currents are maintained lower than few microampere/cm^2 to avoid macroscopic heating of the target.

In table 1, I summarize some of the results obtained so far. It is clear the formation of many newly synthesized species. Of particular relevance to exobiology are the results obtained by irradiating mixtures containing water, methane and ammonia for which a marginal evidence for the formation of amino acids has been inferred by IR spectroscopy (Strazzulla and Palumbo 1997). A much more firm evidence comes from the results by Kobayashi et al. (1995). They showed, by ion mass chromatography, that after irradiation of an ice mixture of propane (or methane, or carbon monoxide), ammonia, and water with 3 MeV protons, a wide variety of amino acids such as glycine, alanine, and amino butyric acids were detected after acid hydrolysis. When propane was the carbon source the G-value for glycine was measured to be 2×10^{-4}.

TABLE 1. List of experiments made to study, by IR spectroscopy, chemical changes induced by ion irradiation of frozen gases

Target	Ion (E in keV)	Major species	Ref.
H_2O:NH_3:CH_4 (1:3:2)	H (1000); Ar (60)	C_2H_6; CO, CO_2; N-N-N; X-CN	(*a*); (*i*)
H_2O:N_2:CO_2 (1:1:1)	H (1000)	CO; NO; CH_4;	(*a*)
H_2O:N_2:CO (5:1:1)	H (1000)	CO_2; CH_4;	(*a*)
SO_2	H (1000)	SO_3; S_8; sulfates	(*b*)
H_2O:CO_2 (1:1)	H (700); He (3)	CO; H_2CO_3;	(*c*); (*d*)
C_6H_6	He (3)	acetylenes; org. residue	(*e*)
C_4H_{10}	He (3)	CH_4; alchenes; org. res.	(*f*)
H_2O:CH_3OH (0:1; 2:1)	H (1000); He (3)	CO; CO_2; H_2CO?	(*g*); (*h*)
CO (1:1)	He, Ar (3-60)	CO_2; suboxides;	(*l*)
CO:O_2 (1:1)	He, Ar (3-60)	CO_2; O_3;	(*l*)
CO:N_2 (1:1)	He, Ar (3-60)	CO_2; suboxides;	(*l*)

(*a*) Moore et al. 1983; (*b*) Moore 1984; (*c*) Moore and Khanna 1991; (*d*) Brucato et al. 1996; (*e*) Strazzulla and Baratta 1991; (*f*) Celi et al. 1995; (*g*) Moore and Hudson 1994; (*h*) Baratta et al. 1994; (*i*) Strazzulla and Palumbo 1997; (*l*) Strazzulla et al. 1997.

Ion irradiation of hydrocarbon containing ices produces also a complex organic, refractory residue. The cross section of the process converting volatile carbon-containing molecules to a refractory residue (we called it IPHAC, Ion Produced Hydrogenated Amorphous Carbon) has been measured for some combinations of projectile-target (Foti et al. 1984; Strazzulla et al. 1984). We have shown (Strazzulla and Baratta 1992) that a general result obtained irradiating benzene, methane, and butane is that up to a total dose of 10 eV/C-atom, some fraction of the deposited frozen film is converted to a molecular solid. In a second region (10-25 eV/C-atom) a strong H loss is observed: the material has evolved towards a polymer-like compound. By polymer like compound we do not mean a single compound but a disordered organic material containing chains of different length but having however a measurable optical gap (i.e. a measurable energy gap between the valence and the conduction bands). In a third region ($\geq$ 25 eV/C-atom) the material has evolved towards a hydrogenated amorphous carbon film.

It is relevant that the IR spectra (in the 3.4 μm region) of some residues obtained from frozen methane and butane reproduce very well those observed in the diffuse interstellar medium and in the organic residue from the Murchison meteorite (Pendleton et al. 1994).

3. Discussion

I have summarized the clear evidences of the formation, by ion irradiation of simple ices, of more complex organic materials, including amino acids, that are very probably relevant for the chemistry of many objects in the Solar System (e.g., Strazzulla 1997). Of course the experimental results have to be applied with caution to the different astrophysical environments (see e.g. sect. 2 in Strazzulla and Johnson 1991).

It is now fascinating to indagate if the ion-induced chemistry had a role in the development of the life on Earth and/or on other objects in our (Mars?) and/or in other planetary systems. In other words ion irradiation of simple ices induces a pre-biotic chemistry but is, at least a portion of the synthesized materials, really essential for the next step towards a biotic-chemistry?

I believe that at present we are far from the possibility to give an answer to that question. Many other studies have to be conducted in several fields of this really interdisciplinary research field. One of the question that the experimentalist could try to answer concerns the maximum concentration of amino acids that it is possible to synthesize in a given target. In fact we have not to forget that when the amino acid concentration is high enough the high rate of their destruction (e.g., Foti et al. 1991) competes with the formation rate.

References

Baratta G.A., Castorina A.C., Leto G., Palumbo M.E., Spinella F. and Strazzulla G. (1994) Ion irradiation experiments relevant to the physics of comets, *PlSpSci* **42**, 759-766.

Brucato J.R., Palumbo M.E. and Strazzulla G. (1996) Ion irradiation of frozen water-carbon dioxide mixtures, *Icarus* **125**, 135-144.

Celi G., Baratta G. and Strazzulla G. (1995) Vibrational spectroscopy of ion-irradiated frozen butane, *Infr Phys Tech* **36**, 995-1001.

Foti G., Calcagno L., Sheng K.L. and Strazzulla G. (1984) Micrometer-sized polymer layers synthesized by MeV ions impinging on frozen methane, *Nat* **310**, 126-128.

Foti A.M., Baratta G., Leto G. and Strazzulla G. (1991) Molecular alteration and carbonization of glycine by ion irradiation, *Europhys Lett* **16**, 201-204.

Johnson R.E. (1990) *Energetic charged particle interactions with atmospheres and surfaces*, Lanzerotti L.J. ed., Springer Verlag Press.

Kobayashi K., Kasamatsu T., Kaneko T., Koike J., Oshima T:, Saito T., Yamamoto T. and Yanagawa H. (1995) Formation of amino acid precursors in cometary ice environments by cosmic radiation, *Adv. Sp. Res.* **16 n.2**, 21-26.

Moore M.H. (1984) Studies of proton-irradiated SO_2 at low temperatures: implication for Io, *Icarus* **59**, 114-128.

Moore M.H. and Hudson R.L. (1994) Far-infrared spectra of cosmic-type pure and mixed ices, *A&AS* **103**, 45-56.

Moore M.H. and Khanna R.K. (1991) Infrared and mass spectral studies of proton irradiated H_2O+CO_2 ice: evidence for carbonic acid, *Spectrochimica Acta* **47**, 255-262.

Moore M.H., Donn B., Khanna R. and A'Hearn M.F. (1983) Studies of proton-irradiated cometary-type ice mixtures *Icarus* **54**, 388-405.

Pendleton Y.J., Sandford S.A., Allamandola L.J., Tielens A.G.G.M. and Sellgren K. (1994) Near-infrared absorption spectroscopy of interstellar hydrocarbon grains, *ApJ* **437**, 683-696.

Strazzulla G. (1997) Ion bombardment of comets, in *From Stardust to Planetesimals* Y. Pendleton, A.G.G.M. Tielens eds., ASP Conf Series Book, S. Frncisco, p. 423-433.

Strazzulla G. and Baratta G.A. (1991). Laboratory study of ion-irradiated frozen benzene. *A&A* **241**, 310-316.

Strazzulla G. and Baratta G.A. (1992) Carbonaceous material by ion-irradiation in space, *A&A* **266**, 434-438.

Strazzulla G. and Palumbo M.E. (1997) Evolution of icy surfaces: an experimental approach, *PlSpSci, submitted.*

Strazzulla G. and Johnson R.E. (1991) Irradiation effects on comets and cometary debris, in *Comets in the Post-Halley Era* R. Jr Newburn, M. Neugebauer, J. Rahe eds., Kluwer, Dordrecht, p. 243-275.

Strazzulla G., Calcagno L. and Foti G. (1984) Build Up of carbonaceous material by fast protons on Pluto and Triton. *A&A* **140**, 441-444.

EXTRATERRESTRIAL DELIVERY OF SIMPLE BIOMOLECULES TO THE EARTH: SURVIVAL OF AMINO ACIDS AND NUCLEIC ACID BASES

V.A. BASIUK and R. NAVARRO-GONZALEZ
Laboratorio de Química de Plasmas y Estudios Planetarios, Instituto de Ciencias Nucleares, Universidad Nacional Autónoma de México, Circuito Exterior C.U., A.Postal 70-543, 04510 México, D.F., MEXICO

1. Introduction

Geological record and photochemical models point to the Earth's early atmosphere to be composed mainly of CO_2 (and probably a small percentage of CO), N_2 and H_2O [1]. It is hard to explain how organic compounds can form in such a non-reducing environment. That is why the idea of extraterrestrial delivery of organic matter, including biologically important compounds, to the early Earth is especially attractive [2-5]. However, the problem is how biomolecules can survive the high temperatures developed during atmospheric deceleration and on impacts of space bodies to the terrestrial surface [2-5].

According to the estimates, organics in comets and asteroids as small as 100 m in radius, which cannot be efficiently aerobraked at present day atmospheric pressure and density, cannot survive the impacts [4]. Catastrophic airbursts of comets with energy dissipation through evaporation of volatile components have been suggested to be another scenario where high degree of survival of cometary organics is possible [2].

Other possible delivery vehicles of organics to the early Earth are meteorites and interplanetary dust particles [3]. Meteorites leave no doubts due to the discovery of various organic compounds, *e.g.* amino acids, purines and pyrimidines, in carbonaceous chondrites [6-10]. Amino acids can be found there in chemically free form and in the form of hydrolyzable derivatives [11]. As regards the temperatures experienced by meteorite interiors, these space bodies seem to be almost perfect to deliver biomolecules to the Earth: pyrolytic temperatures extend to less than 1 mm depth [12]. On the other hand, the contribution of meteoritic organics is estimated to be very insignificant [3].

The situation with interplanetary dust particles is opposite. This is the most abundant source of extraterrestrial matter accreted to the Earth. Contemporary flux of micrometeoroids is of the order of 10^6-10^7 kg y^{-1} [13], where carbonaceous matter constitutes a substantial part, roughly 10% [3]. But their interiors are subjected to very high temperatures during atmospheric passage. Millimeter and submillimeter-size meteoroids are completely evaporated. However, there are some evidences that particles of the size of the order 1-10 μm are subjected to temperatures less than 500 °C [14-18].

J. Chela-Flores and F. Raulin (eds.),
Exobiology: Matter, Energy, and Information in the Origin and Evolution of Life in the Universe, 295–298.

In any event, the temperatures of about 500 °C seem too high for biomolecules to survive. The pyrolytic data available demonstrate amino acids to decompose at that high temperatures to CO_2, H_2O, NH_3, CO, and simple organic compounds, as well as to undergo condensation into piperazine-2,5-diones (PD's) [19]. PD's are very important in our context since they can regenerate amino acids under hydrolysis. At the same time no studies were focused on whether amino acids decompose *completely*, that would have the most direct relation to amino acid survivability in the bodies entering the Earth's atmosphere. For nucleic acid bases, relevant data are unavailable as well.

Our goals were to estimate (1) how much amino acids and nucleic acid bases can survive without chemical alteration being subjected to a rapid heating; (2) what happens to optical activity of chiral amino acids; (3) yields of PD's formed; (4) what other compounds form that can produce amino acids again upon hydrolysis.

2. Experimental

The compounds selected were non-chiral α-aminoisobutyric acid (Aib), optically active L-alanine (L-Ala), L-valine (L-Val), L-leucine (L-Leu), and nucleic acid bases uracil, cytosine, adenine and guanine. In this study we used as a first approach a nitrogen atmosphere at normal pressure and the temperature of 500 °C.

The products of amino acid volatilization (sublimates) were analyzed quantitatively for amino acid and PD content using high-performance liquid chromatography (this technique was also employed to quantify recovery of the nucleic acid bases), and qualitatively for other transformation products by gas chromatography--Fourier transform infrared spectroscopy--mass spectrometry (GC-FTIR-MS) with auxiliary computer simulation of IR spectra [20].

3. Results and Discussion

Percentage of amino acid recovery was 2% for Aib and Leu, 10% for Val and 11% for Ala. PD's formed in yields of 1% for Aib, 3.5% for Leu, 5% for Val and 68% for Ala. PD's derived from optically active amino acids (*i.e.* Ala, Val and Leu) appeared always as diastereomeric mixtures that points to strong (or even total) racemization.

As was expected, the sublimates contained other compounds as well. We extracted them by chloroform and analyzed by means of GC-FTIR-MS. In the case of Aib, we were able to identify methacrylic acid, Aib linear dipeptide and its decarboxylated derivative, 3,3,4,4-tetramethyl-β-propiolactam (resulting from the PD degradation with the loss of HNCO), N-formyl α-aminoisobutyric aldehyde (resulting from the Aib dipeptide degradation) and bicyclic amidine 2,2,5,5,8,8-hexamethylhexahydroimidazo[1,2-a]pyrazine-3,6-dione. Likewise PD's, hexahydroimidazo[1,2-a]-pyrazine-3,6-diones are direct products of amino acid intermolecular condensation [20] and hydrolyze in a strong acidic solution to starting amino acids [21]. Thus in the case of Aib, in addition to PD, derivatives which can be hydrolyzed to the starting amino acid are Aib linear dipeptide, its decarboxylated derivative and the bicyclic amidine.

In the case of Ala, we identified in the extract propionic acid, several primary and secondary linear amides, corresponding PD, its dehydrogenated and dehydrogenated-plus-N-methylated derivatives, two isomeric dimethyl-substituted N-methyl-γ-butyrolactams, 3-ethyl-5-methylhydantoin, singly and doubly dehydrogenated bicyclic amidines (whereas the parent amidine has not been detected). Of them, the compounds preserving at least one Ala residue are the piperazinediones, hydantoin and singly dehydrogenated bicyclic amidine. Similar classes of compounds, but with much more numerous products, formed in the cases of Val and Leu.

For nucleic acid bases we quantified their recovery only. The highest value of 12% was found for adenine, 2.2% for uracil, 0.2% for cytosine, whereas guanine decomposed completely.

The data reported here refer to the specific conditions of our experiments. To do extrapolations to real conditions of atmospheric passage and impacts to the terrestrial surface, a detailed knowledge is needed about thermal history of the organics-bearing space bodies which is not available at present. Some suggestions still can be done. Obviously, large asteroid and comet impacts resulting in complete pulverization and evaporation of the projectiles leave no possibilities for the biomolecules to survive, since the temperatures can exceed 10,000 °C [4]. If atmospheric disruption produces asteroid and comet fragments in centimeter/meter size range that are able to reach the planetary surface, a degree of biomolecules' survival can be as high as in meteorites: in terms of heating experienced by the interiors, such fragments are remnants of meteorites in which pyrolytic temperatures extend to less than 1 mm depth [12].

As we already mentioned, another possibility exists for comets. If they catastrophically airburst in the upper atmosphere, energy dissipation through evaporation of volatile components can also make possible a high degree of survival of cometary organics in final small particles [2], but it is unclear to what extent.

Maximum heating of interplanetary dust particles occurs in the 80-100 km altitude region [22], and this circumstance is especially important to predict the behavior of amino acids and nucleic acid bases in chondritic meteoroids. This altitude corresponds to a very reduced pressure, less than 1 mbar, *i.e.* vacuum. At the same time it is known that amino acids and nucleic acid bases sublime under reduced pressure and temperatures of >200 °C [23,24]. This implies that the biomolecules will sublime from the dust particles as soon as the temperature exceeds 200 °C and their vapors will dissipate in the upper atmosphere. Thus, at least interplanetary dust particles should be able to deliver the simple biomolecules to the terrestrial atmosphere without substantial thermal decomposition. The problem remains, however, with possible photolysis and radiolysis of the biomolecules at the high altitudes.

Acknowledgments

This work was supported by grants CONACyT-4282-E9406 and DGAPA-IN102796.

References

1. Kasting, J.F.: Earth's early atmosphere, *Science 259* (1993), 920-926.
2. Greenberg, J.M.: Synthesis of organic compounds in interstellar dust and theit transport to Earth via comets, *Adv. Space Res. 9* (1989), (6)15-(6)23.
3. Anders, E.: Pre-biotic organic matter from comets and asteroids, *Nature 342* (1989), 255-257.
4. Chyba, C.F., Thomas, P.J., Brookshaw, L., and Sagan,C.: Cometary delivery of organic molecules to the Early Earth, *Science 249* (1990), 366-373.
5. Steel, D.: Cometary supply of terrestrial organics: Lessons from the K/T and the present epoch, *Origins Life Evol. Biosphere 21* (1992), 339-357.
6. Cronin, J.R., and Moore, C.B.: Amino acid analyses of the Murchison, Murray, and Allende carbonaceous chondrites, *Science 172* (1971), 1327-1329.
7. Kvenvolden, K., Lawless, J., Pering, K., Peterson, E., Flores, J., Ponnamperuma, C., Kaplan, I.R., and Moore, C.: Evidence for extraterrestrial amino-acids and hydrocarbons in the Murchison meteorite, *Nature 228* (1970), 923-926.
8. Lawless, J.G., Kvenvolden, K.A., Peterson, E., Ponnamperuma, C., and Moore, C.: Amino acids indigenous to the Murray meteorite, *Science 173* (1971), 626-627.
9. Stoks, P., and Schwartz, A.W.: Uracil in carbonaceous meteorites, *Nature 282* (1979), 709-710.
10. Van der Velden, W., and Schwartz, A.W.: Search for purines and pyrimidines in the Murchison meteorite, *Geochim. Cosmochim. Acta 41* (1977), 961-968.
11. Cooper, G.W., and Cronin, J.R.: Linear and cyclic aliphatic carboxamides of the Murchison meteorite: Hydrolyzable derivatives of amino acids and other carboxylic acids, *Geochim. Cosmochim. Acta 59* (1995), 1003-1015.
12. Sears, D.W.: Temperature gradients in meteorites produced by heating during atmospheric passage, *Modern Geol. 5* (1975), 155-164.
13. Love, S.G., and Brownlee, D.E.: A direct measurement of the terrestrial mass accretion rate of cosmic dust, *Science 262* (1993), 550-553.
14. Bradley, J.P., Brownlee, D.E., and Fraundorf, P.: Discovery of nuclear tracks in interplanetary dust, *Science 226* (1984), 1432-1434.
15. Mackinnon, I.D.R., and Rietmeijer, F.J.M.: Bismuth in interplanetary dust, *Nature 311* (1984), 135-138.
16. Christoffersen, R., and Buseck, P.R.: Epsilon carbide: A low-temperature component of interplanetary dust particles, *Science 222* (1983), 1327-1329.
17. Rietmeijer, F.J.M., and Mackinnon, I.D.R.: Poorly graphitized carbon as a new cosmothermometer for primitive extraterrestrial materials, *Nature 315* (1985), 733-736.
18. Rietmeijer, F.J.M., and Mackinnon, I.D.R.: Layer silicates in a chondritic porous interplanetary dust particle, *J. Geophys. Res. 90* (1985), suppl., D149-D155.
19. Ratcliff, M.A., Medley, E.E., and Simmonds, P.G.: Pyrolysis of amino acids. Mechanistic considerations, *J. Org. Chem. 39* (1974), 1481-1490.
20. Basiuk, V.A., and Navarro-González, R.: Identification of hexahydroimidazo[1,2- a]pyrazine-3,6-diones and hexahydroimidazo[1,2-a]imidazo[1,2-d]pyrazine-3,8-diones, unusual products of silica-catalyzed amino acid thermal condensation and products of their thermal decomposition using coupled high-performance liquid chromatography--particle beam mass spectrometry and gas chromatography--Fourier transform infrared spectroscopy--mass spectrometry, *J. Chromatogr. 776* (1997), 255-273.
21. Basiuk, V.A., Navarro-González, R., and Basiuk, E.V.: Behavior of amino acids when volatilized in the presence of silica gel and pulverized basaltic lava, *Origins Life Evol. Biosphere 28* (1998), in press.
22. Hunten, D.M., Turco, R.P., and Toon, O.B.: Smoke and dust particles of meteoric origin in the mesosphere and stratosphere, *J. Atmos. Sci. 37* (1980), 1342-1357.
23. Gross, D., and Grodsky, G.: On the sublimation of amino acids and peptides, *J. Am. Chem. Soc. 77* (1955), 1678-1680.
24. Yanson, I.D., and Teplitsky, A.B.: Heats of sublimation of crystals of organic compounds, *J. Phys. Chem. USSR 49* (1975), 428-429.

Section 9.
Exobiology on Titan

TITAN 'S CHEMISTRY : EXOBIOLOGICAL ASPECTS AND EXPECTED CONTRIBUTION FROM CASSINI-HUYGENS

F. RAULIN[1], P. COLL[1], Y. BENILAN[1], P. BRUSTON[1], , M-C. GAZEAU[1], P. PAILLOUS[1], N. SMITH[1], R. STERNBERG[1], D. COSCIA[1&2] and G. ISRAEL[2]
[1]*LISA, CNRS & Universités Paris 12 et Paris 7, Avenue du Général de Gaulle, 94010 Créteil Cedex France*
[2]*Service d'Aéronomie du CNRS, B.P. 3, 91371 Verrières le Buisson Cedex France*

Abstract. Because of the presence of a dense atmosphere, of an environment very rich in organics, and of many couplings involved in the various parts of its "geofluid", Titan, the largest satellite of Saturn, is a reference for studying prebiotic chemistry on a planetary scale. By coupling the data obtained from simulation experiments carried out in the laboratory, theoretical modeling and observations, it is already possible to draw a quite precise figure of such organic chemistry, with all its complexity. In particular, new data have been obtained from experiments simulating the organic chemistry of Titan's atmosphere (gas and aerosol phases), in an open reactor, within the correct range of temperature, pressure, carefully avoiding any chemical contamination. They show a very good agreement with the observational data, demonstrating for the first time the formation of all the organic species already detected in Titan's atmosphere. In particular, C_4N_2, never detected before in this kind of laboratory experiment, although it has been detected in Titan's atmosphere, is now clearly identified, together with many other species not yet detected in Titan. This strongly suggests the presence of more complex organics in Titan's atmosphere and surface, including high molecular weight polyynes and cyanopolyynes. The NASA-ESA Cassini-Huygens mission was successfully launched in October 1997. The Cassini spacecraft will reach the Saturn system in 2004 and will become an orbiter around Saturn and fly-by Titan, while, simultaneously it will send the Huygens probe into Titan's atmosphere. Both will systematically study organic chemistry in Titan's "geofluid". In situ measurements, in particular from Huygens GC-MS and ACP instruments, will provide detailed analysis of the organics present in the air, aerosols, and surface. This very ambitious mission should provide much information of crucial importance for our knowledge of the complexity of Titan's chemistry, and, more generally for the field of exobiology.

J. Chela-Flores and F. Raulin (eds.),
Exobiology: Matter, Energy, and Information in the Origin and Evolution of Life in the Universe, 301–311.

1. Introduction

Exobiology is not only the search for life elsewhere, it also includes the study of the origins of Life on Earth, and the study of extraterrestrial chemistry, with, in particular, the search for extraterrestrial organic molecules and the study of the processes involved in their formation and evolution. In this respect, Titan is probably, with the comets, one of the most exobiologically interesting bodies in the Solar system (Carle et al, 1992).

In this paper, we present the exobiological aspects of Titan's atmosphere : both in the gas and aerosol phases. We report new laboratory data supporting such aspects and briefly describe some of the scientific investigations of the Cassini-Huygens mission which could provide scientific return of important exobiological interest.

2. Titan's observational data

Titan is the largest satellite of Saturn and the second satellite by size in the Solar system. But, above all, it is the only satellite in the Solar system having a dense atmosphere. It is also the only satellite whose surface is not visible, due to the presence of dense aerosol layers which entirely mask it from the outside of the satellite atmosphere.

Titan's environment has been already explored in detail , mainly thanks to the Voyager mission in 1980 and 1981. In particular, the Infrared spectrometer IRIS, the UV instrument UVS and the radio-occultation experiment have been able to determine the main chemical composition of Titan's atmosphere, to identify several trace atmospheric species, and to determine the vertical pressure/temperature profile. Many observations and models of Titan, after Voyager, have provided complementary data on the chemical composition, as well as on the surface state [see for instance Owen et al, 1992; ESA, 1992; Lebreton, 1997; Gautier, 1997, Gautier and Raulin, 1997].

In spite of noticeable differences such as size and distance to the sun (Table 1), the current vision of this satellite often induces a comparison with the primitive environment of the Earth, because of several similarities between both planetary bodies. As recently emphasized by Clarke and Ferris (1997a), both (Table 1) have a dense atmosphere. Titan's one is even much denser than the present atmosphere of the Earth: the near surface atmospheric density on Titan, taking into account the temperature (3 times lower on Titan) and the pressure differences, is 4.5 times higher than on the Earth today. Both have several energy sources : mainly solar UV and electrons (lightning and aurorae on the Earth, Saturn magnetospheric electrons on Titan). Both have N_2 as the main atmospheric constituent, with the presence of greenhouse gases, condensable in the corresponding environment (methane on Titan, water on the Earth), and non-condensable (molecular hydrogen on Titan, equivalent to carbon dioxide on the Earth), and anti-greenhouse species (hazes and clouds in Titan's atmosphere play the same role than aerosols and clouds in the Earth's atmosphere). One of the main consequences of this is that Titan and the Earth atmosphere have similar temperature vertical profiles, with a troposphere, a tropopause and a stratosphere.

TABLE 1. Titan versus the early Earth : analogies and differences (adapted from Clarke & Ferris, 1997a).

	Titan	Early Earth
Dimension (mean radius)	2575 km	6378 km
Mean distance to the sun	9.5 au	1 au
Atmosphere surface pressure	1.5 bar	$\geq$ 1 bar
main constituents	N_2, CH_4	CO_2, N_2, CH_4 (?)
minor constituents	many organics	H_2O, organics
Temperature : surface	94 K	273 - $\leq$ 373 K
tropopause	70 K	?
stratosphere	> 70 - 180 K	?
Liquid bodies on the surface	CH_4 - C_2H_6 lakes	water oceans

Furthermore, Titan's environment is very rich in organics, as was probably the primitive Earth. Indeed, many organics have already been detected in Titan's atmosphere, mostly in the gas phase, and many others are supposed to be present, both in the gas and in the aerosol phases, and on the surface. On Titan, the organics are mainly produced by photochemistry and electron impact transformation of the atmosphere, starting with the dissociation of methane and molecular nitrogen, and the coupling of the chemistry of these main constituents (Yung et al, 1984; Toublanc et al, 1995). On the primitive Earth, the organics may not have been produced in the atmosphere, since models of the early Earth atmosphere suggest that it was not chemically reducing, contrary to Titan's atmosphere; it is now suggested that they were produced in the submarine hydrothermal vents, or, more likely, imported from extraterrestrial materials (meteorites, micro-meteorites, cometary dust or other cometary importation). However, the nature of the organics already detected in Titan's atmosphere is similar to those involved in the terrestrial prebiotic chemistry : nitriles (mainly HCN and HC_3N), acetylenic compounds, (mainly acetylene and diacetylene), have been identified and their oligomers are supposed to be also present.

The analogies between Titan and the Earth can be pursued even further. Modeling of the behavior of CH_4 in the atmosphere and of its photochemical products on the surface strongly suggests the presence of liquid bodies, mainly made of methane and ethane, on Titan's surface (Lunine et al, 1983). Although oceans, globally or even partially covering the surface seem now fully excluded, from the recent observations of Titan's surface in the near IR (Smith et al, 1996), the presence of lakes, small seas or subsurface oceans is still possible (Lunine et al, 1993; Lunine, 1997).

Of course, there are limitations in such analogies. The strongest is due to the temperature differences. This causes the full absence of liquid water on Titan. The resulting consequence is that Titan's prebiotic-like chemistry is currently mainly limited to the atmosphere, and is mainly a C-H-N chemistry (Raulin et al, 1992; Raulin, 1997). Such chemistry, however, has so many links with the terrestrial Earth prebiotic chemistry that is essential for exobiology to study it. Because of its occurrence now, with all its diversity, with the many couplings, chemical as well as physical, between the three parts of what can be called « Titan's geofluid » (gas phase, aerosol phase and - if liquid bodies are there - surface), studying Titan today is a way to go back to the early Earth, while keeping in mind the liquid water and temperature limitations.

3. Titan's organic chemistry simulations

An efficient way to study such chemistry is to try to simulate it in the laboratory. Indeed, a wide range of experiments have already been carried out (Toupance et al, 1975; Raulin et al, 1982; Sagan et al, 1984; Thompson et al, 1991; McDonald et al, 1994; de Vanssay et al, 1995; Coll et al, 1995; McKay, 1996; Coll, 1997; Coll et al, 1997a, 1997b, and Refs. included) to mimic the chemical evolution of Titan's atmosphere, by irradiating with different energy sources gas mixtures of N_2 and CH_4. Irradiations have been performed with UV light, as well as high energy particles, and electrons (spark and silent discharges). As seen on Table 2, during those experiments, all the organics already observed in the gas phase in Titan's atmosphere have been detected in such simulation experiments. Furthermore, many others are produced in the laboratory which could also be present on Titan. Until very recently, only one organic compound remained detected in Titan and not in laboratory simulation : C_4N_2. A full program of experimental research has been developed at LISA to study this problem, and to reassess the quality of Titan's experimental simulation. It concerns not only the gas phase but also the solid products obtained in such experiments, usually called « tholins » (Sagan and Khare, 1979) and which are supposed to mimic Titan's aerosols.

The main idea which was driving these new sets of experiments was to use conditions closer to those prevailing on Titan, first of all concerning the low temperature, but also - more generally - to develop experimental tools, including experimental protocols compatible with the search and qualitative analysis of products of low thermal stability, as C_4N_2.

This program started with the use of spark discharge systems. It was first able to evidence the formation of organic compounds never detected before in such simulation experiments : triacetylene, C_6H_2, and cyanobutadiyne, HC_5N (de Vanssay et al, 1995; Coll et al, 1995). Such discovery is of great importance, since it strongly suggests that, in parallel to the already proposed pathways of complexification of the chemistry in Titan's atmosphere, through polyynes, one should also consider other pathways involving cyanoplolyynes. This also suggests strongly that the macromolecules which are the end-products of such chemical evolution, are not only made of C and H atoms, but also include N atoms (Coll et al, 1995; Coll, 1997, Coll et al, 1997b).

TABLE 2. Titan's atmosphere composition and organic compounds found in simulation experiments. M : major products, < : less abundant by one order of magnitude, << by two orders or more; * : all isomers. (a) : Gupta et al, 1981; (b) : Toupance et al, (1975), Raulin et al, (1982), Thompson et al, (1991); (c) : de Vanssay et al, (1995), Coll et al, (1995); (d): Coll et al, (1997c).

COMPOUNDS	**TITAN**	**UV** (a)	**Arc & Silent Discharge** (b)	**Low Temp. Arc** (b)	**Low Temp. Silent** (c,d)
N_2 (%)	90-98	90	90	90	98
CH_4 (%)	2-5	10	10	10	2
C_2H_6	M	M	<	<	<
C_2H_4	<<	<<	<	<	<
C_2H_2	<		M	M	M
C_3H_8	<	<<	<<	<<	<
C_3H_6			<	<	<<
CH_3C_2H	<<		<	<	<
CH_2CCH_2			<<	<<	<<
C_4H_{10}*			<<	<<	<<
C_4H_2	<<		<<	<<	<<
C_6H_6			<<	<<	<<
C_6H_2, C_8H_2				<<	<<
other hydrocarbons			many, <<	many, <<	many, <<
HCN	<		M	M	M
C_2N_2	<<		<<	<<	<<
CH_3CN	<<		<	<	<<
CH_2CHCN			<	<	<<
CHCCN	<<		<	<	<<
C_3H_5CN*			<<	<<	<<
C_4N_2	**solid**			?	<<< (d)
other nitriles			many, <<	many, <<	many, <<
CH_3N_3			<< ?		<< ?
HC_4CN				<<	<<
other N-organics			<< ?		several <<
CO	5x10-5				
CO_2	10-8-10-9		<< ?		

More recently, by using silent discharge and flow systems, able to operate at low temperature, at last, the formation of C_4N_2 was clearly demonstrated (Coll, 1997, Coll et al, 1997a, 1997c). But these latest data also evidence the very low mole fraction of this compound (of the order of a fraction of ppb).

This discovery strongly supports the idea that such experiment is a good simulation of Titan's organic chemistry and that many of the other compounds obtained during the experiment and not yet found in Titan's atmosphere should be searched for. This is the case of many gas phase products, including polyynes such as C_6H_2 and C_8H_2, and cyanopolyynes, such as HC_5N. Furthermore, such validation also supports the hypothesis that the tholins obtained during these recent experiments are good analogues of Titan's aerosols. Now the availability of such analogues is essential to interpret the obtained data of Titan's hazes and to develop haze models. More precisely, this requires the availability of the spectra of the tholins in a wide range of wavelengths, from the UV to the IR, with the values of the index of refraction (real and imaginary parts), together with information on the size distribution and morphology of the particles.

Until now such data were available only from one type of tholins, produced in the Cornell group (Khare et al, 1984, McDonald et al, 1994). However, chemical analysis of such tholins showed a contamination of the products by oxygen atoms. The experimental programs developed at LISA use conditions which avoid such a contamination (irradiation system in a glove box under nitrogen atmosphere, allowing the recovery and sampling of tholins under inert atmosphere).

Furthermore, Cornell's tholins were produced within room-temperature conditions, far from Titan's ones. Now, as shown on Table 3, clearly, the chemical composition of the tholins strongly depends on the conditions (pressure, temperature, inert atmosphere). The C/N ratio varies from 1 to 11. Khare's et al (1984) tholins which were used for systematic spectroscopic analyses have a C/N ratio of 1.9 (Sagan et al, 1984), but with more than 10% contamination by O atoms. In the case of the experimental conditions which seems to be the more representative of Titan's atmosphere (low temperature, low pressure, absence of oxygen), the value is 2.8 (Coll, 1997; Coll et al, 1997c).

TABLE 3 : Titan's tholins C/N ratio from various simulation experiments

Experiment	Temperature	Pressure	Sampling solid Products	C/N ratio
Sagan et al, 1984	Room T	Low P	O-contamination	1.9
Coll et al, 1995	Low T	High P	Airtight	11
McKay et al, 1996	Room T	High P	O-contamination	5.5
Coll et al, 1997a	Room T	Low P	Airtight	1.7
Coll , 1997; Coll et al, 1997c	Low T	Low P	Airtight	2.8

Preliminary data on the IR spectra and on the index of refraction of these more representative tholins show that such differences in elemental composition induce important differences in spectral values. In particular, the variation of the imaginary part of the index of refraction with wavelength in the 0.3 - 0.9 micron range, is totally different (Coll, 1997). There is now a need for systematic studies of the spectral data of these new tholins. Such studies are in progress at LISA, in collaboration with Dr; Laffay's group (University of Paris 6) .

4. The Cassini-Huygens mission : expected exobiological returns.

In fact, we must admit that we have so far no direct information on the chemical composition of Titan's haze particles. Is our assumption that they are made of organic tholins right ? Is Titan's chemistry even more complex than we expect ? Are O atoms involved in it, in particular through the atmospheric CO organic chemistry (Clarke and Ferris, 1997b) ? Are there purine or pyrimidine bases, amino acids or analogues present ? To corroborate such hypotheses and try to answer the many interrelated questions, it is essential to study in detail Titan's environment. This is one of the main objectives of the Cassini-Huygens mission (Raulin et al, 1992; ESA, 1992; Lebreton, 1997; Owen et al, 1997; Raulin, 1997).

The spacecraft Cassini, built by NASA, was successfully launched on October 15, 1997 from Cape Canaveral. It carries the Titan's atmospheric probe Huygens, built by ESA. When it reaches the Saturnian system in 2004, Cassini will become an orbiter around Saturn, fly-by Titan, and will release the Huygens probe which will penetrate into Titan's atmosphere and slowly descend to the surface. Cassini carries twelve scientific experiments which will allow a detailed exploration of the environments of Saturn and Titan (Table 4). Six interdisciplinary scientists (IDS) will study different aspects of these planets by using the data derived from several of the scientific experiments of the mission in a correlative way. The Huygens probe carries six scientific instruments (Table 4). Three IDS programs are associated to Huygens, including an exobiological one.

Several instruments of the orbiter and most of the instruments of the probe will provide data of exobiological interest on Titan. The spectrometers on the orbiter (first of all CIRS and UVIS) will determine the chemical composition of different zones of Titan's atmosphere and should be able, in particular, to detect new organic species. The Cassini Radar will be able to see and map Titan's surface through the haze layers, and to evidence eventually the presence and distribution of liquid bodies.

On Huygens (Lebreton, 1997), the GC-MS instrument (Niemann et al, 1997) is a gas chromatograph with three GC columns in parallel, coupled to a quadrupole mass spectrometer with five ion sources. It will perform a detailed molecular analysis of the atmosphere along the descent trajectory of the probe, including the analysis of trace organic constituents with concentrations as low as 10 to even 1 ppb. The ACP experiment (Israel et al, 1997) will collect the atmospheric aerosols and, after transfer into an oven, will heat them at different temperatures (up to the pyrolysis temperature

TABLE 4 . Cassini -Huygens Science Instruments and IDS's

Cassini Instruments and InterDisciplinary Programs	P.I., Team Leader or IDS		Exobiological Return
Optical Remote Sensing Instruments			
Composite Infrared Spectrometer (CIRS)	V. Kunde	USA	+++
Imaging Science Subsystem	C. Porco	USA	+
Ultraviolet Imaging Spectrograph (UVIS)	L. Esposito	USA	+++
Visual & I.R. Mapping Spectrometer	R. Brown	USA	++
Fields Particles and Waves Instruments			
Cassini Plasma Spectrometer	D. Young	USA	+
Cosmic Dust Analysis	E. Grün	Germa.	+
Ion & Neutral Mass Spectrometer	H. Waite	USA	+
Magnetometer	D. Southwood	U.K.	
Magnetospheric Imaging Instrument	S. Krimigis	USA	
Radio & Plasma Wave Spectrometer	D. Gurnett	USA	
Microwave Remote Sensing			
Cassini Radar	C. Elachi	USA	+++
Radio Science Subsystem	A. Kliore	USA	+
Interdisciplinary Scientists			
Magnetosphere and Plasma	M. Blanc	France	+
Rings and Dust	J.N. Cuzzi	USA	+
Magnetosphere and Plasma	T.I. Gombosi	USA	+
Atmospheres	T. Owen	USA	++
Satellites and Asteroids	L.A. Soderblom	USA	+
Aeronomy & Solar Wind Interaction	D.F. Strobel	USA	+

Huygens Instruments and InterDisciplinary Programs	P.I. or IDS		Exobiological Return
Gas Chromatograph-Mass Spectrometer	H. Niemann	USA	+++
Aerosol Collector & Pyrolyser	G. Israël	France	+++
Huygens Atmospheric Structure Instrument	M. Fulchignoni	Italy	++
Descent Imager/Spectral Radiometer	M. Tomasko	USA	+++
Doppler Wind Experiment	M. Bird	Germa.	+
Surface Science Package	J. Zarnecki	U.K.	+++
Interdisciplinary Scientists			
Aeronomy	D. Gautier	France	++
Atmosphere/Surface Interactions	J.I. Lunine	USA	++
Chemistry and Exobiology	F. Raulin	France	+++

of 600°C). The resulting gases will be chemically analyzed by the GC-MS instrument. ACP will thus provide the first direct in situ molecular and elemental analysis of Titan's hazes. HASI will measure, in particular, the pressure and temperature vertical profiles, essential data for modeling the behavior of organics in Titan's atmosphere. DISR will determine photon fluxes, cloud structure and images of the surface. SSP will also provide important information on the physical state and chemical composition of the surface, all data of exobiological importance .

5. Conclusions

For a better understanding of the terrestrial prebiotic chemistry, it is now crucial to find extraterrestrial places where analogous organic processes are currently going on. Titan, a planet-size prebiotic laboratory, similar in many aspects to the primitive Earth, but with the absence of liquid water, is such a place.

Studying its complex organic chemistry, with all the physical and chemical couplings which are involved in the different parts of its geofluid (gaseous atmosphere, aerosols and surface), should give us many new insights, new ideas and new avenues on the early Earth prebiotic chemistry. Clearly, the Cassini-Huygens mission is going to provide a scientific return essential for the field of exobiology.

Acknowledgments

We wish to thank prof. Guy Cernogora for hhis essential help in the development of the new cold plasma reactor.. This review work has been supported by grants from the French Space Agency (CNES: Centre National d'Etudes Spatiales).

References

Carle, G., Schwartz, D., and Huntington, J. (eds) (1992) Exobiology in Solar System Exploration, *NASA SP* **512.**

Clarke D.W. and Ferris J.P. (1997a) Chemical evolution on Titan : comparisons to the prebiotic Earth, *Origins of life and Evol. Biosphere*, 27, 225-248.

Clarke, D.W. and Ferris, J.P. (1997b) Titan haze : structure and properties of cyanoacetylene and cyano-acetylene-acetylene photopolymers, *Icarus* **127**, 158-172.

Coll, P. (1997) Modélisation expérimentale de l'atmopshère de Titan : production et caractérisations physico-chimiques d'analogues des aérosols et de la phase gazeuse enfin représentatifs, *Thèse de Doctorat de l'Université Paris 12.*

Coll, P., Coscia, D., Gazeau, M.-C., de Vanssay, E., Guillemin, J.-C., and Raulin, F. (1995) Organic chemistry in Titan's atmosphere: new data from laboratory simulations at low temperature, *Adv. Space Res.* **16** (2), 93-104.

Coll, P., Coscia, D., Gazeau, M-C.G. and Raulin, F. (1997a) New planetary atmopshere simulations : application to the organic aerosols of Titan, *Adv. Space Res.*, **19** (7) 1113-1119.

Coll, P., Coscia, D., Gazeau, M-C.G. and Raulin, F. (1997b) Review and latest results of laboratory investigation of Titan's aezrosols, *Origins of Life and Evol. Biosph.*, in press.

Coll, P., Coscia, D., Smith, N., Gazeau, M.C., Israel, G., Cernogora, G. and Raulin, F. (1997c), Experimental laboratory simulation of Titan's atmopshere in the frame of the Cassini-Huygens space mission, in preparation.

European Space Agency (1992) Symposium on Titan, *ESA SP*- **338**.

Gautier, D. (1997) The aeronomy of Titan, *ESA SP* **1177**, 199-209.

Gautier, D., and Raulin, F. (1997) Chemical composition of Titan's atmosphere, *ESA SP* **1177**, 359-364.

Gupta, S. , Ochai, E. and Ponnamperuma, C. (1981) Organic synthesis in the atmosphere of Titan, *Nature*, 273, 725-27.

Israel, G., Niemann H., Raulin, F., Riedler, W., Atreya, S., Bauer, S., Cabane, M., Chassefiere, E., Hauchecorne, A., Owen, T., Sablé, C., Samuelson, R., Torre, J.P., Vidal-Madjar, C., Brun, J.F., Coscia, D., Ly, R., Tintignac, M., Steller, M., Gelas, C., Condé, E. and Millan, P. (1997) The aerosol collector pyrolyser (ACP) experiment for Huygens, *ESA SP*- **1177,** 59-84.

Khare, B.N., Sagan, C., Arakawa, E.T., Suits, F. Callicott, T.A. and Wilaims, M.W. (1984), Optical constants orf organic tholins produced in a smulated Titanian atmopshere : from soft X-rays to microwave frequencies, *Icarus* 60, 127-137.

Lebreton, J.P., European Space Agency (1997) Huygens : Science, Payload and mission, *ESA SP*- **1177**.

Lunine, J.I., Stevenson, D.J., and Yung, Y.L. (1983) Ethane Ocean on Titan. *Science* **222**, 1229-1230.

Lunine, J.I.. (1993) Does Titan have an ocean ? A review of current understanding of Titan's surface, *Rev. Geophys.* **31** (2), 133-149.

Lunine, J. (1997) Titan surface-atmopshere interactions, *ESA SP*- **1177**, 211-218.

McDonald, G.D., Thompson, W.R., Heinrich, M., Khare, B.N., and Sagan, C. (1994) Chemical investigation of Titan and Triton tholins, *Icarus* **108**, 137-145.

McKay, C.P. (1996) Elemental composition, solubility, and optical properties of Titan's organic haze, *Planet.. Space Sci.* **44** (8), 741-747.

Niemann, H., Atreya, S., Bauer, S., Biemann, K., Block, B., Carignan, G., Donahue, T., Frost, L., Gautier, D., Harpold, D., Hunten, D., Israel, G., Lunine, J., Mauersberger, K., Owen, T., Raulin, F., Richards, J., and Way, S. (1997) The gas chromatograph mass spectrometer aboard Huygens, , *ESA SP*- **1177,** 85-107.

Owen, T., Gautier, D., Raulin, F., and Scattergood, T. (1992) *Titan*, in *NASA SP* **512**, 127-143.

Owen, T., Raulin, F., McKay, C.P., Lunine, J.I., Lebreton, J.P. and Matson, D. (1997) The relevance of Titan and Cassini/Huygens to Pre-biotic chemùistry and the origin of Life on Earth,, *ESA SP*- **1177, 231-233.**

Raulin, F. (1997) Titan's organic che'muistry and exobiology, *ESA SP*- **1177, 219-229.**

Raulin F., Mourey, D. and Toupance, G. (1982) Organic Syntheses from CH_4-N_2 Atmospheres : Implications for Titan, *Origins of Life* **12**, 267-279.

Raulin, F., Frère, C., Paillous, P., de Vanssay, E., Do, L.,, and Khlifi, M.(1992) Titan and exobiological aspects of the Cassini-Huygens mission, *J. British. Interplanet. Soc.* 45, 257-271.

Sagan, C. and Khare, B.N. (1979) Tholins : organic chemistry of interstellar grains and gas, *Nature*, **277,** 102-107.

Sagan, C., Khare, B.N. and Lewis, J.S. (1984) Organic matter in the Saturn system, in *Satrun*, University of Arizona Press, Tucson, pp. 788-807.

Smith, P.H., Lemmon, M.T., Lorenz, R.D., Sromovsky, L.A., Caldwell, J.J. and Allison, M.D. (1996) Titan's surface revealed by HST imaging, *Icarus* **119**, 336-349.

Thompson, W., Todd, H., Schwartz, J., Khare, B., and Sagan, C. (1991) Plasma Discharge in N_2 + CH_4 at low Pressures : experimental Results and Applications to Titan, *Icarus*, **90**, 57- 73.

Toublanc, D., Parisot, J.-P., Brillet, J., Gautier, D., Raulin, F., and McKay, C.P. (1995) Photochemical modeling of Titan's atmosphere, *Icarus* **113**, ,2-26.

Toupance, G., Raulin, F. and Buvet, R. (1975) Formation of prebiological compounds in models of the primitive Earth's atmosphere. I : CH4-NH3 and CH4-N2 atmospheres". *Origins of Life,* **6,** 83-90.

de Vanssay, E., Gazeau, M.-C., Guillemin, J.-C., and Raulin, F. (1995) Experimental simulation of Titan's organic chemistry at low temperature, *Planetary Space Sci.* **43**, 25-31.

Yung, Y.L., Allen, M. and Pinto, J.P. (1984) Photochemistry of the atmosphere of Titan: comparison betwen model and observations, *Astrophys. J. Suppl. Ser.* **55**, 465-506.

PRODUCTION OF ORGANIC GASES AND AEROSOLS BY ELECTRICAL ACTIVITY IN TITAN'S TROPOSPHERIC CLOUDS

RAFAEL NAVARRO-GONZALEZ
Laboratorio de Química de Plasmas y Estudios Planetarios, Instituto de Ciencias Nucleares, Universidad Nacional Autónoma de México, Ciudad Universitaria, México D.F. 04510, MEXICO

Abstract. The corona discharge chemistry of a Titan simulated atmosphere was investigated by GC-FTIR-MS techniques. The main products are hydrocarbons (ethane, ethyne, ethene, propane, 2-methylpropane, n-butane, n-pentane, and n-hexane), nitriles (hydrogen cyanide, ethanenitrile, propanenitrile, and butanenitrile) and tholins (composed by a complex mixture of highly branched hydrocarbons). Because little UV penetrates to the lower atmosphere to destroy the molecules formed there, the corona-produced species may be long-lived and contribute significantly to the composition of the lower atmosphere and surface.

1. Introduction

The presence of clouds and convective activity in the lowest region of the atmosphere suggest the possibility of electrical activity in Titan's troposphere. An opportunity to search for evidence of lightning activity on Titan occurred during the Voyager 1 encounter with Saturn on November 12, 1980, when the spacecraft passed within 4.4×10^3 km of Titan's cloud tops. Because optically thick cloud and haze layers prevented lightning detection at optical wavelengths, Desch and Kaiser (1990) searched for lightning-radiated signals at radio wavelengths during about 1 hr of the closest encounter. About 98% of Titan's surface was monitored during this interval and the instrument failed to detect any lightning-associated spherics. This negative result cannot completely dismiss the possibility of a weak electrical activity in Titan's clouds. Titan's atmosphere appears to contain no polarizable gas that could lead to generation of a gross electrical structure of clouds by precipitation mechanisms (Navarro-González and Ramírez, 1997). Instead the electrical dipole within Titan's clouds could result by convective motions to bring externally derived ions into the cloud, where they are attached to cloud particles. These ions and precipitating particles are continuously being generated by cosmic ray and Saturnian magnetospheric interactions with Titan's atmosphere. The net dipole within the cloud would result from updrafts transporting the positively charged particles to its upper portions while the downdrafts deliver negative charged aerosols from above the cloud top to the base. It is expected that weak electric fields would be developed within Titan's clouds that would result in corona activity.

J. Chela-Flores and F. Raulin (eds.),
Exobiology: Matter, Energy, and Information in the Origin and Evolution of Life in the Universe, 313–316.

This paper summarizes some recent findings from this laboratory on the corona chemistry of a Titan simulated atmosphere.

2. Experimental Conditions

An atmosphere composed of 88% N_2, 10% CH_4 and 2% Ar (used as an internal standard) was prepared with ultra-high purity gases using a Linde mass flow measuring and control gas blending console. A coaxial corona discharge was generated in a Pyrex tube with a central tungsten electrode and an externally grounded copper plate. The discharge was induced with a high frequency Tesla coil with an output power of about 2 W (Navarro-González *et al.*, 1998). The samples were irradiated in a closed-system from 2.5 min up to 1.5 days. Immediately after irradiation, the samples were analyzed using a gas chromatograph interfaced in parallel with a FTIR-detector and a quadrupole mass spectrometer. The columns used were PoraPlot Q and Ultra 1 for gases and organic aerosols, respectively.

3. Results and Discussion

Methane and nitrogen linearly decompose with increasing irradiation time at comparable rates over a wide irradiation interval in the corona discharge of N_2-CH_4-Ar. The major products are molecular hydrogen, ammonia, a variety of hydrocarbons and several nitriles. Figure 1 shows a typical chromatogram of the products formed when a mixture is exposed to corona discharges for 30 min at 500 Torr and at 298 K. Table 1 summarizes the techniques used to identify them and also lists their production yield relative to ethane. Ethane is the dominant product; ethene and ethyne co-elute in the chromatographic separation. Their combine yield is comparable to that of ethane. The mass and infrared data of peak 1 in Fig. 1 indicate that ethyne is the main product where the C_2H_2 / C_2H_4 ratio can be from 3 to 8, respectively, depending upon the analytical technique used and assuming similar responses. The yields of higher unsaturated hydrocarbons decrease as the hydrocarbon length increases. Another interesting feature of the corona discharge is the production of branched hydrocarbons and linear nitriles.

The yields of the majority of the products increase linearly with irradiation and then reach steady-state values at about 15 to 30 min. Only for the case of C_2H_2/C_2H_4, their yields reach a maximum and then gradually decrease. At higher irradiation times (*e.g.*, > 2hrs.), a pale yellowish film is gradually deposited on the walls of the reactor. Analysis of these tholins by GC-MS-FTIR was complicated by their low yield of formation. Preliminary data indicates that they consist of a complex suite of highly ramified hydrocarbons ranging from C_8 to C_{20} containing NH_2 and CN groups. There was no detection of aromatic hydrocarbons.

It is not possible to estimate the electric field potential in Titan's methane clouds; however, it seems likely that weak electric fields as those used in the present study could be achieved and, therefore a variety of hydrocarbons and nitriles can be expected to

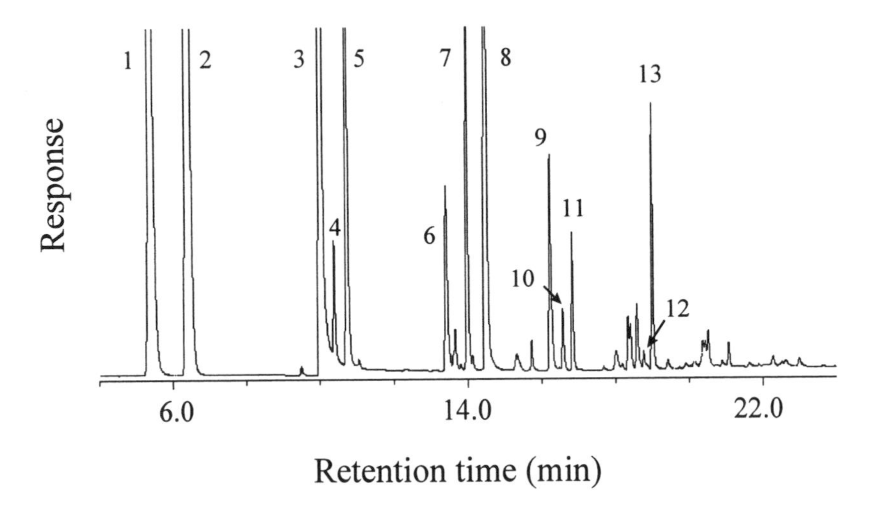

Fig. 1. Gas chromatogram of a gas mixture of hydrocarbons and nitriles produced by 30 minute corona discharge irradiation of a N_2-CH_4-Ar atmosphere: 1. Ethene/ethyne; 2. Ethane; 3. Hydrocyanic acid; 4. Propene/propyne; 5. Propane; 6. 2-Methylpropane; 7. n-Butane; 8. Ethanenitrile; 9. 2-Methyl-2-butane; 10. n-Pentane; 11. Propanenitrile; 12. n-Hexane; 13. Butanenitrile.

Table 1. Production yields of organic gases in the corona discharge irradiation of a mixture of N_2, CH_4 and Ar

Compound	Identification Technique	% Yield*
Hydrocarbons:		
Ethane	MS, IR	100
Ethene / Ethyne	MS, IR	93
Propane	MS, IR	19
2-Methylpropane	MS, IR	8
Propene / Propyne	MS	7
n-Butane	MS, IR	14
2-Methylbutane	MS, IR	8
n-Pentane	MS, IR	2
n-Hexane	MS	11
Nitriles:		
Hydrogen cyanide	MS, IR	49
Ethanenitrile	MS	78
Propanenitrile	MS	2
Butanenitrile	MS	8

*Relative to ethane.

form in Titan's troposphere. Although photochemistry and charged particle chemistry occurring in the stratosphere can account for many of the observed hydrocarbon species in Titan, the predicted abundance of ethene is to low by a factor of 10 to 40 (Yung, *et al.*, 1984; Borucki *et al.*, 1988). While some ethene will be produced by charged-particle chemistry, its production by corona processes and its subsequent diffusion into the stratosphere appears to be an adequate source. Furthermore, because little UV penetrates to the lower atmosphere to destroy the molecules formed there, the corona-produced species may be long-lived and contribute significantly to the composition of the lower atmosphere and surface.

This study has shown that a complex organic chemistry can be expected in the lower atmosphere initiated by corona discharges on surfaces of methane-rain particles within Titan's clouds. Even with the development of weak electric fields, it is still possible to form a variety of hydrocarbons and nitriles. An important feature in corona discharge processes is the formation of mostly saturated nitriles and hydrocarbons (for C_3 and up).

A more thorough investigation is needed to adequately assess the importance of corona processes in Titan's troposphere. In particular, it is important to derive energy yields for the organic gases and aerosols and conduct the experiments at low temperatures.

4. Acknowledgments

The assistance of Miss Sandra I. Ramírez and Miss Graciela Matrajt Arbertman is greatly appreciated. This work was supported by grants from the National Council of Science and Technology of Mexico (CONACyT-4282-E9406) and the National Autonomous University of Mexico (DGAPA-IN102796).

5. References

Borucki, W.J., L.P. Giver, C.P. McKay, T. Scattergood, and J.E. Parris, Lightning Production of Hydrocarbons and HCN on Titan: Laboratory Measurements, *Icarus*, **76**, 125 (1988).

Desch, M.D., and M.L. Kaiser, Upper Limit set for Level of Lightning Activity on Titan, *Nature*, **343**, 442 (1990).

Navarro-González, R., and S.I. Ramírez, Corona Discharge of Titan's Troposphere, *Adv. Space Res.* **19(7)**, 1121 (1997).

Navarro-González, R., A. Romero, and C. Honda, Power Measurements of Spark Discharge Experiments, *Origins Life Evol. Biosphere* (in press).

Yung, Y.L., M. Allen, and J.P. Pinto, Photochemistry of the Atmosphere of Titan: Comparison between Model and Observations, *Astrophys. J. Suppl. Ser.*, 55, 465 (1984).

Section 10.
Extrasolar Planets

THE SEARCH FOR LIFE OUTSIDE THE SOLAR SYSTEM

J. SCHNEIDER
CNRS - Observatoire de Paris, 92195 Meudon, France
e-mail: schneider@obspm.fr

1 Introduction

The question of the existence of Life elsewhere in the Universe occupies a peculiar position in science: it belongs to the 'avant-garde' of well formulated scientific problems, and nevertheless can be understood by layman. This is because the root of the question is essentially subjective: 'Do poeple ressembling ourselves exist and live among the stars we see at night?'. Today, particularly with the discovery of extra-solar planets, the search for Life outside the Solar System is becoming a scientific reasonable goal. Nevertheless one prerequisit is to have an idea of what we want to search for, i.e. an idea of what we mean by 'Life'. In order to miss the fewest conceivable forms of Life, we should start with as few *a priori* assumptions as possible. Let us briefly sketch the argument in 2 steps:

1./ The very essence of Life is a subjective notion. It is, at first, only recognized as such on *a priori subjective* grounds. In terms of modern psychology, 'Life' is, to begin with, an 'object-relation' built on affect and various forms of identifications, introjection and incorporation.

2./ Scientific analysis then recognizes that living systems are based on extreme complexity. The search for living systems is thus the search for cery complex systems.

The above requirements are too general to serve as guidelines for actual searches of Life outside the solar system. We need more specific hypotheses on the physical constitution of living systems. There are several possibilities: chemical (organic) systems, electromagnetic plasmas, solid state physics, liquid electronics, liquid crystals, others ? ...

2 Remote detection of Life

2.1 Remote signatures of Life

Once Life has been objectivated as a complex, out of equilibrium, system, its remote detection leads to the search for out of equilibrium samples of particles. This can be done in two different ways:

J. Chela-Flores and F. Raulin (eds.),
Exobiology: Matter, Energy, and Information in the Origin and Evolution of Life in the Universe, 319–326.

1./ One can search remotly for out of equilibrium populations of constituents (atoms, molecules, ions, ..) present in situ where the Life has developed. I will come back to this strategy in section 5.

2./ One can search for out of equilibrium populations of particles reaching the observer, or 'signals'. It is the strategy usually called SETI (Search for Extra-Terrestrial Intelligence). Because the word 'Intelligence' refers to some kind of intention, it poses yet unsolved philosophical problems [10], [11]. I therefore prefer to just call this strategy 'Search for Complex Signals'. These particles could *a priori* be of any sort. Because of the magnetic field present in the interstellar space, charged particles are not suited for that purpose. Presently, only two kinds of stable neutral particles are known: the photon and the neutrino. The search for electromagnetic signals in the radio or optical domain has led to several projects such as Phoenix or OSETI. They are reviewd by J. Tarter in these Proceedings. To give an open minded touch to the present contribution, let me remark that the yet unexplained ultra-high energy ($\geq 10^{20} eV$) cosmic rays, whose origin is not explained by any known natural acceleration mechanism (photons are exluded and new exotic objects such as 'vortons' have been invoked), could well result from accelerators built by a technologically advanced 'civilization': these accelarators would produce some 'leaks' in the form of ultra-high energy neutrinos, which we detect on Earth [2].

2.2 Habitability Zones

Once a complex process called 'Life' has been chosen from the list above, a 'habitable zone' (HZ) is a region of the Universe where the physical conditions (temperature, density, pressure, radiation content, matter content etc) are favourable for the development of that form of Life. One can schematically suggest the following HZ for each category of 'Life':

(a) chemical (organic) systems: planets at 300 K (see section 3)
(b) electromagnetic plasmas: interstellar medium
(c) solid state physics: solid planets, asteroïds, ... [9]
(d) liquid electronics, liquid crystals: planets at 200 - 250 K.

3 Chemically-based Life.

A standard conservative choice is to restrict the search for life to chemical systems. There is no real scientific reason for such a restriction other than strategic prudence. We can then specify further more restricting hypotheses:

1./ Consider only *carbon-based organic chemistry*

2./ Require the presence of *liquid(s)*: favors the convective and hydrodynamical mixing of molecules

3./ The liquid must be *water*: it is a very good dissolvent and is abundant in nature.

4./ Require the existence of a *solid/liquid interface* to enhance the exchanges between molecules.

From the above 'decision tree', the safest environment which our imagination has yet found is a solid 'habitable' planet or satellite. It is a planet or a satellite having a temperature of about 300 K, i.e. approximately at the Sun-Earth distance (= 1AU) from its parent star.

4 Detection of habitable zones.

The potential success of a given detection method depends naturally on its technological limitations, but also on the different characteristics of the planets and satellites: their mass M, radius R and distance a from the parent star (I assume circular orbits). I will describe some of the methods considered today.

4.1 Habitable single planets.

1./ Gravitational perturbation of the star:

When a planet is in orbit around a star at a distance a, the star makes around the center of mass of the system an orbital revolution with the same period P on a circle with a radius $a_* = a(M_P/M_*)$. This motion leads to the periodic modulation of three observables of the star with a period P and an amplitude respectively given by:

1.1/ Radial velocity:

$$\Delta V_R = \frac{M_P}{M_*} \times \frac{GM_*}{a} = 10 \left(\frac{M_P}{M_\oplus}\right) \left(\frac{M_*}{M_\odot}\right)^{-1/2} \left(\frac{a}{1\mathrm{AU}}\right)^{-1/2} \mathrm{cm/s} \quad (1)$$

1.2/ Astrometric position:

$$\Delta\alpha = \frac{M_P}{M_*} \times \frac{a}{D} = 0.3 \left(\frac{M_P}{M_\oplus}\right) \left(\frac{M_*}{M_\odot}\right)^{-1} \left(\frac{a}{1\mathrm{AU}}\right) \left(\frac{D}{10\mathrm{pc}}\right)^{-1} \mu\mathrm{arcsec.} \quad (2)$$

where D is the distance of the planetary system to the observer.

1.3/ Time of arrival of periodic signals

$$\Delta T_A = \frac{M_P}{M_*} \times \frac{a}{c} = 0.015 \left(\frac{M_P}{M_\oplus}\right) \left(\frac{M_*}{M_\odot}\right)^{-1} \left(\frac{a}{1\mathrm{AU}}\right) \mathrm{sec.} \qquad (3)$$

where c is the speed of light. The best targets for having periodic signals are pulsars, for which several planets have been found (see the table; updates are on the World Wide Web at the URL: http://www.obspm.fr/planets).

As is developed in more details by M. Mayor in these Proceedings, the resolution achievable for radial velocity (1 m/sec) and astrometric (a few μarcsecs.) measurements does not allow to detect terrestrial planets in their habitable zone. There is nevertheless a hope in an extreme but plausible situation: for a 'super-Earth' of say 5 Earth masses in the habitable zone of a $M_* = 0.5 M_\odot$ K5-type star (having thus a temperature of $\approx$ 4,400 K), the amplitude of the V_R modulation is about 1.2 m/s with a period of 100 days. With a precision of 1 m/s for individual measurements, such a modulation can be detected at a 4 sigmas level after 4 orbital periods, i.e. within about 1.1 year.

2./ Direct imaging.

This method faces a difficulty: due to the wave-like nature of light, a stellar image always makes a diffraction halo having an opening angle λ/B where B is the telescope aperture. To separate the planet from this stellar halo, one must use a telescope (or a combination of telescopes) with an aperture (or a baseline) B given by

$$B \geq \lambda \frac{D}{a} = 20 \left(\frac{\lambda}{10\mu}\right) \left(\frac{D}{10\mathrm{pc}}\right) \left(\frac{a}{1\mathrm{AU}}\right)^{-1} \mathrm{m} \qquad (4)$$

where λ is the observation wavelength. In order to avoid redhibitory complications due to atmospheric turbulence, it is necessary to go to space.

3./ Transit of the parent star.

If the orbital plane of the planet is correctly oriented, it produces a drop in the star light during transits of the star disk by the planet. The detection of a transit in the star lightcurve requires three conditions:

3.1/ The orbital plane of the planet must be correctly oriented: for random orientations, the geometric probability is $p = R_*/a$. For an Earth around a 1 $R_\odot$ star, this probability is 0.5%. Since, in addition, the star must be photometrically monitored continuously over at least one entire orbital revolution of the planet, this makes the transit method very inefficient for large a and favors small a since then p is larger and the required time base is shorter.

3.2/ The duration of the transit is $D_T = (P/\pi) times (R_*/a)$, *i.e.* 13 h for an Earth. This duration is not very sensitive to a.

3.3/ The relative brightness drop $\Delta F/F$ is

$$\Delta F/F = \left(\frac{R_P}{R_*}\right)^2 \tag{5}$$

The photometric precision of the lightcurve must be better than $\Delta F/F$. For a 1 $R_\oplus$ planet the drop is 10^{-4}. In ground-based observations, the photometric precision is at best 0.1% [5]. But in space, one can reach a precision of the order of the photon noise, *i.e.* $\approx 10^{-4}$ for a magnitude 15 star with a 50 cm aperture telescope for a 1 h. exposure. The COROT space mission, to be launched in 2002 and mainly devoted to stellar sismology, will have the capability, as a secondary objective, to detect by this method a few tens of telluric planets [3].

4./ Gravitational lensing of a background star.

The planet can produce a gravitationnal amplification A_G of the light of background stars with a duration T_G depending on its transverse velocity V [4]. Once the parent star makes an amplification, the probability that the planet amplification exceeds 5% is 20% [4] for a Jupiter-mass planet. For an Earth-mass planet, the amplification exceeds 1% only in $\sim$ 3% of cases [1]. Unfortunately, one cannot only detect planets in the HZ of their parent star. Furthermore, a lensing event is seen only once and it is not possible to investigate the planet at 4 kpc any further by any other method. This makes the lensing method less attractive.

4.2 Habitable moons of giant planets.

The detection of these objects is impossible by any method perturbationg the star's observables [12]. But they can be detected by the transit method.

1./ Profile of the light curve:

Like the standard detection of transits of single terrestrial planets, a satellite of a giant planet can be detected by the superposition of the transits of the giant planet and of the satellite

2./ Timing of the giant planet transits:

Suppose a giant planet with a mass M_P has been detected by the transit method at a distance a of the star. Then, when a satellite makes an orbital revolution around the giant planet at a distance a_S with a period $P_S = \sqrt{a_S^3/GM_P}$, the latter makes around the center of mass of the system an orbital revolution with the same period P_S on a circle with a radius $a_P = a_s(M_S/M_P)$. This motion leads to a periodic modulation of the time of

Table 1. Catalog of extrasolar planets (January 1998)

Star distance	$M_P[.\sin i_P]$ Jup. Mass (J) Earth Mass (E)	a_P AU	P_P years (y) days (d)	eccentricity
PSR 1257+12 300 pc	3.4 (E) 2.8 (E) 0.3 (J)	0.36 0.47 40	66.54 (d) 98.22 (d) 170 (y)	0.0182 0.0264 ?
51 Peg 13.7 pc	0.47 (J)	0.05	4.2293 (d)	0.0
ups And 16.5 pc	0.68 (J)	0.057	4.611 (d)	0.109
55 Cnc 13.4 pc	0.84 (J) > 5 (J)	0.11 > 4	14.648 (d) > 8 (y)	0.051 -
rho Crb 18 pc	1.1 (J)	0.23	39.645 (d)	0.03
16 Cyg B	1.5 (J)	1.72	804 (d)	0.67
47 Uma 13.4 pc	2.8 (J)	2.11	2.98 (y)	0.03
tau Boo 15 pc	3.87 (J)	0.0462	3.3128 (d)	0.018
70 Vir 22 pc	6.6 (J)	0.43	116.6 (d)	0.4
PSR B1620-26 3.8 kpc	< 10 (J)	20	100 (y)	-
HD 114762 28 pc	10 (J)	0.3	84.5 (d)	0.25

transits of the giant planet with an amplitude ΔT_T given by $DeltaT_T = (a_S/V_P) \times (M_S/M_P$ where $V_P = \sqrt{GM_*/a}$ is the giant planet velocity. For a Saturn-mass planet at 0.5 AU from a 1 $M_\odot$ star, the numerical value of ΔT_T is:

$$\Delta T_T = 8 \left(\frac{a_S}{10^{-2}\mathrm{AU}}\right) \left(\frac{M_S}{M_\oplus}\right) \mathrm{min.} \tag{6}$$

The COROT space mission [3] will, as for single planets, be able to make such detections.

5 Remote detection of a possible signature of chemically-based Life

Suppose an earth-like planet is discovered in the habitable zone of its parent star. Is it possible to detect the presence of Life on it? The answer is that it is possible to detect spectroscopic signatures of what is perhaps dues to complex organic chemistry. The argumentaion rests on the observation that on Earth all the molecular oxygen and ozone in the atmosphere are of biogenic origin [7]. The detection of ozone may be easier than to detect oxygen. Oxygen can be detected by its absorption band at 760 nm [8].

For technical reasons, it is presently more easy to detect the ozone absorption band at 9.6 μ than to detect oxygen in the visible. It is necessary for that to separate the planet from its parent star. Such a separation requires, at 9.6 μ, an interferometer in space with a baseline of at least 20 m [6]. This has led to the IRSI project (Infra-Red Space Interferometer) which is one of the two potential interferometric Cornerstones of the European Space Agency, to be launched (if approved) in 2015.

References

[1] Afonso C., Gaucherel C. & Schneider J., 1997, Earth-mass planets detection probability by gravitational microlensing (submitted)

[2] The Pierre Auger Project Design Report, 1995, Fermilab

[3] Deleuil M. Barge P., Leger A. & Schneider J., 1997 Detection of Earth-like Planets with the COROT Space Mission in *Planets Beyond the Solar System and the Next Generation of Space Mission* Soderblom Ed. Astron. Soc. of Pacific Conference Series **119**, 259

[4] Gould A. & Loeb A., 1992, Discovering Planetary Systems through gravitational microlenses. *Astrophys. J.*, **396** 104

[5] Henry G., Baliunas S., Donnahue R., Soon W. & Saar S., 1997, *Astrophys. J.*, **474**, 503

[6] Léger A., Mariotti J.M., Puget J.L., Rouan D. & Schneider J. 1996, How to Evidence Primitive Life on an Exo-Planet? - The DARWIN Project -, in *Proceedings of the First International Circumstellar Habitable Zones Conference* L. Doyle ed., Travis Publ. Co.

[7] Lovelock J.E. 1975, Thermodynamics and the Recognition of Alien Biospheres, *Proc. Roy. Soc. Lond.* **B 189**, 167

[8] Owen T. 1980, The Search for Early Forms of Life in Other Planetary Systems: Future Possibilities Affored by Spectroscopic Techniques, in *Strategies for the Search of Life in the Universe*, ed. Papagiannis, Reidel., p. 177

[9] Schneider J., 1977, A model for non-chemical form of Life: crystalline physiology. *Origins of Life*, **8**, 33

[10] Schneider J., 1994, The Now, Relativity Theory and Quantum Mechanics. in *Time, Now and Quantum Mechanics* M. Bitbol and E. Ruhnau Eds. Editions Frontieres.

[11] Schneider J., 1997 Time and the Mind/Body Problem: a Quantum Perspective. *American Imago*, **54**, 307

[12] Schneider J., 1998, A next step in extrasolar planets search: the detection of their moons and rings (submitted).

EXOCHIRALITY IN THE SOLAR SYSTEM AND BEYOND

A.J. MACDERMOTT
Department of Chemistry
University of Cambridge
Lensfield Road
Cambridge CB2 1EW, UK

Animals are made of only L-amino acids and not their D mirror images — biology is *homochiral*, in contrast to non-living systems which are *racemic* (i.e. contain equal numbers of L and D molecules). A search for extra-terrestrial biology or pre-biotic chemistry can therefore be approached as a **S**earch for **E**xtra-**T**errestrial **H**omochirality, **SETH** (MacDermott, 1995, MacDermott *et al.*, 1995). Recent discoveries of excesses of L amino acids in the Murchison meteorite represent the first definitive identification of *exochirality* (chirality outside the Earth) and demonstrate for the first time the operation of a pre-biotic chiral influence.

The chiral influences that may have produced biomolecular homochirality fall into three categories: *local* chiral influences include circularly polarized sunlight (Wolstencroft, 1984) and magnetic and electric fields (Barron, 1996); a *solar-system-wide* chiral influence is provided by circularly polarized radiation from neutron stars (Greenberg *et al.*, 1994); and the *universal* chiral influence of the weak force (mediated by the Z° boson) produces a very small *parity-violating energy difference* or *PVED* between enantiomers (Hegstrom *et al.*, 1979). Calculations of the PVED (MacDermott, 1995a) show that the natural enantiomers, e.g. L-amino acids, D-deoxyribose, and right-hand DNA and RNA are indeed more stable than their "unnatural" mirror images in most cases. Most of these chiral influences are small, but their effect could be amplified by autocatalytic amplification mechanisms (Kondepudi, 1997). It actually does not matter *which* hand is found by SETH: *any* deviation from the racemic state implies at the very least the operation of a chiral influence, and in the right context could imply life itself. But which hand is found may reveal something about the *origin* of homochirality. Chance or local chiral influences will produce different hands on different planets; the solar-system-wide effect of neutron stars will give the same hand throughout any one solar system but different hands in different solar systems; but the universal effect of the weak force will give the same hand in all solar systems.

Turning now to the Murchison enantiomeric excesses, Cronin & Pizzarello (1997) looked at α-methyl amino acids (which are non-terrestrial, to avoid contamination) and found excesses of the L form ranging from 2 to 10%. Engel and Macko (1997) found D/L ratios of 0.5 to 0.6 in terrestrial amino acids including alanine, but terrestrial contamination was excluded because δ^{15}N isotope ratios were non-terrestrial and the same for both enantiomers. The Murchison amino acids may be showing a primordial enantiomeric excess from irradiation of the

J. Chela-Flores and F. Raulin (eds.),
Exobiology: Matter, Energy, and Information in the Origin and Evolution of Life in the Universe, 327–332.

pre-solar cloud by a neutron star, or alternatively an amplified electroweak excess (MacDermott, 1997). Calculations of the PVED show that L-alanine is definitely more stable than D (Tranter, 1985), and our own preliminary calculations suggest that this electroweak stabilization of the L form extends also to Cronin's α-methyl amino acids. The Murchison enantiomeric excesses are taken as evidence of a pre-biotic chiral influence rather than life itself (Cronin & Pizzarello, 1997, MacDermott, 1997) because of the *total lack of fractionation* of molecules useful to life: all isomers, including branched chains, are present in the ratios to be expected from lab syntheses. There have been reports of sausage-like "microfossils" in Murchison (Hoover, 1997) similar to those found by McKay *et al.* (1996) in Mars meteorite ALH84001 — although many may consider their morphology less convincing, and in contrast to the ALH84001 case they are not accompanied by much other evidence for life. However, an open mind should be kept about Murchison — in particular there should be a search for life-like δ^{13}C isotope ratios as found by Pillinger in Mars meteorites ALH84001 and EETA79001.

The COSAC GC-MS experiment aboard the Rosetta mission to Comet Wirtanen will attempt to detect (pre-biotic) homochirality using pairs of chiral GC columns. One of the pair will have a homochiral coating, the other a racemic coating of the same substance. Enantiomers will produce two peaks through the homochiral column where there was only one through the racemic column. If the two peaks also have identical MS they are very likely enantiomers — but one cannot be absolutely sure without chiroptical detection. A possible optical rotation (OR) detector for future missions is the SETH Cigar (MacDermott *et al.*, 1996), a cigar-sized space polarimeter with no moving parts in which the normal rotating analyzing polarizer is replaced by multiple fixed polarizers at different angles, used with a diode array detector. Direct polarimetric detection is ideal with liquid chromatography (HPLC) — output from the columns could go straight into the SETH Cigar — but impossible with gas chromatography (GC) because of the small size of the samples. HPLC is not realistic for space use at present, but improvements in miniaturization and handling of liquids in space could make chiral HPLC-OR (with the SETH Cigar as detector) the technique of the future for Mars missions. In the meantime the SETH Cigar could be used in parallel with GC-MS (i.e. independently, not as the GC detector) on nearer-term Mars missions.

The ideal Mars exobiology package would be a drill plus GC-MS plus the SETH Cigar. A drill is essential to get below the surface to where traces of extinct or extant life might be found. The GC-MS should have columns to detect isotopic signatures (similar to Rosetta's MODULUS), and others to detect organics, including chiral columns (as in Rosetta's COSAC). For only minimal extra mass (0.8 kg), a Cigar could be added to clinch identification of chirality. Different molecules need different GC columns for optimal resolution, so it is possible that chirality would be missed by the GC due to incomplete resolution of unknown target molecules — but the Cigar could pick up an enantiomeric excess through optical rotation. Furthermore, the non-volatility of amino acids makes them difficult in GC, but they are ideal in the Cigar (complexing agents could be used in the solvent to enhance the amino acids' optical rotation). In particular, *homochirality* plus *non-equilibrium isotope ratios* found *together* would be an extremely powerful indicator of life that is difficult to mimic abiotically.

What are the prospects of finding homochirality on Mars as a relic of past life? It is generally thought that any life which evolved on Mars would have gone extinct when Mars went cold and dry 3.5 billion years ago. But Bada & McDonald (1995) have shown that although racemization half-lives of amino acids are typically 10^6 years in wet conditions, this rises to 10^{13} years or more in dry and/or frozen conditions. The homochiral signature of biology could therefore be preserved beneath the Martian surface even if the extinction occurred 3-4 billion years ago, provided there were no protracted periods of re-melting. Racemization would be even less problematic if the extinction occurred more recently, as suggested by Pillinger's recent discovery of life-like $\delta^{13}C$ isotope ratios in Mars meteorite EETA79001, which is only 180 million years old.

The life-like signs in EETA79001 of course beg the question of whether life may have persisted on Mars even up to the present day. It is therefore timely to reappraise the results of the famous Labelled Release (LR) experiment on the 1976 Viking mission. This experiment appeared to test positive for life on Mars, but the result was later dismissed because another Viking experiment tested negative, and so the "no extant life on Mars" paradigm became set. The LR experiment (Levin 1997, Digregorio, 1997) took Mars soil and dissolved it in solutions of ^{14}C-labelled nutrients (amino acids, esters, etc.). If the Martian soil released $^{14}CO_2$ from the labelled nutrients (but not when sterilized at 160°), this would indicate microbial activity — and this test was positive on Mars. Furthermore 60% less $^{14}CO_2$ was evolved at 46°C, and 90% less at 51°C: such large changes with slight temperature differences are very characteristic of microbial activity, and very difficult to explain by ordinary chemical reactions. But although the Viking LR tested positive for life on Mars, the Viking GC-MS detected no organics. An inorganic explanation was therefore devised — the famous soil oxidant, most popularly H_2O_2. But the oxidant theory does not explain the unusual temperature behaviour of the LR.

Furthermore it is most significant that the Viking GC-MS was much less sensitive than the LR: whereas the LR could detect as few as 50 cells, the GC-MS needed 10^6 cells to produce enough organics to detect. The Viking instruments were tested on Antarctic soil sample #726, and again the LR tested positive for life (Levin & Straat, 1976) while the less sensitive Viking GC-MS found no organics. A wet chemistry analysis of soil #726 *did* find organics (Lavoie, 1979), although a recent study (Cronin, 1995) suggested that the "organics" may in fact have been coal rather than fresh biological material — but this study involved digesting the sample with concentrated HF-HCl, which might have destroyed delicate biomolecules. So the evidence is far from clear-cut, but there is clearly a strong possibility that the Viking GC-MS was too insensitive to detect the rather low levels of biological activity that might be present in marginal environments such as Antarctica and Mars. Significantly it is now known that there *is* life in the Antarctic Dry Valleys, although this was not known at the time of Viking. New evidence since Viking has found life thriving in a variety of extreme environments (Flam, 1994): it is now known that organisms can survive prolonged desiccation, e.g. lichens inside rocks in the Antarctic Dry Valleys where precipitation rarely occurs (and even then only as light snow), and also bacteria in salt blocks, bacteria revived after millions of years' dormancy, etc..

It is certainly true that there can be no liquid water on the surface of Mars. Although the total atmospheric pressure is 6.65mbar (which is above water's triple pressure of 6.11mbar), the atmosphere contains only 300ppm of H_2O (the rest being mainly CO_2), so the partial pressure of water is far below the triple pressure and any liquid water on the surface would instantly evaporate. But just below the surface, in enclosed pores of the soil, it is possible that the partial pressure of water could transiently rise above the triple pressure as frost sublimes in the morning sun and diffuses to the surface (Carr, 1996). Transient liquid water could allow microbes to get a "quick drink", and indeed microbes have adapted to environments on Earth where moisture is similarly transient, e.g. in Death Valley (Levin, 1997). Indeed, it is worth noting (Digregorio, 1997) that although the bacteria in Death Valley are well documented (Levin & Heim, 1965) and were easily detected by the LR, their density is below the sensitivity limits of the Viking GC-MS (Digregorio, 1997). It is therefore conceivable that life-forms may have gradually adapted to the increasing dryness on Mars, possibly even learning to survive on water vapour alone. An open mind should therefore be kept on the results of the Viking LR. In particular the time is ripe for Levin's proposed up-dated version of the LR (Levin, 1997) that would resolve the issue by presenting L and D nutrients separately (in contrast to Viking which used racemic mixtures): evolution of $^{14}CO_2$ from one enantiomer but not the other would be conclusive for microbial rather than surface oxidant activity.

An ideal Mars experiment should give useful data even if there is no life on Mars, while at the same time not losing the opportunity to look for extant as well as extinct life. To this end we propose a "triple" package combining the proposed University of Leicester experiment (Sims *et al.*, 1997) with the SETH Cigar and the up-dated LR. Sims' Leicester experiment dissolves crushed Mars rock in a carousel of half a dozen buckets containing solvent or different combinations of solvent plus nutrients. It aims to measure parameters such as pH, conductance and turbidity, which might be relevant to possible Martian biology but are at the same time useful data to have *even if there is no life.* The experiment is greatly improved by making some of the buckets SETH Cigars to detect optical rotation as the signature of *extinct* (or possibly *extant*) life. Finally, just in case of *extant* life, a $^{14}CO_2$ detector can be added for only a tiny extra mass, and ^{14}C-labelled L and D nutrients can be offered in separate buckets. If $^{14}CO_2$ were evolved from the bucket with one hand of nutrient but not the other (and stopped by heating to 160°C), this would be a conclusive indication of life — and even more so if Mars soil dissolved in solvent alone gave optical rotation with the SETH Cigar. The "triple" package, lightweight (2kg) and simple, could be an ideal *in situ* pre-screener for sample-return missions: it would clearly be irresponsible to return to Earth optically active samples or samples that responded to only one enantiomeric nutrient.

Irrespective of what technique is used to search for life, a marginal environment such as Mars presents a "needle in a haystack" problem: caches of fossils or pockets of extant life are likely to be highly *localized* and therefore easily missed. But Earth-like biospheres will show *global* signatures. The Darwin space interferometer (Leger *et al.*, 1996) (an array of five 1.5m telescopes under consideration by ESA for launch between 2009 and 2017) will detect global infra-red signatures on Earth-like extra-solar planets: CO_2 indicates a telluric planet; H_2O as well indicates it is habitable; and finding O_3 in addition should be a signature of life itself. A follow-up Darwin II mission between 2015 and 2025, with larger mirrors to give higher resolution, would be able to detect fainter signatures of life such as methane and CFCs. To these signatures we would like to add *bulk homochirality*. Light reflected from an Earth-like planet will show a slight circular polarization of order 0.1% to 1% due to reflection circular dichroism (differential absorption of left and right circularly polarized light) from the highly chiral chlorophyll molecules in vegetation (or plankton in the oceans) in the red visible at 6700 Å. Looking for circular polarization on Darwin II could thus enable **SETH**, the **S**earch for **E**xtra-**T**errestrial **H**omochirality, to become **SEXSOH**, the **S**earch for **EX**tra-**SO**lar **H**omochirality. For Darwin II to detect a 1% circular polarization it needs to gather 250 times as many photons (MacDermott, 1997) as Darwin I (5 x 1.5m telescopes), which would require a single 53m dish or 28 x 10m mirrors or 1250 x 1.5m mirrors. But these figures take no account of future improvements in technology, e.g. smoother mirrors might make a 20m dish sufficient. The O_3 signature of life from Darwin I could conceivably be mimicked by hitherto unsuspected elaborate inorganic chemistry, but the case for life would be greatly strengthened if Darwin II found that the planet was also optically active. *Bulk* optical activity on extra-solar planets with SEXSOH can *only* come from extant, global life, not localized pre-biotic chemistry or fossils. Furthermore, finding the same hand in many different solar systems would tend to support the weak force for the origin of chirality — and the Darwin missions will examine several hundred 300 nearby stars. But even if we find the "wrong hand", bulk optical activity is still the chiral signature of life!

Acknowledgements

I would like to acknowledge the contribution of Mark Sims (U Leicester), Gil Levin (Biospherics Inc.), and my Co-Investigators in the SETH and SEXSOH Consortia: L.D.Barron (U Glasgow), A.Brack (CNRS, Orleans), T.Buhse (Wake Forest U, NC), J.R.Cronin (Arizona State U), A.F.Drake (U London), R.J.Emery (RAL, UK), G.Gottarelli (U Bologna), J.M.Greenberg (U Leiden), R.Haberle (NASA Ames), R.A.Hegstrom (Wake Forest U, NC), K.Hobbs (Glaxo Wellcome), B.T.Jones (Wake Forest U, NC), D.K.Kondepudi (Wake Forest U, NC), C.N.Matthews (U Illinois), C.P.McKay (NASA Ames), S.Moorbath (U Oxford), A.J.Penny (RAL, UK), F.Raulin (U Paris 12), M.C.W.Sandford (RAL, UK), D.W.Schwartzman (Howard U), W.Thiemann (U Bremen), G.E.Tranter (Glaxo Wellcome), R.D.Wolstencroft (ROE, UK), J.C.Zarnecki (U Kent).

References

Bada, J.L. and McDonald, G.D. (1995) Amino acid racemization on Mars: implications for the preservation of biomolecules from an extinct Martian biota *Icarus* **114**, 139-143.

Barron, L.D. (1996) Symmetry and molecular chirality *Chem. Soc. Rev.* **15**, 189-223.

Carr, M.H. (1996) *Water on Mars*, Oxford University Press.

Cronin, J.R. (1995) Organic carbon of Antarctic soil #726: a Viking footnote, Dept. Chem. and Biochem., Arizona State Univ., Nov. 16, 1995.

Cronin, J.R. and Pizzarello, S. (1997) Enantiomeric excesses in meteoritic amino acids *Science* **275**, 951-955.

Digregorio, B.E., with Levin, G.V. and Straat, P.A. (1997) *Mars: The Living Planet*, Frog Ltd, Berkeley, California.

Engel, M.H. and Macko, S.A. (1997) Isotopic evidence for extraterrestrial non-racemic amino acids in the Murchison meteorite *Nature* **389**, 265-268.

Flam, F. (1994) The chemistry of life at the margins *Science* **265**, 471.

Greenberg, J.M., Kouchi, A., Niessen, W., Irth, H., van paradijs, J., de Groot, M. and Hermsen, W. (1994) Interstellar dust, chirality, comets and the origins of life: life from dead stars? *J. Biol. Phys.* **20** 61-70.

Hegstrom, R.A., Rein, D.W. and Sandars, P.G.H. (1979) Parity non-conserving energy difference between mirror-image molecules *Phys. Lett. A* **71** 499.

Hoover, R.B. (1997) Meteorites, microfossils, and exobiology *Proc. SPIE* **3111**, 115-136.

Kondepudi, D.K. (1987) Selection of molecular chirality by extremely weak chiral interactions under far from equilibrium conditions *BioSystems* **20**, 75-83.

Lavoie, J.M., Jr. (1979) Support Experiments to the Pyrolysis/Gas Chromatograph/Mass Spectometric Analysis of the Surface of Mars *Ph.D. Thesis, MIT.*

Leger, A. *et al.* (1996) Could we search for primitive life on extra-solar planets in the near future? The Darwin Project *Icarus* **123**, 249-255.

Levin, G.V. (1997) The Viking Labeled Release Experiment and Life on Mars *Proc. SPIE* **3111**, 146-161.

Levin, G.V. and Heim, A.H. (1965), in *Life Sciences and Space Research III*, M.Florkin, ed., North Holland.

Levin, G.V. and Straat, P.A. (1976) Labeled Release - An experiment in radiorespirometry *Orig. Life* **7**, 293-311.

MacDermott, A.J. (1995) The weak force and SETH: the Search for Extra-Terrestrial Homochirality *AIP Conf. Proc.* **379**, 241-254.

MacDermott, A.J. (1995a) Electroweak enantioselection and the origin of life *Orig. Life Evol. Biosphere* **25**, 191-199.

MacDermott, A.J. (1997) Distinguishing the chiral signature of life in the solar system and beyond *Proc. SPIE* **3111**, 272-279.

MacDermott, A.J. *et al.* (1996) Homochirality as the signature of life: the SETH Cigar *Planet. Space Sci.* **44**, 1441-1446.

MacDermott, A.J. *et al.* (1997) Homochirality as the signature of extra-terrestrial life *IAU Colloquium* **161**, 505-510.

McKay, D.S. *et al.* (1996) Search for past life on Mars: possible relic biogenic activity in Martian meteorite ALH840001 *Science* **273**, 924-930.

Sims, M.R., Cole, R.E., Grant, W.D., Mills, A.A., Powell, K. and Ruffles, R.W. (1997) Simple techniques for detection of Martian microorganisms *Proc. SPIE* **3111**, 164-174.

Tranter, G.E. (1985) The parity violating energy differences between the enantiomers of α-amino acids *Mol. Phys.* **56**, 825-838.

Wolstencroft, R.D. (1984) *IAU Symposium* **112**, Boston.

DRVS AND THE SEARCH FOR LIFE ON EXTRASOLAR PLANETS

RAJENDRA VIKRAMSINGH
B-10 Westend, New Delhi 110021, India
3611 Lupine, Palo Alto, California 94303
E-mail via: psteiner@ix.netcom.com

Abstract: Specific molecules of interest in the search for life on extrasolar planets may be expected to have a large number of spectral lines. Correlation techniques applied to a sufficiently large set of spectral lines will greatly enhance the detection of the specific molecule. Stellar emissions are orders of magnitude brighter than planetary emissions and interferometric techniques have been proposed to reduce this contrast. Additional contrast reduction may be achieved by using Differential Radial Velocity Spectroscopy (DRVS), since the spectrum from the planetary molecules will vary periodically in relation to the stellar spectrum due to a difference in their radial velocities towards the observer.

1. Introduction

Differential Radial Velocity Spectroscopy (DRVS) may be used to observe the spectral lines of a faint planetary source even in close proximity to its extremely bright star. DRVS filters the spectra according to their differential Doppler shift, and is essentially unaffected by the magnitude of the signals. DRVS can identify the signature of a designated molecule if it is present in the extrasolar planetary atmosphere.. Molecules composed of H ,C, N, and O are prime candidates in the search for life on extrasolar planets. Carbon dioxide has been detected in the spectra of Earth, Venus, and Mars, but ozone and water have been detected only in Earth's atmosphere.

2. Planetary Spectra and Correlation Gain

A planet in orbit around its star produces a periodically varying positive and negative Doppler shift of the planetary spectral lines relative to the stellar spectrum. During one orbit the planetary radial velocity towards the observer varies over tens of kilometres per second while the stellar radial velocity varies due to reflex motion by only a few metres per second. The stellar spectrum shifts a little but the Doppler shift of and

J. Chela-Flores and F. Raulin (eds.),
Exobiology: Matter, Energy, and Information in the Origin and Evolution of Life in the Universe, 333–336.

change in the planetary spectrum are several thousand times greater. The differentialDoppler shift of each planetary spectral line is proportional to the difference at the particular moment between the radial velocities of the planet and its star towards the observer. The separation (frequency difference) between these spectral lines also keeps changing proportionately. The Doppler shifted spectrum is thus distinct for each moment of the planetary orbit.

The shifting planetary lines of a specific molecule in the shifting planetary spectrum can be tracked either with a comb of matched filters or better with a dynamic matched filter that Doppler shifts the laboratory spectrum of the specific molecule by the radial velocity (which is changing from moment to moment) of the planet. Substantial correlation gain can be achieved for the planetary spectrum and the larger amplitude stellar signal is rejected as uncorrelated "noise".

3. Detection of Giant Planets

Recent discoveries of massive planets orbiting close to several stars are based on precise measurements of the Doppler shift of the stellar spectrum which disclosed small periodic variations of only a few metres per second in the observed radial velocity of the star that are believed to be due to stellar reflex motion caused by one or more orbiting planets. This indirect planetary detection technique requires reference spectra with a large number of precisely calibrated line and excellent long term stability.

Large radial velocity variations, believed to be due to stellar reflex motion,have been measured; +/-56.83m/s in 4.231 days for 51 Pegasi by a >0.45 Jupiter mass planet at 2.1 AU, and +/-47.3 m/s in 1092 days for 47 Ursae Majoris by a >2.3 Jupiter mass planet at 2.1 AU. Less massive planets in larger orbits should be detected soon. Stellar radial velocity variations can be measured to better than 15 m/s at several observatories and almost 3 m/s at Lick in California. Reflex motion of the Sun due to Jupiter at 5.2AU changes the Sun's radial velocity by +/-13 m/s in 11.9 years, but the Earth creates a currently unmeasurable change of only +/-0.09 m/s in one year. Even if radial velocity changes of a few centimetres could be measured the presence of a small planet would be ambiguous because of intrinsic stellar phenomena that create greater radial velocity variations. Unfortunately, the detection of a planet, large or small, by stellar reflex motion does not provide enough clues to its atmosphere and the possibility of life on it.

4. Direct Detection of Planetary Emissions

There are two major problems, based on the Solar system as the model, which have

J. Chela-Flores and F. Raulin (eds.),
Exobiology: Matter, Energy, and Information in the Origin and Evolution of Life in the Universe, 339–346.

weighed heavily against a direct search for extrasolar planets and a study of theirspectra. Firstly, the flux expected from the planets is very low and it would require large apertures and long observation times to collect enough photons. Since planetary photons are necessary to study planetary atmospheres there is no getting around this problem which is well understood by astronomers studying faint objects. The second problem is that the parent star's (Sun) flux is a million times the flux of the planet (Earth) in the infrared and a billion times in the visible spectrum.These values refer to average flux over relatively broad bandwidths and overstate the stellar/planetary spectral lines amplitude ratio. Anyway, to overcome the high flux ratio problem spatial techniques, such as interferometric nulling, have been devised and hardware implementation is in progress. The interference null will be placed on the star thereby reducing its flux by a factor of a thousand or more. Adaptive optics may be used to further improve the performance of large aperture ground based interferometers. Space borne interferometers with smaller apertures are also planned.

Differential Radial Velocity Spectroscopy tackles the high stellar/planetary flux ratio problem by focusing on the significant difference in the Doppler shifts of the stellar and planetary spectra. During one orbit the radial velocity variation of the planet is several thousand times greater than that of the star due to its reflex motion. For example, the radial velocity of the Earth varies by +/-30 km/s and the Sun by only 0.09 m/s, 51Pegasi planet by +/-130 km/s and the star only 56.83 m/s, and 47Ursae Majoris planet by +/-21 km/s and the star only +/-47.3 m/s. The Doppler shift of the stellar spectrum is therefor negligible in comparison with the Doppler shift of the planetary spectrum. Precise measurement of the spectral line positions is not required but accurate tracking of the shifting and changing planetary spectrum is necessary in order to achieve maximum correlation gain.

One way to realize the matched filtering is to begin with the zero Doppler shift spectrum of the specific molecule. Laboratory spectra are available for a number of molecules of interest in the search for life. The Doppler shifted spectrum, due to the known raial velocity of the star, can be calculated. The difference in the radial velocities of the planet and its star can vary from zero to a maximum of +/-orbital velocity (v), thus defining the widest frequency bands around each planetary spectral line that may have to be searched by the tracking matched filter. For planetary orbits inclined at an angle i the maximum radial velocity difference will be v sin i, and the width of the frequency band to be searched will be reduced.

Large correlation gains are realized with matched filters in covert radar and communication systems because the spectral signature of the signal buried deep under noise is known. Similarly, if a specific molecule is present in the planetary atmosphere under observation, and its Doppler shifted spectrum is reasonably well postulated, correlation techniques will disclose it even in the presence of uncorrelated "noise" of the stellar spectrum with the much larger amplitude.

5. Conclusion

DRVS analysis can be applied to data already collected in the visible spectrum to search for specific molecules around the extrasolar planets discovered so far. Infrared and millimeter wave spectra will be better suited for molecules of interest in the search for life. Theoretical models of extrasolar atmospheres and spectra indicate exciting possibilities of reasonable temperatures for extrasolar planets. Habitable zones may be expected on some of these extrasolar planets. Initial studies can be conducted with existing infrared and millimeter wave observatory facilities. Software for data analysis may need to be developed. Hardware implementation of DRVS, as sketched in poster papers presented in Denver (April 1995), Capri (July 1996), and Trieste (September 1997), could be examined and implemented at a later date.

6. References

Baranne, A. et al (1996) A spectrograph for accurate radial velocity measurements, *Astron. Astrophys. Suppl. Ser.* **119**, 373-390.

Burrows, A. et al (1997) Advances in the Theory of Brown Dwarfs and Extrasolar Giant Planets, *Proceedings of Workshop held in Tenerife*, Spain, March 17-21, 1997.

Butler, R.P. and Marcy, G.W. (1996) A Planet orbiting 47 Ursae Majoris, *Ap.J.***464**, L153-L156.

Butler, R.P. et al (1996) Attaining Doppler Precision of 3 m/s, *ASP* **108**, 500-509.

Jorgensen, U.G. (1995) Molecular Databases, *ASP Conf. Ser.* **78**, 179-203.

Mayor, M. and Queloz, D. (1995) A Jupiter-mass companion to a solar-type star, *Nature* **378**, 355-359.

Vikramsingh, R. (1993) Differential Radial Velocity Spectrometry for Detection of Earthlike Extrasolar Planets, *ASP Conf. Ser.* **74**, 237-243.

Vikramsingh, R. (1995) Differential Radial Velocity Spectrometer for Extrasolar Planetary Studies, *Earth, Moon and Planets* **70**, 213-220.

Section 11.
Search for Extraterrestrial Intelligence.

A FULL-SKY SURVEY FOR ULTRA-NARROWBAND ARTIFICIAL SIGNALS

GUILLERMO A. LEMARCHAND
Instituto Argentino de Radioastronomía (CONICET) &
Centro de Estudios Avanzados (Universidad de Buenos Aires)
C.C. 8 - Sucursal 25, (1425) Buenos Aires, ARGENTINA
E-Mail: lemar@seti.edu.ar

1. Introduction

Almost forty years of SETI studies have repeatedly indicated that searching in the microwave spectral range is a very promising approach. Existing terrestrial radio-astronomical technology is sufficiently sensitive to detect extraterrestrial artificial signals, no much stronger than some leaving Earth. Although it is perhaps natural and seems reasonable to many to assume that planetary systems and life are typical, we realize that there may be enough variety in the Milky Way and enough cosmic time to lead the evolution of beings, civilizations and technological levels far different from us. It could be technically possible that a few of those hypothetical civilizations are transmitting beacons many orders of magnitude stronger than the rest of the existing galactic civilizations. The rationality of making full-sky surveys for ultra-narrowband artificial signals, compared with the target-search strategies (e.g. Project Phoenix) is based in the fact that it is much more easier to detect those beacons that are intrinsically strongest, but remote, than those that are nearer but weakest. Drake [1] showed that this is true even when for every 300 civilizations transmitting at certain radio power, there is but one civilization which transmits signals ten times more powerful. Under this fact it is probably not right to point "only" to the nearby stars. The full-sky survey take this advantage of scanning in all sky directions trying to get the strongest signals at the so-called "magic frequencies". Table 1 shows the main technical characteristics of the on-going SETI full-sky surveys. All of them are economically sponsored by *The Planetary Society*. The *SETI Institute* also provides funds for the SERENDIP program.

Project BETA, that operates from the 26-meter radiotelescope at Agassiz Station, covers the "waterhole" of 1.4 to 1.7 GHz, using 250 million-channel spectrometer, in conjunction with a 3-beam antenna system, producing a 250 Mbytes/s output stream [2].

The Berkeley SETI effort has been ongoing for more than twenty

J. Chela-Flores and F. Raulin (eds.),
Exobiology: Matter, Energy, and Information in the Origin and Evolution of Life in the Universe, 339-346.

years. The SERENDIP IV spectrometer has 168 million channels of 0.6 Hz each and is in operation from Arecibo in Puerto Rico [3]. An eight-million channel version of the SERENDIP system will be installed at the 64-meter Parkes radiotelescope in Australia to make an extragalactic full-sky southern survey.

Table 1. Technical characteristics of the main on-going microwave SETI sky surveys

Characteristics	BETA	META II	SERENDIP IV
Site	Oak Ridge	BuenosAires	Arecibo-Pto.Rico
No. of channels (millions)	250 x 8	8.4	168
Antenna diameter (meters)	26	30	305
Spectral Resolution (Hz)	0.5	0.05	0.6, 1.2, 2.4,...600
Instantaneous Bandwidth (MHz)	40	0.4	100
Total Bandwidth Coverage (MHz)	320	1.2	180
Sensitivity ($W.m^{-2}$)	$3x10^{-24}$	$8x10^{-24}$	$\sim10^{-24}$
Sky Coverage (% of 4 π)	70	50	30
Types of Signals	C, SC	C	C, CH, P

2. The Southern Hemisphere Full Sky-Survey for Narrowband Signals : META II

The META II system, an 8.4 million channel spectrometer to performed a full sky survey for artificial narrowband signals, has been used from one of the two 30-meter radio-telescopes of the Instituto Argentino de Radioastronomía (IAR) for 5 years. After more than 14,000 hours of observations and $\sim10^{13}$ spectral channels already analyzed, we got 29 extra-statistical events or "alerts". Unfortunately, the software and hardware available with the original system did not allow us to determine the real origin of those "alerts". With the sponsorship of *The Planetary Society* a new data acquisition system for the META II spectrometer was developed by IAR's engineers. Some of the advantages of the new system are the following: the possibility of integration times from fraction of seconds to hours in a specific high resolution frequency domain window, the possibility to use it as a high resolution astronomical analyzer (e.g. the study of high resolution OH masers profiles), the use of lower thresholds with the correlation analysis of the interesting candidate signals found in previous observations. The new system will also be able to control the antenna movements, local oscillators, frequencies, frames, polarizations, IFs, post detection data analysis, etc. It will also be possible to switch between any of the two 30-meter radiotelescopes and to expand our search to other frequencies.

The original META system has been described in detail elsewhere [4-5]. The main characteristics of the IAR's facilities are the following: the radiotelescope is a 30-meter parabolic reflector with an equatorial mount, but has operated in a meridian transit mode for the SETI survey, being moved in declination by one half-power beamwidth every day. It's operation was done 12 hours a day, while during nights the same dish was used for conventional astronomy. The radiotelescope can make a complete sky survey between ($-90^o \leq \delta \leq -10^o$) in a period of 500-700 days.

The original system first IF was centered at 126.5 MHz and the second IF was centered at 30 MHz. The software from the old *Wicat* platform computed ephemeris information and controls the local oscillators (LO), *i)* setting the first LO and polarization switch at the beginning of each 20-second integration interval τ_i (where $\Delta B \tau_i=1$) such that the center of the received band corresponds to the chosen frequency as observed in the chosen reference frame, and *ii)* sweeping the second LO to compensate for the Doppler *chirp* caused by the Earth's rotation. The control computer also *iii)* orchestrates overall timing (frame buffering and initiation of Fourier transforms), *iv)* checks for processor errors using 16 redundant processors (1 million channels), *v)* examines the 8.4 million channel spectra for unusual signal features, and *vi)* archives numerical and graphical information connected with unusual signals or system malfunction.

In order to receive transmissions at λ=21 cm. referenced to the Galactic Barycenter or Cosmic Blackbody rest frames, a total bandwidth of at least 300 kHz was needed, because of the uncertainties of ± 30 km/sec in our knowledge of those frames. However the resolution bandwidth should be kept matched to cover the natural bandwidth of carriers propagated through the interstellar medium, for optimum detection of interstellar carriers. META II has a maximum 2^{23} = 8,388,608 channels, ΔB = 0.05 Hz per channel, resulting in an overall instantaneous bandwidth of 400 kHz.

The dual polarization single-conversion receiver, with programmable first LO, fixed second LO and swept third LO (for Doppler Effect compensation of Earth rotation) feeds a 128 point DFT implemented with a A41102 Austek Microsystems Frequency Domain Processor. It performs a multiplication between the signal and a selected window in the time domain, so the signal is convolved with the window in the frequency domain. This constitutes a filter bank pre-processor for an array of 128 FFT processors of 64 K points. The complex amplitudes that appear in any given DFT channel, considered as a time series feeds a corresponding ultra-narrow band 64 K complex FFT (128 in total), resulting in 8,388,608 channels covering 400 kHz. Each Fourier processor computes its complex FFT and resultant power spectra. The whole system performs an 8 million point complex FFT in 15 seconds.

The META strategy was to look for properly chirped narrowband signals at guessable rest frames. The chirp was necessary to compensate the Doppler shift produced by the Earth rotation and avoid spreading the signal over many channels. We use three rest frames of recognized universality: the Local Standard of Rest (LSR), the galactic Barycenter, and the Cosmic Blackbody. Table 2 shows a list of the available feeds and their characteristics from IAR's radiotelescopes that were used or will be used in the future for SETI observations.

Table 2. Physical parameters of the available feeds.

	Antenna I		Antenna II	
Central Frequencies [MHz]	1420 (H)	1667 (OH)	1420 (H)	3300 (CH)
T_{sys} [K]	35	45	100	60
Δ B [MHz]	10	150	10	150

3. Observational Results and Post-Detection Bayesian Analysis

Lemarchand *et al*· [6] showed the results of the first five years of observations of the META II system. After analyzing more than 14,000 hours of observations and ~10^{13} different signals, we were unable to explain 29 ultra-narrowband (ΔB~0.05 to 0.1 Hz) events that were over 30 times the average mean power. If we assume that none of the *extra-statistical events* represent narrowband sources of extraterrestrial origin, we are still faced with the problem of their origin. The absence of "chirp broadening" or "marching signals" behavior in these events seems to rule out their explanation as fixed terrestrial carriers, but still allows several mundane possibilities:

(1) If terrestrial microwave carriers display an ensemble of drift rates at a level of parts in 10^9 per second, these events could represent the enhanced with approximately the correct drift rate.

(2) It is possible that intermittent and undetected processor error could be responsible for some of the extra-statistical events. One plausible mechanism is occasional "soft" errors in dynamic RAM memory (which comprise 6912 of META's 20,000 integrated circuits) caused by a particles. These occur in the kind of memory chips used by META (64 K DRAM) at a rate of ~500 FITs ("failure in time", defined as 1 error per 10^9 chip hours). Using this rate, one can estimate that there ought to be ~42 soft errors during 14,000 operating hours of META II. To this count we apply factors of 1/2, 1/2 and 1/8 respectively, to account for the fact that only half act in the direction to increase the size of the signal, that the error must be in the upper half of the integer to create an isolated spectral feature, the error must occur during the last two of the 16 "butterflies" that comprise the Fourier transform. Thus we estimate that soft memory errors might account for ~3 peaks during the whole set of observations.

(3) A more mundane source of processor error is socket and connector degradation as a source of processor unreliability. This is not surprising, given the fact that there are ~2 x 10^5 socket pins and 2 x 10^4 backplane pins in the META system. This source is difficult to estimate but it might be possible that some of the events have this origin.

Recently, Cordes, Lazio and Sagan [7] derived likelihood and Bayesian tests to analyze the extra-statistical events according to three models: (I) radiometer noise fluctuations; (II) a population of constant galactic sources which undergo deep fading and amplification due to interstellar scintillations, consistent with ETI transmissions, and (III) real, transient signals of either terrestrial or extraterrestrial origin. The application of these models to META "events" shows that:

- If any of the events in META is due to scintillating celestial source, it likely results from a modest to large scintillation gain combined with favorable noise fluctuations.
- Existing re-observations of META candidate signals (follow the source for periods of ~1 hour) were incapable of ruling out the case where a real ETI source with constant, intrinsic signal strength underlies the measured candidate signal. This conclusion holds even for the case where the scintillations remain correlated between the time of an initial detection and prompt re-observations. Future re-observations are certainly capable of ruling out a constant source model for the META detections.
- To test the "real" origin of the signals, much lower thresholds and larger number of re- observations are needed than have been performed to date.

We are planning to perform this new re-observations strategies, in order to follow-on the candidate signals. To do this, we made an important improvement in the computer system (hardware and software), to be able to analyze the huge amount of information that will be produced, setting up the observations at a lowest threshold. In this way, we will have the capacity to track each candidate source, for at least four hours per run, at a threshold of 5-10 times the mean power P_0. The new software and hardware capacity will distinguish the presence of any CW signal at a level of 5-10 P_0 that can correlates -in frequency and sky position- with a previous more intense signal.

4. The System Upgrade

The requirements for the new data acquisition system were the following:

- To use a PC platform in order to get a data format compatible with several post-detection processing systems.
- To allow the system integration times from seconds to hours in the same fixed frequency window.
- To control the antenna movements, local oscillators, frequencies, frames, polarizations, IFs, post detection data analysis.
- To switch between antenna I and II and their corresponding different feeds and frequencies.
- To make the system compatible for other high resolution astronomical uses (e.g the study of high resolution profiles).
- To expand our search to other frequencies (e.g. 1665, 1667 and 3300 MHz).

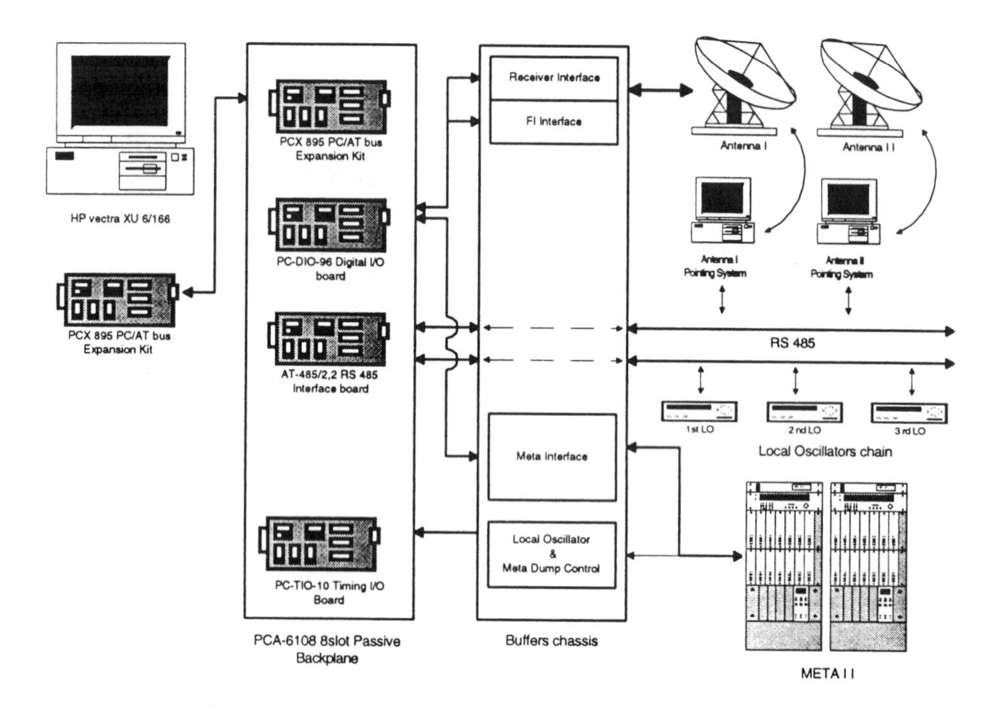

Figure 1 : Block diagram of the new META II acquisition system.

In Figure 1 we present a block diagram of the new hardware upgrade. The new system is based in a PC platform with MS-Windows NT. The choice of a general purpose desktop PC's was based upon the fact that they constitute one of the most universal and easy to up-date platforms for an acquisition system. This standard computer platform constitute the heart of a high open capability system architecture. The new PENTIUM-PRO system has a high computational performance, state-of-the-art video, large memory capacity, network connectivity and ever decreasing prices with both low cost maintenance and easy-to-find spare parts in countries like Argentina.

The operating systems of MS Windows family, offers intuitive GUI's access to more memory, multitasking and inter-process communication. NT 4.0 operating system constitutes a powerful platform for client-server computing. It combines the ease for use and broad-based application capability with the new multi-threading capability. Most of the hardware at IAR is based in different routines written for these types of systems and we decided to concentrate our limited human resources to up-grade all the new systems with this type of platform.

The whole set of low-level functions (those that interact directly with the META hardware and currently coded in 68000 assembler language) were re-written into new **C** functions to work with the new interface (National I/O boards).

The tools of the new high-level commercial software (LabWindows/CVI 4.0) reduce or eliminate many of the time-consuming programming tasks of traditional instrumentation system development. GUI design and control, instrument programming, plug-in DAQ board control, and system trouble-shooting were simplified significantly.

We hope we will be able to expand the RF hardware in order to increase the META II instantaneous bandwidth from 400 kHz to 1 MHz. With the new commands we will be able to control the spectral resolution directly from the PC and to decide what kind of sky survey we would like to make (e.g. only the galactic plane), or the automatic observation of a big list of targets, or to change the observing antenna and feed. All these things were impossible with the original system.

With the new system in operation we will be able to make a relative sensitive full sky survey of the 1667 MHz OH maser lines. As was already showed by Cordes [8] the interstellar and circumstellar masers are natural amplifiers of high gain, so long as they are not saturated. Cordes showed that weak ETI sources, otherwise detectable at only parsec distances, may be rendered detectable across the Galaxy if viewed through a cosmic maser. This program will probably start by mid-1998.

5. References

1. Drake, F. D. (1973) Techniques of Contact, in C. Sagan (ed.), *Communication with Extraterrestrial Intelligence* (CETI), MIT Press, Cambridge.
2. Leigh, D. and Horowitz, P. (1997) Millions and Billions : The META and BETA Searches at Harvard, in C.B. Cosmovici, S. Bowyer and D. Werthimer (eds.), *Astronomical and Biochemical Origins and the Search for Life in the Universe,* Editrice Compositori, Bologna, pp.

601-610.

3. Bowyer, S.; Werthimer, D. ; Donnelly, C. ; Cobb, J , Ng, D. and Lampton, M. (1997) Twenty Years of SERENDIP, The Berkeley SETI Effort : Past Results and Future Plans in C. B. Cosmovici, S. Bowyer and D. Werthimer (eds.), *Astronomical and Biochemical Origins and the Search for Life in the Universe,* Editrice Compositori, Bologna, pp. 667-676.
4. Horowitz, P. and Sagan, C. (1993) Five Years of Project META: An All-Sky Narrow-Band Radio Search for Extraterrestrial Signals, *Astrophysical Journal,* **415** (1), 218-235.
5. Colomb, F.R., Hurrell, E.E. ; Lemarchand, G. A. and Olalde, J.C. (1995) Results of Two Years of SETI Observations with META II, in Seth Shostak (editor), *ASP Conference Series,* vol.74, pp. 345-352.
6. Lemarchand, G. A., Colomb, F. R., Hurrell, E. E., and Olalde, J. C. (1997) Southern Hemisphere SETI Survey: Five Years of Project META II, in C.B. Cosmovici, S. Bowyer and D. Werthimer (eds.), *Astronomical and Biochemical Origins and the Search for Life in the Universe,* Editrice Compositori, Bologna, pp. 611-621.
7. Cordes, J. M., Lazio, T. J. W. and Sagan, C. (1997) Scintillation-Induced Intermittency in SETI, *Astrophysical Journal,* **487** (2), 782-808.
8. Cordes, J. M. (1993) Astrophysical Masers as Amplifiers of ETI Signals, in G. Seth Shostak (ed.), *Third Decennial US-USSR Conference on SETI, APS Conference Series,* vol.47, pp. 257-266.

Acknowledgments: META II was built at Harvard University by E.E. Hurrell and J.C. Olalde with the supervision of Prof. Paul Horowitz in 1989. The Planetary Society (TPS) provided the funds for the construction, while the CONICET from Argentina provided the funds for five years of observations. For the new system upgrade TPS has provided the funds for the equipment and the new observations and analysis expenses. We would like to thanks the great help we have received from Tom McDonough, Dan Werthimer, Paul Horowitz and Jonathan Weinroub. The new design was made by E.E. Hurrell, J.J. Larrarte, with the help of J.C. Olalde, D. Perilli, Abel Santoro and C. Cristina. The new front-end design was made by A. J. Bava and J. Sanz. GAL want to express his acknowledgments to the organizers for providing the economical support to participate in this conference.

SETI THESE LAST FEW YEARS

An Interdisciplinary Overview

J. HEIDMANN
Observatoire de Paris
92195 MEUDON, FRANCE[1]

Abstract

We present a swift survey of the recent aspects, and prospects, of SETI from diverse viewpoints such as bioastronomy, strategy, technology, searches,radio frequency interference, space law, diplomacy, sociology, culture, education.

1. Introduction

It may seem akward to pretend to give news of SETI, the Search for ExtraTerrestrial Intelligence, while everyone knows already that the Big News, the very discovery of an artificial signal emanating from an extraterrestrial technology, did not yet hit the front lines of human preoccupations. Though SETI started its operations nearly four decades ago, without success, its stakes are so huge that tremendous efforts are conducted, with success, to broaden its scope and cutting edge in every domain which learned people think advisable to tackle.

After the cultural blow delivered by the United States Congress in 1993 by cutting all public funding to the then larger effort conducted for years by the National Aeronautics & Space Administration, SETI, already expanding all over the world at a large number of universities and institutes, brightly recovered. It is now a fundamental scientific and technical research.

[1] internet: HEIDMANN@OBSPM.FR

J. Chela-Flores and F. Raulin (eds.),
Exobiology: Matter, Energy, and Information in the Origin and Evolution of Life in the Universe, 347–353.

2. Bioastronomy

Bioastronomy is this new field of research in which astronomers, really the persons on the front line, look for signs of life in the cosmos. The most valuable contribution they brought lately is the discovery of planets around other stars of solar type. Though most are as massive as Jupiter, because Earth-like ones are still too difficult for our technologies, they demonstrate that our solar system is not unique, increasing our odds to discover life elsewhere, hence conforting SETI.

The second contribution was derived from the detailed analysis of a meteorite ejected from Mars, ALH 84001, in which carbonate grains are indicative of a martian ground activity comparable to the one in the hot springs of Iceland, a wet environment favorable for the appearance and development of microbial life. The possibility that an other planet, Mars, in our own solar system, may have also witnessed the apparition of life would indeed be a tremendous push for the universality of life and an encouragment for SETI.

Note also that the latest observations of the Galileo space probe indicate that the Europa satellite of Jupiter may have a global liquid ocean, under an ice crust, which might be an interesting place where to look for some eventual form of life.

3. Strategy/Technology

The main strategical advance is that in the bolt and nuts field, SETI is pursued in a large number of flexible institutes: SETI Institute (ex-NASA), Berkeley, Ohio and Harvard Universities, Argentina and Bologna Radio Institutes, Parkes Australian Radio Facility and Paris- Meudon- Nançay Observatory, while the main technological progress is the tremendous rise in the number of simultaneous "listening" radio channels used: from 1,000 15 years ago to 250,000,000 now. The billion is in sight!

4. Radio Frequency Interference

The real SETI killer problem is the tremendous increase of human-made interferences, emanating mainly from constellations of telecommunication satellites beaming down radio links into our skyward-looking delicate radiotelescope. The rise of hand-held phones, direct-to-home TVs, mobiles, based on tens of billions of dollars enterprises, cannot be checked. The humankind radio window to the universe from down here is doomed by man himself!

It is why a few years ago I started an interdisciplinary effort to protect for the next 20 or 30 years a well singled out minimal locale on the farside of the Moon: the 100 km diameter Saha crater. Very strong support was readily obtained from lawyers of the International Institute of Space Law (IISL). In the words of one of its directors, A.A.Cocca: "The reservation of a lunar zone for scientific activities, and its further utilization aiming to the common good of humanity must be recognized and constitute a precedent (...) Nothing opposes this project to receive the academic, legal and political support."

5. Space Law

International law regulates the relations between states through customs, treaties, UN resolutions, institutes. Its space part comprises five treaties: Outer Space (1967), Rescue (1968), Liability (1972), Registration (1975) and Moon (1979). The rules are: not to damage others, enforcement to be made by states, what is not prohibited is permitted, space exploration to be made for the general benefit of humankind, and to inform UN, public and scientists of *organic* life discovery. From this, SETI should be aided and protected.

The space radio aspect is regulated by the International Telecommunication Union which insures that a registered activity can claim for protection. SETI, part of Radioastronomy, is considered as a passive service and then qualify for protection. However the only 'protection' it has is just a footnote or two in the ITU World Administration Radio Conferences. Though the Outer Space treaty insures free exploration and scientific use of the Saha crater, this is insufficient for protection. More should be done by member states.

Recently, F. Lyall, also director of the IISL, stressed the need to set aside a specific lunar site and work for its radio protection by creating a lunar regime. For that purpose the Antarctica and Deep Sea-Bed regimes could be used as models by a review conference to flesh the Moon treaty when its exploitation appears as feasible. In its words: "Care should be taken to ensure that the fullest protection is given to the requirements of a far-side lunar radio observatory", and, in order to insure that negociators are conversant, "approval by space relevant organisations such as Committee of Space Research (COSPAR), the International Academy of Astronautics (IAA) and the International Astronomical Union (IAU) would be useful".

6. Diplomacy

The detection of an artificial signal would give a new perspective to diplomacy, above national interests, helping to "Think Big", according to M.A.G. Michaud, an US Ambassador. Since the end of half a century of WWII and Cold War, we have a lack of vision, strifing in selfish, ethnic violent nationalisms and economic competition. We look for a set of global shared agendas on environment (Rio), population growth (Cairo), Moon ?, Mars ??

An external incentive arising from SETI together with a positive signal could be decisive. Already an agenda is taking shape, first with the Declaration of Principles about a detection which calls for verification and announcement through the United Nations. The IAA and IISL worked out a document endorsed by non-governmental organisms. Though not legally binding,.it is an enterprise undertaken by our Species in its interest.

A second step is taken with a Reply Protocol: in case we get a signal, should we answer, yes or not?; if yes, when?, and what?; as a short-cut to obtain consensus, I suggested to send a complete universal encyclopedia.

We are preparing a White Paper for intergovernmental level and we need support from at least one national government. Australia volunteers in case it is backed up by an other one. Such a support would allow to push it to the Committee on the Peaceful Uses of Outer Space, and then to the UN General Assembly. This would lead to an international behaviour equivalent to the Outer Space Treaty of 1967, then legally binding. According to M.Michaud, humankind would then enter interstellar diplomacy in the interest of life survival in the universe.

7. Culture

What could be the consequences of a detection on our culture? Whether we just get a "dial tone" or an information flow, the effect would be intellectual, not physical because of the distances involved. On the short-term the consequences would be funneled by the media and the political worlds. But on the long-term it would compare to the diffusion of ideas across terrestrial cultures which is best documented by the history of sciences. According to S.J.Dick, historian at the Naval Observatory, a good case is provided by the Greek science. It was lost A.D.500 at the end of the Roman Empire, then went through the Arabs, to be translated in the 12th century in Spain; "First a trickle, eventually a flood, it radically altered the intellectual life in the West". During the 13th century, it was worked for a reconciliation by the best scholars; Aquinas, Alexander the Great, Bacon. For SETI, it could be also that legions of scientists, cryptographers...work for years on the information received, opening a new field of science. Let us hope that the views would shift from exclusion ,to coexistence to blending, to, at last, reach "objective knowledge".

Other exemples are provided by cosmology which shifted our views from geo- to helio- to galactocentric human situation. This lead to a scientific revolution, to new physics with Galileo, Newton, to other potential Earths and theological controversies. The extraterrestrial idea was accepted as soon as mid-18th century, then elaborated in a midst of oppositions and implications.

Darwinism also is enligthning. Its idea had a long history, then got a strong immediate widespread reaction following Darwin's Origins and Wallace's Natural Selection. With strong followers like Huxley, it spurred biology, with genetics giving a second Darwinian revolution, followed by a long theological debate (creationism, Simpson's uniqueness). Man's place among the apes can be paralleled to the concept of the Plurality of Worlds. Even terrestrial Darwinism can be considered as a subset of an ET world view. Over a lengthy period, the biological world will have its Galileo, Huxley and may overhelm its skeptics. According to S.J.Dick: "It is up now to SETI to determine whether we will make the shift to a new scientific world view: from the *physical* world to the *biological* universe."

8. Sociology

The sociological impact of a signal detection has been completely distorted by UFOs reporting by unresponsible media: ETs are already here for most people! In order to cool down this popular enthusiasm, it is enough to recall a 1997 CIA study in which it appears that the Air Force mislead the public with UFOs in the 50s and 60s to protect the extraordinarily sensitive security project constituted by the very high altitude U2 and SR-71 planes spying over the USSR.

Pools were conducted by D.Tarter among responsible media and SETI workers in order to assess the possible sociological impacts of a detection. They widely range all over from panic, exploitation, credibility concerns to just simply rapidly waning interest.

As a matter of fact the sociological aspect could be applied to the ETs themselves. If we discover a signal, even just one, this will be immediate proof that advanced technologies are plenty in the universe, just for simple statistical reasons, as we would then have two cases of previously suspected rare occurences: theirs and ours. We would then have a strong drive to discover more of them and would be helped by the finesse aquired by the first unraveling. So it is expected that within a few years we would uncover dozens of them. Furthermore ETs discovered in any direction and in the opposite one, having originated and evolved independently, should be different, still sharpening more our scientific drive. It is why I think that the S in SETI, after meaning Search, will come to mean Survey, then Study and at last Sociology of extraterrestrial civilizations.

9. Education

SETI being widely multidisciplinary, it has been tested in the tertiary education as an incentive for rousing the interest of students in physics. Its range encompasses astronomy, cosmology, particle and nuclear physics, quantum mechanics, relativity, nuclear technology, environmental and human issues, communications, intelligences, ethics, culture. Roberta (Bobbie) Vaile, Professor at Western Sydney University, produced a 'Contemporary Physics' course with two 1h lectures and one 2h workshop per week. As soon as the second lecture the number of students jumped to 110, achieving success in research, analysis, decision making, presentation, communication, expression, debates, leading to final written examinations overwhelmingly positive. SETI is thus a powerfull method to excite students and public interest when properly used.

Coordination of efforts is made through the SETI Institute, the American Astronomical Society Committee on Education and the Astronomical Society of the Pacific (390 Ashton Ave., San Francisco CA 94112).

10. General bibliography

Cosmovici, C.B. & al., eds (1996) *Astronomical & biochemical origins & the search for life in the universe*; 5th International conference on bioastronomy & IAU Colloquium No. 161, Editrice Compositori, Bologna

Heidmann, J., ed. (1997) *SETI-5, the search for extraterrestrial intelligence*, Acta Astronautica Special Issue, Journal of the International Academy of Astronautics, Elsevier, Oxford, in press

Heidmann, J.(1997) *Extraterrestrial intelligence*, Canto Series, Cambridge University Press, Cambridge (wide public)

Shostak, G.S., ed. (1993) *Progress in the search for extraterrestrial life*, 4th International Bioastronomy symposium, Astronomical Society of the Pacific Conference Series, 74, San Francisco

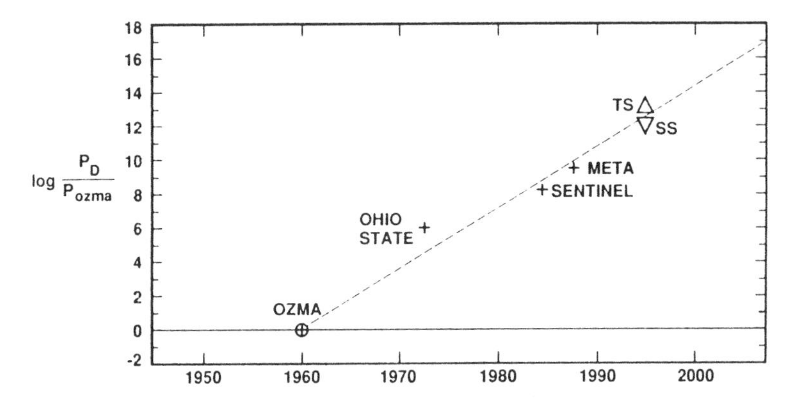

Presently, SETI is essentially aimed at stars closer than a hundred light-years, a very small distance compared to the 100,000 ligth-years diameter of our galaxy! However its efficiency is doubling every eight months, as shown in this diagram by Klein and Gulkis which plots it over the 40 years since the first Drake's search OZMA in 1960. Thus the future prospects are high.

SETI FROM SPACE AND THE MOON FAR SIDE IN THE NEXT CENTURY

C. MACCONE

"G. Colombo" Center for Astrodynamics,
Via Martorelli 43, I-10155 Torino (TO), Italy
E-mail: cmaccone@to.alespazio.it

ABSTRACT

It is currently possible to envisage a space mission, hereafter called SETIMOON, to take place early in the next century with the goal of doing SETI from the far side of the Moon. This space mission is intended to setup and operate a SETI antenna inside the SAHA farside Moon crater, as recently proposed by Jean Heidmann. A preliminary study of the SETIMOON mission is presented herewith. One important feature is that no astronaut's work is requested. On the contrary, the setup of two antennas (one inside SAHA and one on the nearside Moon plain called Mare Smythii), could be achieved in one of the following, alternative ways:

i) by virtue of a tethered system, gradually deployed in orbit around the Moon until the two antennas land softly. The tether itself also is the cable connecting the two antennas.

ii) by landing the main radiotelescope inside SAHA first, and then shooting, from SAHA up to Mare Smythii, the folded Earth antenna tied up by a tether to the radiotelescope already positioned inside SAHA. Again, the tether would then become the cable connecting the two antennas.

An analysis of the tradeoffs between the pro-s and con-s of these possibilities is timely, since the SETIMOON space mission can probably be expected to become operative earlier than any other SETI space mission.

One more SETI space mission envisaged for the next century is the FOCAL deep-space mission. It is intended to let a large space antenna reach the focus of the gravitational lens of the Sun at the distance of 550 Astronomical Units (15 times the Pluto orbit) and beyond, up to 1000 Astronomical Units. This would take full advantage of the huge gains provided by the Sun's focussing effect, reaching about 100 dB at 1420 MHz. A detailed study about the FOCAL space mission was already submitted by this author to ESA as an official Proposal back in 1993.

Finally, large space antennas will soon be orbiting the Earth. Though they are originally designed for doing space VLBI and radioastronomy, they could possibly be used for SETI also.

J. Chela-Flores and F. Raulin (eds.),
Exobiology: Matter, Energy, and Information in the Origin and Evolution of Life in the Universe, 355–360.

1. INTRODUCTION

SETI, the Search for ExtraTerrestrial Intelligence, has always been conducted from the Earth's surface up to now. Since the beginnings of experimental SETI in 1960, in fact, the only needed instrumention to do experimental SETI research was largely that requested for doing basic radioastronomy, i.e. :

1) A radio antenna with the highest possible collecting area, to pick up the weakest possible signals; plus just one new and important addition:
2) An ultra-powerful spectrum analyzer capable of resolving the smallest possible bandwiths (nowdays much smaller than 1 Hz) in the highest possible number (millions or billions of channels searched at the same time by virtue of today's dedicated computers).

As long as this equipment was already available on the surface of the Earth, there was clearly no need to go out to space. Still, one special problem started plaguing experimental SETI more and more since 1960: the Radio Frequency Interference (RFI) produced by all sorts of electronic devices. On the surface of the Earth, RFI is progressively polluting just those radio channels above 1 GHz, like the "magic" frequency of the hydrogen line at 1.420 GHz, where possible signals are expected to be trasmitted by ETI and are capable of reaching the Earth across the vast distances of interstellar space. To face the RFI problem, efficient techniques of RFI rejection were developed and applied, especially for the currently ongoing "Phoenix Project" [1] ("NASA SETI Project" until 1983) and "Meta Project" (run by the Planetary Society). Yet, the RFI problem has gradually become more and more "unsolvable" because of the current enormous increase of electronic equipment of all kinds on Earth.

A radical solution to the RFI problem for SETI must thus be planned, and we would like to show that this may by achieved only in space and by virtue of new space missions.

2. DOING SETI FROM THE FAR MOON SIDE: THE SETIMOON-TETHER SPACE MISSION

A whole set of different space missions would enable Humankind doing SETI from the Moon far side, where RFI simply does not exit because Earth-originated radio waves are halted by the Moon's mass. Here we'll confine ourselves to describe only one such mission, called SETIMOON-TETHER. The goal is to set up a large radio-telescope inside the Saha crater on the Moon far side, plus a radio antenna in the Mare Smythii plain, located on the near Moon side, that would serve for the radio link with the Earth. The Saha Crater and the Mare Smythii were selected by the French radioastronomer, Jean Heidmann [2], [3] because:

1) they are both located only a few degrees away from the Moon equator, so they are virtually in the Moon equatorial plane. This would facilitate the landing manoeuvers of the two spacecrafts carrying the radiotelescope and the folded up antenna, respectively, and tied to each other by a TETHER.

2) the distance between Saha and Mare Smythii is about 350 km, i.e. the minimal distance between one spot on the near and one on the far side of the Moon that can be found by taking into account the phoenomenon of the Lunar Librations. These are the "rocking motion" of the Moon due to two causes. The geometrical libration in latitude occurs because the Moon equator is inclined 6.5 degrees to the plane of its orbit. At one time during the month the Moon's north pole is tipped toward the Earth and half a month later the south pole is tipped toward us allowing us to see slightly beyond each pole in turn. The geometrical libration in longitude is due to the eccentricity of the Moon orbit. The rotation of the Moon on its axis is uniform, but its angular velocity around its orbit is not, since it moves faster near periapsis and solwer near apoapsis. This permits us to see 7.75 degrees around each limb of the Moon, and this angle of 7.75 degrees in in the equatorial Moon plane accounts just for the minimal distance of 350 km on the Moon surface between Crater Saha and Mare Smythii.
3) because of the above distance, the Tether length for the SETIMOON-TETHER spacecraft must be at least 350 km long. This is a considerable length, but it should be reminded that the gravitational field of the Moon is only about 1/6 of the terrestrial one, thus the Tether weight is six times smaller there. Also, there is no magnetic field on the Moon: thus the Tether is a mechanical device only, rather than an electrically conductive cable as it is around the Earth.

The critical Phase SETIMOON-TETHER Mission clearly is the Tether Deployment Phase. The author of this paper is not going to disclose now any numerical value obtained by virtue of the Tether computational codes, because that would be too premature and easily-criticizable at the moment. However, it may be said that the envisaged basic manoeuver implies a Moon approach by the Spacecraft in the Moon equatorial plane. Then, in the equatorial plane, the Tether is gradually deployed in such as way as to let the Earth-link antenna be landed first on its location in the Mare Smythii plain. Next, the tensed Tether makes the radiotelescope land inside crater Saha. And finally the Tether itself touches down on the Moon surface being tense and rectilinear. The deployment of both the antenna and the radiotelescope then completes the SETIMOON-TETHER space mission, the Tether now acting the communication link between the radiotelescope and the antenna.

3. DOING SETI FROM THE SUN'S FOCUS: THE 'FOCAL' SPACE MISSION

The gravitational lensing effect of the Sun for electromagnetic waves is one of the most amazing discoveries produced by the general theory of relativity. The first paper in this field was published by Albert Einstein in 1936 [4], but his work was virtually forgotten until 1964, when Sydney Liebes of Stanford University [5] gave the mathematical theory of gravitational lensing by a galaxy located in between the Earth and a very far cosmological object, like a quasar. In 1978 a 'twin quasar' image, caused by the gravitational field of an intermediate galaxy, was spotted for the first time by the British astronomer Dennis Walsh and his colleagues, and several more examples of gravitational lenses were discovered in the following years. Thus, no more doubts could possibly exist about gravitational lensing as predicted by general relativity. Von R. Eshleman of Stanford University went on to apply this theory to the case of the Sun in 1979 [6]. His paper discussed for the first time the possibility of sending a spacecraft to 550 AU from the Sun, where the nearest focal point for electromagnetic waves lies, to exploit the enormous magnifications provided by the gravitational lens, especially at microwave frequencies, like the hydrogen line at 1420 MHz (21 cm wavelength). This is also the frequency that all SETI radioastronomers regard as 'magic' for interstellar communications, and thus the tremendous potentials of the gravitational lens of the Sun for getting in touch with alien civilizations became obvious. The first SETI radioastronomer in history, Frank Drake (Project Ozma, 1960) of the SETI Institute in Mountain View, California, presented a paper on the advantages of using the gravitational lens of the Sun for SETI at the Second International Biostronomy Conference held in Hungary in 1987 [7], and so did Nathan 'Chip' Cohen of Boston University [8]. Non-technical descriptions of the topic were also given by them in their popular books [9] and [10]. However, planning and then having funded a space mission to 550 AU to exploit the gravitational lens of the Sun is a difficult task to achieve. This was faced by space scientists and engineers for the first time at the 1992 Conference on 'Space Missions and Astrodynamics' organized in Turin, Italy, by the author of this paper [11], who, in 1993, submitted a formal Proposal to ESA to have the space mission study funded [12]. The most rewarding direction of space where to launch our spacecraft was also discussed by Jean Heidmann and the author [13], but it seems clear that a demanding space mission like the one to 550 AU shouldn't be devoted to SETI only: the computation of the parallaxes of stars in the Galaxy and the detection of gravitational waves, plus other plasma physics experiments, are goals that necessarily would side the SETI utilization of the space mission to 550 AU, dubbed 'SETISAIL' in other papers [14], and 'FOCAL' in the Proposal submitted to ESA in 1993.

4. DOING SETI FROM EARTH-ORBITING ANTENNAS

Very Large Baseline Interferometry (VLBI) is a radioastronomical technique that enabled radioastronomers to reach unprecedented resolutions of the order of one tenth of a thousandth of an arcsecond in the observation of radio sources. In consists in having two or more radiotelescopes, greatly separated from each other, to look at the same source at the same time, and then having their respective data "synthesized" by a computer. VLBI has a drawback, however: the best angular resolution achieved on Earth is clearly constrained by the linear separation between any pair of observing antennas, that may not exceed the diameter of the Earth. The only way to overcome this difficulty is to use one radiotelescope on the ground and one antenna in space. Thus the concept of space VLBI is born, and the design of spacecrafts made up by a large radio antenna (10 to 15 meters in diameter) picking up signals from outer space and sending them to Earth has kept space experts busy over the last fifteeen years. While the history of the space VLBI Projects started around 1980, the first experimental step towards a space VLBI system was made by the United States between 1986 and 1988, when trials of the orbiting VLBI concept were made using a Transfer and Data Relay Satellite System (TDRSS) antenna. Actually, using a 4.9 meter TDRSS orbiting antenna and the 64 meter ground radio telescopes at Usuda (Japan) and Tidbinbilla (Australia), fringes were successfully detected at 2.3 GHz and 15 GHz for a number of sources.

In the 1980's Russia started planning its RadioAstron satellite: a solid-panel 10 meter dish to observe at 0.3, 1.6, 5 and 22 GHz bands. With a perigee of 5,000 km and an apogee of 80,000 km, it would be ideal to probe sources with high brightness temperature, and it has been under construction for years, but has not been launched yet.

In the 1980's Japan also entered the competition by her VSOP (VLBI Space Orbiting Project). The official proposal was made to ISAS in March 1987, and VSOP was then designed and constructed during the next ten years. On February 12, 1997, at 13.50 Japanese Standard Time, VSOP was finally launched succesfully from the Kagoshima Space Center in Japan, by the newly-designed Japanese launcher M-V-1. Its highly elliptical orbit has a 1000 km perigee, a 20,000 km apogee, an inclination of 31° and a period of 6 hours and 11 minutes. VSOP carries a 8-meter antenna with 1.6, 5, and 22 GHz band receivers. It is now imaging compact and active celestial radio objects with best resolution of 0.0001 arc second, and also galactic nuclei, star-forming regions and other stellar objects.

SETI radioastronomers started considering the possible use of space VLBI antennas for SETI since the early days of space VLBI in the 1980's (for instance, [15]). Doing SETI in space has several advantages over doing it on the ground, the most important of which may be summarized as follows:

1) RFI can be highly reduced by using two largely separated antennas rather than just one. This is already true for ground-based SETI, as recent progress made in the development of the Phoenix Project shows ([1]), but would be highly increased by using one antenna in space rather than on the ground.
2) For SETI, using an array is better than a single dish, because the array has a narrow beamwidth and the overall sensitivity can be increased by adding collecting area. Thus, one can better determine the incoming direction of a transient signal, such as an ETI signal would be expected to be.
3) Two of the frequencies "magic" to SETI, i.e. the water maser frequency at 22 GHz (or 1.35 cm wavelegth in the K-band) and the OH maser frequency at 1.6 GHz (or 18 cm, in L-band) are going to be observed by both VSOP and RadioAstron, that are thus ideal SETI spacecrafts from this point of view.

Further advantages of SETI from space could obtained by (I.Almar and J. Fejes, 1994):

4) A ground VLBI antenna, plus a SERENDIP-type MCSA (i.e. a spectrum analyzer of the "piggyback type" used at Berkeley by S. Bowyer, D. Werthimer and co-workers (1997)), plus an orbiting VLBI satellite could not only lead to terrestrial RFI reduction, but more reliable identification of transient signals also.
5) H maser frequency standards sent up to orbiting VLBI satellites are an active beacon from the Earth, detectable by ET's from anywhere within the Oort Cloud.

As a final remark about the near-future possibilities of doing SETI from space, however, it is rather sad to admit that not many members of the international community of radioastronomers took SETI seriously thus far. They are more simply "classical" radioastronomers, interested in the exploration of the far-away universe rather than in the exploration of the very near one proposed by SETI. And, unfortunately, even smaller is the interest for space-borne SETI in the international space community, that is an aggregation of science, industrial corporations, and politically-made decisions. Therefore, many years will have to pass before the utter importance of making contact with some other civilization in the Galaxy will be perceived as reality even by the most open-minded part of the world countries.

ACKNOWLEDGEMENTS

The author is indebted to Prof. Julian Chela-Flores for inviting him to give this talk at the Trieste Conference on Chemical Evolution - V - Exobiology. He also would like to thank Prof. Jean Heidmann for coordinating his own talk with the author's talk.

REFERENCES

1. Tarter, J.C.: Results Project Phoenix: Looking up from down under, *Proceedings of the Bioastronomy '96 Conference* held in Capri, Italy, July 1-5, 1996, Editrice Compositori, Bologna, 1997, 633-643.
2. Heidmann J.: Saha crater: a candidate for a SETI lunar base, *Acta Astronautica* ***32*** (1994), 471.
3. Heidmann J.: SETI Programmes All Over the World (and further out), *Journal of the British Interplanetary Society*, ***48*** (1995), 447-452.
4. Einstein A,: Lens-like Action of a Star by the Deviation of Light in the Gravitational Field, *Science*, ***84*** (1936), 506-507.
5. Liebes, S. Jr.: Gravitational Lenses, *Physical Review* ***133*** (1964), B835-B844.
6. Eshleman, V: Gravitational Lens of the Sun: Its Potential for Observations and Communications over Interstellar Distances, *Science*, ***205*** (1979), 1133-1135.
7. Drake F.: Stars as Gravitational Lenses, *Proceedings of the 99th Colloquium of the International Astronomical Union* "Bioastronomy - The Next Steps", held in Balatonfüred, Hungary, June 22-27, 1987, G. Marx editor, Kluwer Academic Publishers, Norwell, MA (1988), 391-394.
8. Cohen N.: The Pro's and Con's of Gravitational Lenses in CETI, *Proceedings of the Bioastronomy International Conference* held in Balatonfüred, Hungary, June 22-27, 1987, G. Marx editor, 395.
9. Drake F. and Sobel D.: *Is Anyone Out There ?*, Delacorte Press, New York, 1992, in particular 230-234.
10. Cohen N., *Gravity's Lens*, Wiley Science Editions, New York, 1988.
11. Maccone C.: Space Missions Outside the Solar System to Exploit the Gravitational Lens of the Sun, *Proceedings of the International Conference on 'Space Missions and Astrodynamics'* held in Turin, Italy, June 18, 1992, C. Maccone editor, *Journal of the British Interplanetary Society*, ***47*** (1994), 45-52.
12. Maccone C.: FOCAL, A New space Mission to 550 AU to Exploit the Gravitational Lens of the Sun, *A Proposal for an M3 Space Mission submitted to the European Space Agency (ESA)* on May 20, 1993, on behalf of an international Team of scientists and engineers. Later (October 1993) re-considered by ESA within the 'Horizon 2000 Plus' space missions plan.
13. Heidmann J. and Maccone C.: AstroSail and FOCAL: two extraSolar System missions to the Sun's gravitational focuses, *Acta Astronautica*, ***35*** (1994), 409-410.
14. Maccone C.: The SETISAIL Project, *Proceedings of the 1993 Bioastronomy Symposium* 'Progress in the Search for Extraterrestrial Life' held at the University of California at Santa Cruz, 16-20 August 1993, G. Seth Shostak editor, *Astronomical Society of the Pacific Conference Series, Volume* ***74*** (1995), 407-417.
15. Maccone C.: The QUASAT satellite and its SETI Applications, *Proceedings of the 99th Colloquium of the International Astronomical Union* "Bioastronomy - The Next Steps", held in Balatonfüred, Hungary, June 22-27, 1987, G. Marx editor, Kluwer Academic Publishers, Norwell, MA (1988), 343-349.

AUTHOR INDEX

SUBJECT INDEX

LIST OF PARTICIPANTS

ADEGBOYEGA Adebayo Gabriel
Obafemi Awolowo University
Dept.Electronic & Electric.Eng
Ile-Ife
Nigeria

ADJEPONG Samuel Kwasi
University of Cape Coast
Department of Physics
Faculty of Science
University P.O.
Cape Coast
Ghana

AKINOLA Ade Peter
Obafemi Awolowo University
Department of Mathematics
(Osun State)
Ile-Ife
Nigeria

ALLOTEY Francis K.A.
University of Science and
Technology
Department of Physics
Faculty of Science
University P.O.
Kumasi
Ghana

BABICH Alexander Antonovich
Gomel Polytechnical Institute
Department of Mathematics
October Avenue 48
246 746 Gomel
Belarus

BALOGUN Fatai Akintunde
Obafemi Awolowo University
Centre For Energy Research
and Development
Oyo State
Ile-Ife
Nigeria

BALTSCHEFFSKY Herrick
University of Stockholm
Arrhenius Laboratory
Department of Biochemistry
S-106 91 Stockholm
Sweden

BALTSCHEFFSKY Margareta
University of Stockholm
Arrhenius Laboratory
Department of Biochemistry
S-106 91 Stockholm
Sweden

BASIUK Vladimir
Universidad Nacional Autonoma
de Mexico
Instituto de Ciencias Nucleares
Apdo. Postal 70-543
Circuito Exterior. C.U., Df
04510 Mexico City, Mexico

BELLOPEDE Arianna
Universita' di Trieste
Dipartimento di Elettrotecnica.
Elettronica e Informatica - Deei
Facolta' di Ingegneria
Via Valerio 10
I-34127 Trieste, Italy

BONU Adriana
Universita' di Trieste
Dipartimento di Elettrotecnica.
Elettronica e Informatica - Deei
Facolta' di Ingegneria
Via Valerio 10
I-34127 Trieste, Italy

BRESSAN Mario
Universita' degli Studi
'G. D'Annunzio'
Dipartimento di Scienze
Viale Pindaro. 42
65127 Pescara, Italy

BRUSTON Paul
Universites Paris 7 and 12. of Cnrs
Lisa
Faculte' Des Sciences
61 Ave. General de Gaulle
94010 Creteil
France

CAPRARA Giovanni
Corriere della Sera
Via Solferino 28
20121 Milano, Italy

CHACON Elisabeth B.
Universidad Nacional Autonoma
de Mexico
Institute of Nuclear Science
Department of Chemistry
A.P.70-543
04510 Mexico City
Mexico

CHADHA Mohindra S.
Bhabha Atomic Research Centre
Bio-Organic Division
400 085 Bombay-Mumbai
India

CHELA-FLORES Julian
The Abdus Salam International Centre for
Theoretical Physics,
Miramare P.O. Box 586; 34100 Trieste,
Italy, and
Instituto de Estudios Avanzados
(Universidad Simon Bolivar)
Apartado 17606;Parque Central
1015-A Caracas, Venezuela.
Venezuela

COMINI Sandro
Il Gazzettino
Corrispondent
Via Torino 110
Mestre (Venezia)
Italy

CORELLI Paolo
Messeggero Veneto
V. Palmanova 209
33100 Udine
Italy

COSMOVICI Cristiano
Consiglio Nazionale Delle Ricerche
Istituto di Fisica Dello Spazio
Interplanetario
Via G. Galilei
Casella Postale 27
00044 Frascati
Italy

COYNE George Vincent
Specola Vaticana
Citta' del Vaticano
00120 Roma
Italy

DASGUPTA Brahmananda
Saha Institute of Nuclear Physics
Sector 1. Block - 'Af'
Bidhannagar
700 064 Calcutta
India

DAVIES Paul
P.O. Box 389
Burnside
South Australia 5066
Australia

DELCOLLE Fiorenzo
Telepadova - Italia 7
Via Colloredo 148
Pasian di Prato (Udine), Italy

DEVINCENZI Donald
Nasa Ames Research Center
Space Science Division Ms 245-1 -
94035-1000 Moffett Field
(California)
United States of America

DRAKE Frank
Seti Institute
2035 Landings Drive
94043 Mountain View (California)
United States of America

EASWARAN Kalpathy R.K.
Indian Institute of Science
Molecular Biophysics Unit
560012 Bangalore
India

EROKHIN Alexander S.
Moscow State University
Chemical Faculty
Gsp
119899 Moscow
Russian Federation

FALASCHI Arturo
Area di Ricerca
International Centre For Genetic
Engineering & Biotechnology
Padriciano 99
34100 Trieste
Italy

FERRARA Giorgio
Consiglio Nazionale Delle Ricerche
Istituto di Geocronologica e
Geochimica Isotopica
Via Cardinale Maffi 36
56100 Pisa
Italy

FORTUNA Gabriella
Radio Televisione Italiana (Rai)
Sede Regionale - Trieste
Via Fabio Severo 7
34133 Trieste
Italy

FOX Sidney W.
University of South Alabama
Lscb 33
36688 Mobile (Alabama)
United States of America

FRAIOLI Luca
Galileo
Corso Trieste 95
00198 Roma
Italy

FRONTE Margherita
Zadig
V. Lanino 5
Milano
Italy

GREENBERG J. Mayo
Leiden Observatory
Sterrewacht
Postbus 9513
NL-2300 Leiden
Netherlands

GRUBBS Randall O.
the University of South Alabama
Dept.of Marine Sciences
Lscb 25
36688-0002 Mobile (Alabama)
United States of America

GUCLU Mehmet Cem
Istanbul Technical University
Department of Physics
Fac. of Science & Letters
Maslak
80626 Istanbul
Turkey

GUIMARAES Romeu Cardoso
Universidade Federal de Minas Gerais
Instituto de Ciencias Biologicas
Depto. de Biologia Geral
Mg 31270-901 Belo Horizonte
Brazil

HACK Margherita
Osservatorio Astronomico di
Trieste
Via G.B. Tiepolo 11
34131 Trieste
Italy

HEIDMANN Jean
Observatoire de Paris
92195 Meudon
France

HOGBOM Jan Arvid
Stockholm Observatory
Saltsjobaden
133 36 Stockholm
Sweden

HORNECK Gerda
Deutsche Forschungsanstalt Fuer
Luft-Und Raumfahrt E.V (Dlr)
Institut Fuer Luft- Und Raumfahrtmedizin
Abt. Strahlenbiologie
Linder Hoehe 51147 Cologne
Germany

IVANOV Mikhail V.
Russian Academy of Sciences
Institute of Microbiology
Prospekt 60-Letiya Oktiabrya. 7-2
117 312 Moscow
Russian Federation

JAIN Sanjay
Indian Institute of Science
Centre For Theoretical Studies
560 012 Bangalore
India

KEYNES Richard D.
Physiological Laboratory
Downing Street
CB2 3EG Cambridge
United Kingdom

KOBAYASHI Kensei
Yokohama National University
Dept. of Physical Chemistry
Faculty of Engineering
79-5 Tokiwadai
Hodogaya-Ku
240 Yokohama
Japan

KRITSKY (KRITSKI) Mikhail S.
Russian Academy of Sciences
A.N. Bach Institute of Biochemistry
Leninsky Prospekt 33
117071 Moscow
Russian Federation

KUKU Aderemi Oluyomi
University of Ibadan
Postgraduate School
Ibadan
Nigeria

LANCET Doron
the Weizmann Institute of Science
Department of Membrane Research & Biophysics
76100 Rehovot
Israel

LAZCANO Antonio
Universidad Nacional Autonoma de Mexico
Facultad de Ciencias
Apdo. Postal 70 407
Cd. Universitaria
D.F. 04510 Mexico City
Mexico

LEMARCHAND Guillermo Andres
Universidad de Buenos Aires
Centro de Estudios Avanzados
C.C.8 - Sucursal 25
1425 Buenos Aires
Argentina

LOU Sen-Yue
Ningbo Normal College
Institute of Modern Physics
315211 Ningbo
People's Republic of China

MACCONE Claudio
Centro di Astrodinamica
'G. Colombo'
Via Martorelli. 43
10155 Torino
Italy

MACDERMOTT Alexandra Jeanette
University of Cambridge
Department of Chemistry
Lensfield Road
CB2 1EW Cambridge
United Kingdom

MASLIKOV Alexandre Albertovich
Institute For High Energy Physics
(Moscow Region)
142284 Protvino
Russian Federation

MAYOR Michel
Observatoire de Geneve
Chemin Des Maillettes 51
CH-1290 Sauverny
Switzerland

MCKAY Christopher P.
Nasa Ames Research Center
Ms 245-3
94035 Moffett Field (California)
United States of America

METRANGOLO Pierangelo
Universita degli Studi di Milano
Istituto di Chimica Organica
Facolta' di Farmacia
V. Venezian 21
20133 Milano
Italy

MILEIKOWSKY Curt
Avenue de Rochetlaz 14A
Ch-1009 Pully
Switzerland

MOORBATH Stephen
University of Oxford
Department of Earth Sciences
Parks Road
OX1 3PR Oxford
United Kingdom

NAKAMURA Hakobu
Konan University
Department of Biology
Faculty of Science
8-9-1 Okamoto, Higashinada
658 Kobe, Japan

NAVARRO-GONZALEZ Rafael
Universidad Nacional Autonoma
de Mexico
Instituto de Ciencias Nucleares
Lab. de Quimica Do Plasmas Y
Estudios Planetarios
A. Postal 70-543
D.F. 04510 Mexico City
Mexico

NEGRON-MENDOZA Alicia
Universidad Nacional Autonoma
de Mexico
Instituto de Ciencias Nucleares
Apdo. Postal 70-543
Circuito Exterior. C.U.
Df 04510 Mexico City
Mexico

NGUYEN Ngoc Hai
Academy of Sciences of Vietnam
Institute of Mathematics
P.O. Box 631
Vien Toan Hoc, Bo Ho
10.000 Hanoi, Viet Nam

ORO Juan
University of Houston
Dept. of Biochemistry &
Biophysical Science
77204-5934 Houston (Texas)
United States of America

OWEN Tobias
Institute For Astronomy
2680 Woodlawn Drive
96822 Honolulu (Hawaii)
United States of America

PAGAN Fabio
International Centre For
Theoretical Physics
Press Office
P.O. Box 586
34100 Trieste
Italy

PAPPELIS Aristotel
Southern Illinois University
At Carbondale
Department of Plant Biology
62901-6509 Carbondale (Illinois)
United States of America

PIRRONELLO Valerio
Universita di Catania
Istituto di Fisica
Citta Universitaria
Viale A. Doria 6
95125 Catania
Italy

POIDOMANI Simona
Adn - Kronos
Via di Ripetta 73
00186 Roma
Italy

RAMOS-BERNAL Sergio
Universidad Nacional Autonoma
de Mexico
Instituto de Ciencias Nucleares
Apdo. Postal 70-543
Circuito Exterior. C.U.
Df, 04510 Mexico City
Mexico

RANI Meeta
University of Hyderabad
School of Life Sciences
Department of Biochemistry
500 046 Hyderabad
India

RAULIN Francois
Universites Paris 7 and 12. of Cnrs
Lisa
Faculte' Des Sciences
61 Ave. General de Gaulle
94010 Creteil
France

RAULIN-CERCEAU Florence
Grande Galerie de L'Evolution
Museum Nationale D'Historie
Naturelle
36 Rue Geoffry Saint-Hilaire
75005 Paris
France

RIZZOTTI Martino
Universita' degli Studi di Padova
Dipartimento di Biologia
Via Trieste 75Bassi 58/B
35131 Padova, Italy

SANTORO Diego
L'Indipendente
Corrispondent
Sede Centrale
Milano, Italy

SCHENKEL Peter
Centro Internacional de Estudios
Superiores de Comunicacion
Para America Latina (Ciespal)
Cas. 17-11-6064
Quito, Ecuador

SCHIDLOWSKI Manfred
Max-Planck-Institut Fur Chemie
Postfach 30603
D 55020 Mainz
Germany

SCHNEIDER Jean
DARC
Observatoire de Paris-Meudon
Principal Cedex
92195 Meudon
France

SCHWEHM G.H.
European Space Agency (Esa)
Estec
Space Science Department
Postbus 299
2200 AG Noordwijk
Netherlands

SECKBACH Joseph
the Hebrew University of Jerusalem
P.O. Box 1132
Efrat 90435, Israel

SEGRE' Daniel
the Weizmann Institute of Science
Department of Membrane
Research & Biophysics
76100 Rehovot, Israel

SHAH Tahir K.
Universita' di Trieste
Piazzale Europa 1
Trieste, Italy

SHEN Jun
Tongji University
Department of Physics
1239 Siping Road
200092 Shanghai
People's Republic of China

SIPALA Paolo
Universita' di Trieste
Dipartimento di Elettrotecnica.
Elettronica e Informatica - Deei
Facolta' di Ingegneria
Via Valerio 10
I-34127 Trieste, Italy

SNIDARCIC Vittorio
Radio Televisione Italiana (Rai)
Sede Regionale - Trieste
Via Fabio Severo 7
34133 Trieste
Italy

STADLER Peter
University of Vienna
Institute of Theoretical Chemistry
Wahringerstrasse. 17
A-1090 Vienna
Austria

STRAZZULLA Giovanni
Universita' di Catania
Osservatorio Astrofisico
Citta' Universitaria
95125 Catania
Italy

TARTER Jill
SETI Institute
Project Phoenix
2035 Landings Drive
94043 Mountain View
(California)
United States of America

TEWARI Vinod Chandra
Wadia Institute of Himalayan Geology
33 General Mahadeo Singh Road
248001 Dehra Dun
India

VAAS Ruediger
University of Hohenheim
Institute For Zoology
Hohenheim
Germany

VIKRAMSINGH Rajendra
3611 Lupine Avenue
94303 Palo Alto
(California) , USA

WANG Wenqing
Beijing University
{University of Peking}
Dept. of Technical Physics
100871 Beijing
People's Republic of China

WESTALL Frances
Universita' degli Studi di Bologna
Dipartimento di Protezione e
Valorizzazione
Agroalimentare/Diproval
Via San Giacomo. 7
40126 Bologna
Italy

WHITEHEAD Edward P.
European Commission
Science Research and
Development
- Joint Research Centre
Rue de La Loi 200
B-1049 Brussels
Belgium

WIAFE-AKENTEN John
University of Science and
Technology
Department of Physics
Faculty of Science
University P.O.
Kumasi
Ghana

ZACCOLO Elisabetta
Telepadova - Italia 7
Via Colloredo 148
Pasian di Prato (Udine)
Italy